# GEOLOGY: AN INTRODUCTION

**Robert L. Bates**
**Walter C. Sweet**
**Russell O. Utgard**
*OHIO STATE UNIVERSITY*

second edition

Geology
an introduction

D. C. HEATH AND COMPANY
Lexington, Massachusetts   Toronto   London

Published simultaneously in Canada.

Printed in the United States of America.

International Standard Book Number: 0-669-74328-3

Library of Congress Catalog Card Number: 72-3584

As a new student in introductory geology, you will find these few cautionary remarks and practical suggestions to be of help to you throughout the course, and perhaps beyond.

1. *Study the illustrations and the captions that go with them.* Except for a few decorative sketches in Chapter 1, every illustration in this book is included for a definite purpose. Each drawing, photograph, and map has been chosen to help clarify a specific point, and each merits your close attention.
2. *Expect many new terms and definitions.* Every subject, from auto repair to philosophy, has its specialized vocabulary, and geology is no exception. Remember that the existence of such a vocabulary does not make comprehension of a subject more difficult, but easier. It is simpler, for example, to say "seismic" than "of or pertaining to earthquakes." In addition to meeting new terms, you will find a number of familiar ones that have unfamiliar meanings. Such everyday words as *crust, formation,* and *fault* have specific meanings in a geologic context. A feeling for words and an interest in them will help you to learn a large number of necessary new ones.
3. *Develop precision in use of terms and expressions.* In science, coming close to the mark is not enough. Chlorite is not a chloride; silicon is different from silica, and both of these differ from silicates. A map is one thing; a cross section, another. Do not speak of *consolidation* when you mean *compaction.* Discrimination and accuracy in the use of words is one mark of an educated person.
4. *Don't assume that all the evidence is in.* Every science tends to pass through long quiescent stages and shorter periods of intense activity. Geology has been in an active phase since the early 1960s. New discoveries have been made, new hypotheses have arisen to challenge older ones, and there has not yet been time to sort out and evaluate all the data. You will find that several aspects of geological science today are highly debatable; on these issues we have adopted the points of view that seem most reasonable to us and we have indicated that other interpretations are possible. This book makes no pretense of containing "all the answers."
5. *Don't consider geology as merely an academic subject.* Look for daily news about geology: you have probably read that the first trained scientist to land on the moon was a geologist. Look for political considerations based on geology: the planning commission rejects one site for disposal of solid wastes because the subsurface conditions are found to be unfavorable. Look for geological features on your next vacation trip to a national park, which is almost certain to contain exceptionally

# A WORD TO THE STUDENT

fine examples. There is really no excuse for viewing geology as something isolated from daily affairs.

Geology is a new subject to many readers of this book. We consider this an advantage: there is little to unlearn, or to relearn in a new way. You will enlarge your vocabulary, and you will absorb a considerable array of facts. But the primary object of the book, and of the course that it serves, is neither words nor facts. The purpose is the stretching of your mental muscles, the widening of your intellectual horizons, and the quickening of your interest in the world about you. "The object of education," it has been said, "is to make the mind a pleasant place in which to spend one's leisure time." We hope this book will contribute to this end.

The Authors

The purpose of this edition is the same as that of its predecessor, namely, to provide a text of suitable scope for students taking introductory geology as a nonprofessional course. We aim for accuracy of broad coverage instead of exhaustive treatment of detail, realizing that most students who use the book will be taking a geology course for cultural rather than professional reasons.

We continue to allot as much emphasis to historical geology as to physical. If, as we tell students, the present is the key to the past, it seems unwise to spend more time studying the key than investigating what lies beyond the open door. In presenting the history of the continent, we emphasize chiefly the large units of the sedimentary record rather than rely solely on the time-honored but often irrelevant geologic systems.

# PREFACE

We remain indebted to those who helped with the first edition. W. S. Cole and H. K. Lautenschlager read the entire first manuscript, and J. H. Lehr, A. Mirsky, H. J. Pincus, and C. H. Shultz made useful suggestions. The present edition has been read and criticized in full by R. W. Pierce and J. W. Skehan, and in part by D. W. Byerly. We have taken many of their suggestions. We acknowledge with gratitude the assistance of our colleagues, C. Bull, C. E. Corbató, D. H. Elliot, G. Faure, G. E. Moore, Jr., W. A. Pettyjohn, J. M. Schopf, J. F. Sutter, and S. E. White.

Many friends at Ohio State and elsewhere have provided us with photographs. Acknowledgment is made in the appropriate captions. Photos without a line credit were taken by us. The sketches for Chapter 1 were made by Robert W. Tope.

*Robert L. Bates*
*Walter C. Sweet*
*Russell O. Utgard*

# CONTENTS

**1** GEOLOGY AMONG THE SCIENCES 2

1-1 What Science Is 3
1-2 How the Scientist Works 5
1-3 Special Aspects of Geology as a Science 14
1-4 Summary 16

**2** THE EARTH AND ITS CRUST 18

2-1 The Four Spheres 19
2-2 Interior of the Lithosphere 20
2-3 Crust of the Lithosphere 21
2-4 Materials of the Crust: Elements and Minerals 25
2-5 Materials of the Crust: Rocks 32
2-6 Sources of Earth Energy 38
2-7 Summary 41

**3** SEQUENCE AND TIME 44

3-1 Sequence and Superposition 45
3-2 Determination of Sequence 46
3-3 Uniformity of Process 48
3-4 Geologic Time 49
3-5 The Geologic Time Scale 53
3-6 Relative versus Absolute Time 58
3-7 Time and Probability 59
3-8 Summary 61

**4** IGNEOUS ROCKS AND PROCESSES 62

4-1 Magma and Its Mobility 63
4-2 Extrusive Processes and Products 63
4-3 Intrusive Processes and Products 74
4-4 By-products of Igneous Processes 83
4-5 Summary 87

**5** WEATHERING 88

5-1 New Minerals from Old 89
5-2 The Agents of Weathering 90
5-3 Kinds of Weathering 91
5-4 Differential Weathering 97
5-5 Regolith and Soil 97
5-6 Summary 98

**6** DOWNSLOPE MOVEMENTS 100

6-1 Slopes and Gravity 101
6-2 Types of Downslope Movement 101
6-3 Geological Significance of Downslope Movements 107
6-4 Downslope Movements and the Works of Man 108
6-5 Summary 110

# 7 EROSION AND DEPOSITION BY STREAMS, WAVES, AND WIND   112

7-1   The Significance of Streams   113
7-2   Energy Factors in Streams   114
7-3   Erosion: How Streams Acquire Their Load   117
7-4   Transportation: How Streams Carry Their Load   120
7-5   Deposition: Loss of Mechanical Load   121
7-6   Stream Deposits   121
7-7   A Man-modified River: The Nile   129
7-8   Waves and Currents   130
7-9   Wind   137
7-10  Summary   141

# 8 SEDIMENTARY ROCKS AND PROCESSES   144

8-1   The Nature of Sediments   145
8-2   Clastic Rocks   146
8-3   Organic Rocks   151
8-4   Chemically Formed Rocks   156
8-5   Features of Sedimentary Rocks   159
8-6   The Importance of Sedimentary Rocks   163
8-7   Summary   165

# 9 FOSSILS AND FOSSILIZATION   168

9-1   Conditions Favoring Fossilization   169
9-2   The Preservation of Fossils   171
9-3   The Interpretation of Fossils   176
9-4   The Significance of Fossils   178
9-5   Summary   180

# 10 IDENTIFICATION AND CORRELATION   182

10-1  Twofold Nature of the Sedimentary Record   183
10-2  Identification   184
10-3  Correlation   186
10-4  Lithologic versus Chronologic Equivalence   187
10-5  Cambrian Rocks in Northern Arizona: An Example   188
10-6  Summary   190

# 11 LAND SCULPTURE BY STREAMS   192

11-1  A Contrast in Valleys   193
11-2  Valley Development   193
11-3  Regional Reduction   200
11-4  Interruptions in the Cycle   204
11-5  Drainage Patterns   208
11-6  Summary   208

# 12 THE WORK OF GLACIERS 210

12-1 Ice on the Lands 211
12-2 Glaciers and Glacier Motion 211
12-3 Valley Glaciers 213
12-4 Ice Sheets 222
12-5 Some Effects of Glaciation 228
12-6 Pre-Pleistocene Glaciations 231
12-7 Summary 232

# 13 GROUND WATER 234

13-1 Infiltration versus Runoff 235
13-2 Porosity and Permeability. Aquifers 235
13-3 Distribution 236
13-4 Wells and Springs 237
13-5 Geologic Work of Ground Water 240
13-6 Man and Ground Water 245
13-7 Summary 247

# 14 ROCK DEFORMATION 250

14-1 Evidence in the Crust 251
14-2 Dip and Strike 251
14-3 Folds 253
14-4 Domes and Dome Mountains 256
14-5 Faults 257
14-6 Fault-Block Mountains 262
14-7 Earthquakes 262
14-8 Unconformities 266
14-9 Summary 270

# 15 METAMORPHIC ROCKS AND PROCESSES 272

15-1 Rocks Formed from Other Rocks 273
15-2 Contact Metamorphism 273
15-3 Regional Metamorphism 275
15-4 The Common Metamorphic Rocks 276
15-5 The Rock Cycle 282
15-6 Summary 282

# 16 OCEAN BASINS, CRUSTAL PLATES, AND CONTINENTAL DRIFT 284

16-1 Nature of the Ocean Basins 285
16-2 Sea-Floor Spreading 289
16-3 The Plate-Tectonic Theory 294
16-4 Continental Drift 296
16-5 Stability versus Drift 302
16-6 Summary 303

# 17 MOUNTAINS AND THEIR HISTORY 306

17-1 Features of Continental Mountain Systems 307
17-2 Mobile Belts 309
17-3 Development of Mountain Systems 315
17-4 The Role of Heat in Crustal Development 319
17-5 Summary 320

# 18 GEOLOGIC FRAMEWORK OF NORTH AMERICA 322

18-1 Geographic and Geologic Provinces of North America 323
18-2 Main Events in North American Geologic History 335
18-3 Summary 336

# 19 THE PRECAMBRIAN ERAS 338

19-1 Geologic Features and History of the Moon 340
19-2 Precambrian Rocks 345
19-3 Interpreting the Precambrian Rock Record 346
19-4 Precambrian Mineral-Age Provinces 347
19-5 Precambrian Rocks and History of the Lake Superior Region 348
19-6 Inferences as to Precambrian History of North America 352
19-7 Precambrian Ice Ages 354
19-8 Precambrian Mineral Resources 355
19-9 Summary 356

# 20 ROCKS, FOSSILS, MAPS, AND HISTORY 358

20-1 Stratigraphic Procedures 359
20-2 Principal Features of the Cambrian and Later Rock Record 364
20-3 Summary 367

# 21 ROCKS AND PHYSICAL HISTORY OF EARLY PALEOZOIC TIME 370

21-1 Cambrian and Early Ordovician 371
21-2 Middle Ordovician Through Silurian 375
21-3 Summary 380

# 22 ROCKS AND PHYSICAL HISTORY OF LATER PALEOZOIC TIME 386

22-1 Devonian Through Mississippian 387
22-2 The Transcontinental Arch 393
22-3 Pennsylvanian and Permian 395
22-4 Summary 403

# 23 ROCKS AND PHYSICAL HISTORY OF THE MESOZOIC ERA 406

23-1 Triassic and Early Jurassic 407
23-2 Middle and Late Jurassic and Cretaceous 409
23-3 Summary 414

# 24 ROCKS AND PHYSICAL HISTORY OF THE CENOZOIC ERA 416

24-1 Cenozoic Rocks 418
24-2 The Continental Margins 418
24-3 Glaciation in the Pleistocene Epoch 421
24-4 Geologic History in the Making 423
24-5 Summary 426

# 25 NATURE, ORIGIN, AND EVOLUTION OF THE BIOSPHERE 428

25-1 Fossils as History 429
25-2 Nature of the Biosphere 429
25-3 Organization of the Biosphere 430
25-4 Origin of the Biosphere 436
25-5 Evolution of the Biosphere 441
25-6 Summary 443

# 26 THE PRECAMBRIAN BIOSPHERE 446

26-1 Early Precambrian 447
26-2 Middle Precambrian 448
26-3 Late Precambrian 450
26-4 Shells and Skeletons 453
26-5 Summary 454

# 27 THE EARLY PALEOZOIC BIOSPHERE 456

27-1 The Teeming Seas 457
27-2 The Age of Fishes 463
27-3 Summary 466

# 28 THE BIOSPHERE OF THE LATER PALEOZOIC 468

28-1 Stay-at-Homes in the Sea 469
28-2 Colonization of Land 470
28-3 Summary 476

# 29 GYMNOSPERMS AND REPTILES 478

29-1 The Mesozoic Seas 479
29-2 Mesozoic Land Plants 480
29-3 Mesozoic Tetrapods 481
29-4 A Time of Great Dying 495
29-5 Summary 496

# 30 THE AGE OF MAMMALS 498

30-1 The Origin of Mammals 499
30-2 Mesozoic Mammals 499
30-3 Early Cenozoic Radiation 501
30-4 Modernization of the Placentals 504
30-5 History of the Primates 513
30-6 Summary 522

GLOSSARY 525

INDEX 535

# GEOLOGY: AN INTRODUCTION

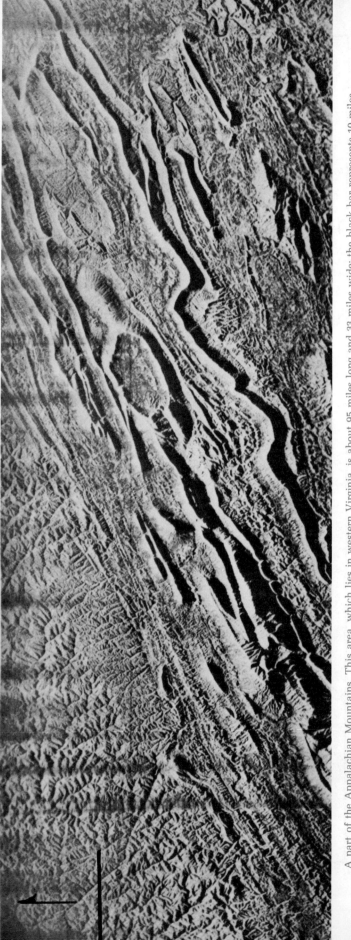

A part of the Appalachian Mountains. This area, which lies in western Virginia, is about 95 miles long and 33 miles wide; the black bar represents 10 miles. The northwestern part is a portion of the Cumberland Plateau. The Valley and Ridge Province occupies all the rest of the area except the extreme southeast corner, which is in the foothills of the Blue Ridge.

The picture was taken by radar, from an airplane at very high altitude. As the radar detector does not utilize visible light, the picture shows no clouds or haze; indeed, it might even have been taken at night. Radar imagery has become a highly useful tool in mapping the earth's surface. (Photo furnished by Corps of Engineers, U. S. Army.)

# 1 GEOLOGY AMONG THE SCIENCES

1-1 What Science Is

1-2 How the Scientist Works

1-3 Special Aspects of Geology as a Science

1-4 Summary

For many readers of these pages, this first course in geology is an introduction to science. We believe that some acquaintance with science is reasonably a part of a liberal education, and that geology is admirably suited to provide it. To set the stage, we begin with a chapter on the nature of science and the place of geology within it.

Science is a way of looking at things, and its object is to discover the order in nature. Though brief, this statement is enormously significant; it assumes that there *is* order in nature and that men have the intellectual capacity and diligence to discover it. In addition, the existence of order implies predictable patterns of behavior, which can be reduced to laws. Such laws explain the ways in which things are put together, how they react with one another, and how they change.

The evidence assembled to date suggests that these assumptions are sound. Scientists have gone far toward finding out how matter is organized, and toward deriving general statements that explain and predict its behavior. Nevertheless, even though we have learned much about the physical world and the forms of life that dwell in it, we are far from knowing all about them, and we shall probably never know as much as we might wish. Our ability to ask questions always seems to outrun our capacity to achieve answers.

To work toward an understanding of the wonderful complexities of our world is an end in itself. Yet science yields discoveries that not only satisfy our curiosity but may be turned to practical advantage. For example, scientific studies on the nature and origin of the many different kinds of rock in the earth's crust have allowed us to predict with some assurance where ore deposits and oil pools may be found, and thus to search for these objectives on a rational basis. Our knowledge of the behavior of underground waters permits us to guard against pollution of valuable water supplies and to correct such pollution if it occurs. Engineers use theoretical information obtained from geological studies when they design highways, tunnels, structures to prevent excessive shore erosion, and all such man-made features that rest on, penetrate, or affect the rocks of the earth.

There is no established pattern through which problems in science are attacked. Some scientists work alone; others, as members of a team or group, sponsored by a university, corporation, or government agency. However, all have a common objective: to discover and codify natural laws, for man's intellectual satisfaction or practical benefit.

## The Divisions of Science

In modern times, especially within the last sixty years or so, science has undergone a tremendous expansion, and this has resulted in its becoming divided and subdivided. This division of science into distinct fields, however, is largely arbitrary and is done for convenience. In Figure 1-1, we show that the principal sciences are in reality closely related by a number of overlapping fields. The existence of these overlapping fields reflects the need for scientists in one discipline to work closely with those in others. In reality, the lines of division between sciences are blurred, and science is approaching the "unity" that it had two centuries ago—although the accumulated knowledge is enormously greater now, and no one person can hope to comprehend more than a fraction of it.

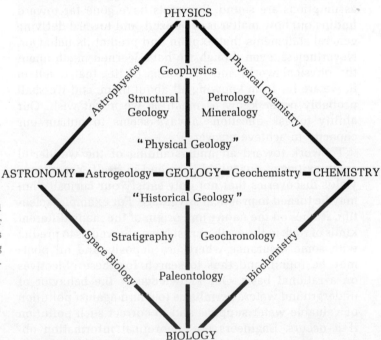

**Figure 1-1**
Geology and its relations with the other sciences. Coverage of the various geological subsciences and overlapping fields is indicated in the text.

## The Place and Scope of Geology

In a general way, we may say that **physics** is concerned with matter and energy and their relation; **chemistry** with the kinds of matter and their changes; **astronomy,** with matter and motion in space beyond the earth; and **biology,** with living things and their development. **Geology** is the study of the earth and its history (the word means "earth discourse"). Since the earth is made up of matter, much of geology consists of studying earth materials (chiefly minerals and rocks) in the light of physical and chemical principles. And since an important part of the earth's history is the development of the plants and animals that have lived on it, geologists collaborate with biologists in the study of ancient life forms. At first glance it may seem that geology is entirely dependent on its fellow sciences. But two themes set geology apart and give it the stature of a science in its own right; these are *time* and *sequence*. The geologist, like his fellow scientists, asks Why? and How? In addition, he asks When? and In what order? The concepts of time and sequence are essential to the science of geology.

Like the other sciences, geology embraces a number of divisions, or subsciences, several of which may be considered as gradational into other fields (Figure 1-1). **Mineralogy** is the systematic study of minerals; **petrology,** of rocks. In **geophysics,** we apply to the earth certain principles and techniques of physics. **Structural geology** deals with the architecture of the earth; and **astrogeology,** with rocks of the moon and with other bodies that we can observe in space. The subject matter of **stratigraphy** is the earth's layered, or stratified, rocks; of **paleontology,** fossils and the history of life; of **geochronology,** the actual age of the earth's rocks, the earth itself, and the solar system. **Geochemistry** is concerned with earth problems in which chemical processes are especially significant. Not shown in Figure 1-1 are **geomorphology,** which is concerned with the nature and origin of landforms, and **economic geology,** the study of ore deposits, oil pools, and other earth materials of economic value.

The first part of this book deals with physical geology and the second part with historical geology. These are not distinct subsciences like the ones just mentioned, but are merely convenient arrangements of the subject matter for study at the introductory level. In physical geology, we discuss chiefly the materials of the earth's crust and the processes that form and modify them; in historical geology, we trace the history of the earth and living things from the earliest times to the present.

## 1-2 HOW THE SCIENTIST WORKS

Science may be most readily distinguished from other ways of looking at nature by the manner in which scientists go about their work, that is, their scientific methods. The workings of the more fundamental of these methods may be illustrated by an example. Since this is a book about geology, we take our example from that field.

### The Birth of a Theory

We shall take for granted that the man involved in the

following imaginary account has the essentials of a scientist: curiosity about nature; keen eyes; the desire to puzzle over his questions, and if possible to find the answers; and finally, the patience to test these answers repeatedly in all manner of situations.

Let us assume that a man lives in a town situated in a valley in the foothills of high mountains. He is interested in the out-of-doors, has an observant eye for natural features, and takes many walks in the vicinity. Before

long, he begins to note things that puzzle him about the landscape and the rocky materials that compose it. For example, just upstream from the town there is a low ridge that extends across the valley. It is more or less irregular, partly covered with soil and grass, and not very conspicuous; but it is nevertheless distinct. He can see that the ridge, where the stream that flows down the valley cuts through it, is made of boulders, pebbles, sand, and clay, all mixed together and without any layering. Picking up some of the larger pebbles, he notes that they are flattened on one or more sides, and scratched, as though they had been rubbed on some rough surface and planed.

His first thought is that loose materials like this must be related in some way to the action of the stream in the valley. But this possibility is ruled out: the ridge extends across the valley, not along it; it consists of rock fragments of all sizes, rather than sand or gravel like a river bar; and it contains scratched pebbles quite unlike

those found in river gravels. How did the ridge get there? For the moment, our friend is quite unable to account for it.

Another feature that interests him is a number of large boulders, some of which are several feet across, lying about singly, partly buried in the soil. Three aspects of these boulders require explanation: they are too large to have been moved by running water, they show scratched and flattened surfaces, and they are all made of a type of rock different from the solid rock (bedrock) beneath them.

Though unable to account for either the ridge of dirt and stones or the big foreign boulders, our man suspects, purely on intuition (a "hunch"), that both may be explained by the same process, whatever that may turn out to be.

When he has a chance, our observant friend takes a trip up the valley to a resort near the lower end of a glacier. He notes with great interest the tongue of ice extending down from the snowfields far above, and observes that the stream that flows through his home town takes its origin in the form of meltwater pouring from the glacier front. Then one day it occurs to him that there is something familiar about the mounds of dirt and stones forming a sort of ridge along the front of the glacier, apparently deposited there by the ice as it melts: this feature looks very much like the ridge he saw several miles down the valley. Further, he is startled to note that the boulders in the ice-front ridge are of the same kind of rock as the foreigners down below, and that this same

rock type makes up the bedrock at this place. It is obvious that the loose boulders being freed by the melting ice are derived from the local bedrock—and some of these boulders are very large.

Suddenly it seems clear that the features he saw down-valley were made by this same glacier, at a time when it extended much farther down than it now does. The ridge and the scattered boulders near his home were simply left behind when the ice front retreated up the valley because of melting.

This seems to our investigator to be an attractive and logical idea; but it occurs to him that certain other factors ought to be checked to see whether the evidence does or does not support it. Are some of the stones in the ice-front ridge scratched and flattened like those down the valley? A few minutes' work turns up half a dozen that are. Are there any features right at the glacial front, and clearly made by the glacier, that might be duplicated down the valley and found if he knew what to look for? Close observation shows that the bedrock immediately in front of the ice, and apparently extending out of sight beneath it, is smoothly polished, evidently by rock fragments frozen in the ice (hence the scratched, planed-off

stones). Clearly the glacier has once stood at least a few feet farther down than it is at present. Polished bedrock should be expected at other places that had been covered by ice.

Not all the evidence must necessarily come from the field. How about the historical records? Do they show that the glacier used to be longer? Unfortunately, the written records go back only a century or so, and are not of much help. But several old-time residents tell of hearing tales passed down of times when the ice stood a mile or more beyond where it stands now.

Our friend puts these lines of evidence together back at his home, and then turns once more to the field. After some searching, he finds an area of smoothly polished bedrock in a farmer's field, plainly like what he saw at the very front of the glacier. Additional field checking, and some rather heated arguments with other men interested in such matters, turn up no facts that cannot be best explained by assuming the presence of glacial ice at some time in the past. And it follows logically that if this conclusion holds true for this valley, it should hold true equally well for other valleys leading out from the nearby mountains. It should apply to every area—valley or plain—that displays ridges of mixed dirt and stones, isolated foreign boulders, smoothed bedrock, and the other features seen in this particular valley.

Although the above account is imaginary, it follows in a general way the work of a Swiss gentleman named

Venetz, who announced in 1821 that the Alpine glaciers had once been much more extensive than they were at the time. This idea, taken up and championed by a better-known countryman named Louis Agassiz, developed into the "Glacial Theory," now universally accepted among geologists.

## The Scientific Methods

Let us now analyze briefly the steps by which the glacial theory was reached in our imaginary example, putting names to some of the mental activities and scientific procedures involved.

The chain of reasoning began with careful **observation** of several puzzling features: the ridge, the large isolated foreign boulders, and the relations of each to the valley and the stream in it. From these observations, our man first guessed (hypothesized) that ridge and boulders were related to the stream. This **hypothesis** was abandoned, however, when **analysis** of the ridge, where it was cut through by the stream, showed that it contained flattened, scratched boulders unlike those in stream deposits. Furthermore, analysis of the local terrain showed that the ridge cut across the stream course rather than lie parallel to it like a sand bar. Finally, it was noted that the large boulders had features in common with those buried in the ridge—that is, in being made of rock different from the local bedrock, and in being scratched or flattened on one or several sides.

Apparently, then, both ridge and boulders, though different things, resulted from the same process or processes. Implicitly, by **synthesis,** our man then brought into use the oldest method of science and one of the most important, namely **classification.** That is, he mentally arranged in a single category two superficially different types of things, related only in the ways noted above. In so doing, he reduced the number of things that required separate explanation, since he could now infer that the same process would explain both ridge and boulders.

At this stage, the investigator probably spent a good deal of time imagining, supposing, or guessing possible ways by which the features he wished to explain might have been produced. For example, he might have imagined that they had been moved into the region by wind, or by currents in the sea. There would then follow a search for features that would support either of these suppositions. Common experience told him that wind cannot move pieces of rock the size of those observed, and careful search of the rubbly materials yielded no facts to support their formation by the sea.

Nature commonly yields answers reluctantly, but in time she does yield them. In our example, the answer was given rather clearly when our observer discovered, at the lower end of a nearby glacier, a feature that indicated a **causal connection** between glacier ice and ridges of scratched boulders. Discovery of this connection led to another hypothesis: the downstream ridge was formed and the boulders were deposited at the end of a glacier at a time when ice covered that part of the valley.

As with previous guesses, however, verification was needed for this one, because guesses, even when based on strong circumstantial evidence, have no special sanctity in science. They must be thoroughly tested, and if necessary, modified, until they explain all the features or phenomena they are supposed to explain. In our example, the hypothesis of glacial origin was tested by finding additional features—the polished bedrock, for example—common to glacier front and downstream area. Thus the hypothesis was strengthened.

As it happened, other observers had noticed bouldery deposits of similar type elsewhere; some explained them in one way, some in another. It was natural that our man would recognize the features common to all these occurrences and **inductively infer** that his glacial hypothesis provided a general explanation for all specific cases. Since additional observation in these other regions supported the glacial hypothesis, it was retained; and gradually, by repeated testing and application, hypothesis became **theory.** Geologists now accept a glacial origin for deposits of the kind described; and by **deductive** use of the glacial theory for specific areas where such deposits are found,

they deduce the former existence there of masses of glacier ice.

None of the processes we have just described is the special and unique property of science or scientists. Keen observation, an active imagination, and the ability to make shrewd guesses at complex relationships are quite as valuable to a detective, say, or to a historian, as they are to a scientist. Perhaps scientific generalization differs most markedly from other kinds in the requirement that it stand up under repeated testing and that it be consistent with other explanations that have been similarly tested. Sometimes, however, a repeatedly tested explanation turns out to be inconsistent with others that seem also to have been thoroughly tested. All will then have to be reexamined, and the generalizations modified until they fit the resulting new knowledge. All scientific explanations that are tested must fit a tailored but constantly remodeled pattern of generalization.

Perhaps the student will be tempted to conclude that to apply the seemingly cold and ponderous methods of science to the world of nature is to take away much of its inherent beauty and grandeur. However, as a scientist, one looks at a landscape from a different point of view from the one he would adopt as a painter; surely there is need for both scientific explanation and work of art. The scientist has found that the methods outlined above must be followed if progress is to be made in the challenging and thrilling business of discovering order in nature.

**1-3
SPECIAL ASPECTS
OF GEOLOGY AS
A SCIENCE**

In addition to the themes of time and sequence, certain special aspects of geology distinguish it from the other sciences. To begin with, there is the matter of scale. Clearly it is impossible to bring into the laboratory a glacier, a volcano, or the Grand Canyon to study them indoors in the manner of physicists, chemists, and biologists with their materials. Yet glaciers, volcanoes, and canyons are the sorts of features the geologist must work with. As a result, most geologic observation is made in the field, on the spot (Figures 1-2, 1-3). There the geologist makes carefully detailed measurements, notes, and sketches; takes photographs; and collects specimens of rocks and fossils. From this material, which *can* be brought back to the laboratory, he constructs as best he can, by maps, cross sections, and the written word, a logical (generally historical) account of the area or feature studied.

Secondly, until recently geology has been distinguished from physics and chemistry (though not from biology) by its qualitative, rather than quantitative, nature. That

**Figure 1-2**
Geologists examine and describe rocks carefully for evidence about their origin. (Photo courtesy of Marathon Oil Company.)

is, it is difficult to apply rigorous mathematical analysis to many of the common geologic features and phenomena, because each is affected by a large set of changeable factors, or variables, or is the result of such factors. Thus measurement and numerical analysis of even the simplest geologic system is exceedingly difficult. At first sight, this qualitative or descriptive aspect of geology makes it look "easier" than its sister sciences; certainly at the introductory level the subject is largely nonmathematical. Yet it would be far more satisfying if we could analyze glacial motion, stream erosion, or the shifting interplay of land and sea with the same precision that a physicist can readily attain under the closely controlled conditions of his laboratory. Since the advent of digital computers, which are able to store and manipulate the sort of data

**Figure 1-3**
Two geologists, one in the foreground and the other atop the hill, collect data for a geologic map. This map will show the nature, distribution, and attitude of the various rocks in the area. (Photo courtesy of Marathon Oil Company.)

that most geologists collect, the science is becoming more quantitative, and the use of statistics and other forms of mathematical analysis is now common in most branches of geology.

Finally, inductive generalizations in science are not all of the same type. What has been called the **concentrative** type prevails in physics and chemistry. In these sciences, a sound theory or law can be established on only a few experiments, a concentrated, local sampling of the evidence, because the uniform and universal nature of physical phenomena has long been known to exist. Much geologic (and also biologic) generalization, on the other hand, is of the **distributive** type. Before it was possible to establish the glacial theory (or the concept of organic evolution) with a high degree of certainty, thousands of facts had to be gathered, from many parts of the world. The evidence is widely distributed, and the geologist must make observations and assemble facts from many places, often over a long period of time, before presenting a scientific generalization with confidence.

**1-4
SUMMARY**

Science is a way of regarding and investigating natural phenomena, with the hope of explaining the patterns of behavior in laws, which describe how matter is organized and how its parts are related and change with time. The lines between scientific disciplines are indistinct, and there are numerous overlapping fields. Geology is the study of the materials that make up the earth, the processes that produce and modify them, and the history of the earth and its inhabitants. Distinctive geologic concepts are those of time and sequence.

The scientist typically uses observation, analysis, synthesis, classification, and inductive inference to arrive at a hypothesis that seems to explain the problem he is investigating. Hypothesis becomes theory if it withstands repeated testing and application. Deductive use of the theory may then explain additional problems of the same general variety.

Because many geological features are too big to be brought into the laboratory geology is fundamentally a field study. It deals with so many variables that it has traditionally been largely descriptive, but the use of computers now makes quantitative studies possible. The generalizations of geology are mainly of the distributive type.

## SUGGESTED READINGS

Matthews, W. H. 1971. *Invitation to Geology: The Earth through Time.* New York: Natural History Press. 148 pp. (Paperback.) An overview of the more fundamental aspects of study of the earth written for the general reader. Provides the essentials of physical and historical geology in addition to insight into geology as a science and the work of the geologist.

Mears, B. 1970. *The Nature of Geology: Contemporary Readings.* New York: Van Nostrand Reinhold. 248 pp. (Paperback.) A collection of readings, prepared for beginning geology students, which demonstrate that geology is a dynamic and relevant science. Writings of Rachel Carson, Hans Cloos, John Muir, John Wesley Powell, and Mark Twain are among those included.

# 2 THE EARTH AND ITS CRUST

2-1  The Four Spheres
2-2  Interior of the Lithosphere
2-3  Crust of the Lithosphere
2-4  Materials of the Crust:
     Elements and Minerals
2-5  Materials of the Crust: Rocks
2-6  Sources of Earth Energy
2-7  Summary

The solid earth is a sphere about 7,900 miles in diameter and a little less than 25,000 miles in circumference. It is one of nine planets that revolve about the sun, which is a star at the center of the solar system. If the sun were represented by a globe one foot in diameter, the earth would be a speck of matter the size of a sand grain, moving about the globe at a distance of 107 feet. (Pluto, the outermost planet, would be a smaller speck, eight-tenths of a mile away.) Since the sun radiates heat and light in all directions, it is clear that the earth intercepts only a tiny fraction of this energy; yet the sun is the source of all the effective heat and light that the earth receives from space. The fundamental importance of solar energy will become clear as we discuss the various geological processes.

The solid earth may be referred to as the **lithosphere** (*lithos* is the Greek word for stone). In its path through space, the lithosphere carries with it an envelope of gases, the atmosphere; and envelopes of water, the hydrosphere, and living matter, the biosphere. (Strictly speaking, the last three are not spheres, but merely shells around the solid earth.)

**2-1 THE FOUR SPHERES**

The **atmosphere** consists chiefly of nitrogen (78 percent) and oxygen (21 percent). The remaining 1 percent is made up of several other gases, including a small but important amount of carbon dioxide. Particles of solid matter, such as dust and smoke, are almost always present. Another constituent is water in its various forms: water vapor, unseen but known to us as humidity; liquid droplets (clouds, fog, rain); and solid particles (high clouds, snow, hail). Most of the atmosphere lies relatively close to the earth's surface: one-half of the total mass is below 18,000 feet, and 99 percent is below 100,000 feet. Beyond 500,000 feet (roughly 100 miles), the extremely rarefied atmospheric gases pass gradually into the virtual emptiness of space. It has been said quite truly that man dwells at the bottom of the ocean of air.

The **hydrosphere** includes the compound $H_2O$ in all its forms, gaseous, liquid, and solid. The oceans constitute the major part of the hydrosphere; streams, lakes, glaciers, and underground water, as well as those forms of water present in the atmosphere, constitute the remainder. Chemically pure water is rare in nature. Most water contains dissolved compounds (salts), especially those of sodium, calcium, magnesium, and iron. "Fresh" water may contain such salts to a total of a few hundred parts per million; seawater has 35,000 parts per million of dissolved salts, and certain brines from underground contain as much as 200,000 parts per million. Other impurities in natural waters include particles of soil and rock, held mechanically in suspension.

19

The **biosphere** is the sum total of living things, both plant and animal. No large part of the earth is entirely biosphere. Many organisms are residents of the hydrosphere, whereas others are found more or less permanently in the atmosphere or lithosphere, or even as occasional inhabitants of all three spheres. Organisms consist dominantly of the elements oxygen, hydrogen, and carbon, and are unique in their ability to assimilate food, grow, and reproduce their kind. On parts of the earth's surface—for example, the tropical rainforest or a coral reef—the biosphere is the dominant feature; in other places it is of moderate or little significance. Especially important to the geologist are the biospheres of the past, parts of which are preserved as fossils. A study of ancient life reveals another unique characteristic of life forms besides those mentioned above: over the tremendous span of geologic time, living things have become progressively larger and more complex.

It will become apparent later that air, water, and organisms are of great importance in many geological processes. For the moment, we merely emphasize the interrelations of these substances with the solid matter of the earth. Deep in caverns (that is, holes in the lithosphere) we find air, moisture, and living creatures such as bats or fish. Dust particles are in the air and sand grains in running water. In the soil of an ordinary pasture, air, water, earth, and organic materials mingle so closely that no one of them can be considered independently of the others.

**2-2**
**INTERIOR OF THE**
**LITHOSPHERE**

Careful measurements, based on experiments using the law of gravity, have shown that the lithosphere as a whole weighs 5.5 times as much as it would weigh if it were made of water; that is, it has a **relative density** of 5.5. The relative density of the rocks at and near the surface is much lower, some 2.2 to 3.0. Hence, to bring the average for the entire lithosphere to 5.5, it follows that the inner part must have a relative density far greater, perhaps as much as 17.

There is strong evidence that the increase in density of the lithosphere from surface to center is not uniform, but takes place by stages. For one thing, in its tidal reactions to the sun and moon the lithosphere acts like a body that has an extremely dense core and a thick outer shell that is less dense. More definite evidence to this effect is provided by the way in which earthquake waves travel through the lithosphere. Most earthquakes (Section 14-7) originate in the outer 50 miles of the lithosphere. Shock waves from the point of origin of an earthquake radiate downward into the solid earth, and if the shock is intense, these waves are picked up on **seismographs** (earthquake-recording instruments) at stations all over the world. Records from major earthquakes show that the shock waves move through the lithosphere at a steadily increasing speed down to a depth of 1,800 miles, at which point there is an abrupt drop in speed of transmission. This speed then picks up gradually to a depth of about 3,100 miles, beyond which it remains constant to the earth's center. The surfaces at 1,800 and 3,100 miles are termed **discontinuities.**

The resulting "picture" of the earth's interior is shown in Figure 2-1. The **mantle,** extending to 1,800 miles, is believed to consist of dense rock; it makes up more than 80 percent of the earth's total mass. The extremely dense **core** probably consists of iron and nickel, like some of the meteorites that fall into the earth from space. Earthquake waves move through the core in such a way as to suggest that the outer part may be liquid and the inner part solid.

Although this picture helps to satisfy our curiosity about the depths of the globe on which we live, and is important to an understanding of the earth as a whole, it deals with regions that are totally inaccessible. Of immensely greater significance to man is the outermost thin shell of the lithosphere.

**Surface Relief**

The scale in Figure 2-1 is so small that the surface of the lithosphere can be shown only as a smooth line, with no irregularities for high or low regions. Although the land surface as we know it appears to be far from smooth, its **relief** (roughness) is negligible compared with the radius of the solid earth. This statement is best demonstrated with actual figures. The highest point on the face of the solid earth is Mount Everest, which projects some 29,000 feet above sea level; the lowest known point is in the Mariana Trench of the northwest Pacific Ocean, which is 36,198 feet deep. The sum of these figures—65,000 feet, or somewhat more than 12 miles—gives us the maximum relief of the surface of the lithosphere. Although 12 miles of vertical relief seems tremendous to a human being, it amounts to only three-tenths of 1 percent of the earth's radius. On a section through a one-foot globe, the resulting roughness would be less than 0.02 inch, or about the thickness of six leaves of this book. From a vantage point in space, the lithosphere appears notably smooth, as shown on photographs taken from rockets and satellites (Figure 2-2).

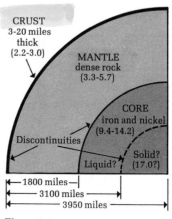

**Figure 2-1**

Zones of the lithosphere. Numbers in parentheses refer to relative density.

**2-3**
**CRUST OF THE LITHOSPHERE**

**Figure 2-2**
The earth from 250 miles up. Looking northeast from the Arabian Peninsula (foreground) across the Persian Gulf toward central Asia. The earth's highest mountains do not show at all on the smooth curve of the planet's profile. (Photo courtesy of NASA.)

### Thickness and Composition

Not only is all the surface relief concealed in the smooth outer line of Figure 2-1, but so is an entire zone—the outermost thin shell of the lithosphere, appropriately termed its **crust.** To man, the crust is all-important. Continents and ocean basins are parts of the crust; lava and big earthquakes originate in it; metals, oil, and other earth materials of value come from it. The crust is the zone in which air, water, and living things mingle with the lithosphere. It is the site of all significant geologic processes and the source of nearly all our information about earth history. For these reasons, we now turn to a description of the principal features of the crust of the lithosphere.

Figure 2-3 represents a vertical slice, or cross section, through a part of the crust. It shows a segment about 400 miles long, extending from mountains on a continent into an ocean basin. The only feature carried over from Figure 2-1 is the mantle, and only the outermost part of the mantle is shown. Most of our evidence about the deeper parts of the crust, like that about mantle and core, comes from interpretation of earthquake waves. The surface of discontinuity between mantle and crust was first defined by Mohorovičić, a student of earthquakes; his name has been applied to the discontinuity and has been conveniently shortened to **Moho.** Like the deeper discontinuities, the Moho is a surface at which the speed of earthquake waves changes markedly.

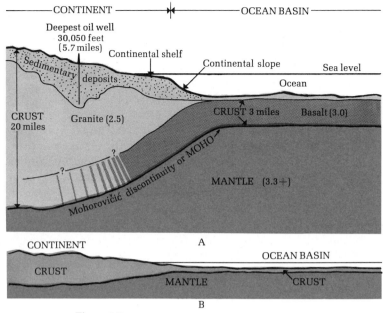

**Figure 2-3**

Cross section of about 400 miles of the earth's crust. Numbers in parentheses refer to relative density. For depiction of detail, the vertical scale in A has been exaggerated 10 times. The lower section, B, is drawn more nearly to true scale.

You will note from Figure 2-3 that the Moho is an approximate mirror image of the surface. Where the surface rises, on continental masses, the Moho is depressed; and where the surface is low, in the oceanic regions, the Moho rises. The continental masses, in addition to being many miles thick, are made of relatively light rock, called granite (relative density about 2.5), and of sediments that are still lighter. The part of the crust beneath the oceans not only is much thinner than the continental masses (it is about three miles thick, or less than 16,000 feet) but is made of heavy rock called basalt, with a relative density of 3.0. Thus the thick, lighter continental segments and the thin, heavy oceanic segments appear to "float" in approximate balance on the surface of the mantle, which behaves very much like a viscous liquid, though it is not one.

Of course we can see continental rocks at first hand, in cliffs, canyons, quarries, and the like. These rocks can also be observed directly in deep mines, to a depth of several thousand feet, and indirectly in cores and by means of other records from oil wells. (The deepest well so far drilled penetrates 30,050 feet, or about 5.7 miles, of rock.) Information from rocks at the surface and at depth gives us a three-dimensional picture of the outer part of the continental masses. For information on deeper zones, we must rely on interpretation of earthquake waves.

## Evidence from the Ocean Basins

The part of the crust beneath the oceans has been out of reach until fairly recently, its thickness and identity being inferred largely from seismic evidence. Today, however, the oceans and the crust that underlies them have become the subjects of intensive study. Exploration of oceanic areas is being conducted by investigators using a variety of newly developed instruments and techniques.

In the late 1950s, scientists initiated an ambitious project: to drill a hole entirely through the crust, penetrate the Moho, and sample the outer mantle. No such plan could be put into effect on land, as we have no equipment capable of drilling a hole 15 or 20 miles deep. Hence it was decided to attempt the Mohole, as it naturally came to be called, in the ocean, where the crust is much thinner (Figure 2-3). Much special equipment was assembled and tested, and new positioning devices were developed to keep a drilling barge directly above a spot on the ocean floor at a distance of some 12,000 feet below. Although the project was discontinued in 1967 for lack of funds before the full Mohole could be drilled, a number of preliminary holes were drilled in the sea bottom to depths of several hundred feet. Most of the cores that were recovered consisted largely of gray clayey sediment, but some included samples of the underlying sea-floor rock. This rock turned out to be basalt, confirming predictions from seismic and other evidence.

In 1964, several leading research institutions set up a program called the Joint Oceanographic Institutions for Deep-Earth Sampling, or JOIDES, the object of which was to formulate and recommend projects for drilling into the oceanic crust. An early JOIDES project resulted in the drilling of six core holes into the continental shelf and slope off the Atlantic coast of Florida. Cores as much as 400 feet long were recovered from several locations, in water ranging up to 3,500 feet in depth. These cores were carefully examined to determine their composition, fossil content, and age, with the object of deciphering the geologic history that they recorded.

A program for wide-ranging exploration of the sub-oceanic crust was initiated by JOIDES in 1968. The objective of this program, the Deep Sea Drilling Project, is to gather information that may help in determining the age and development of the ocean basins. Deep core-holes are drilled in the ocean floor, and information is obtained both from the cores recovered and from in-hole measurements of the physical properties of the materials encountered. Drilling is done from a specially adapted vessel, the *Glomar Challenger,* which is capable of drilling in as much as 20,000 feet of water. All cores and records recovered are intensively studied by scientists of various specialties.

Drilling from the *Glomar Challenger* is done along a series of oceanic traverses, or "legs," plotted out carefully in advance. For example, in February and March 1971, the ship sailed from the Panama Canal to Hawaii, occupying eight sites and drilling holes ranging up to 1,600 feet in depth. In the first three years of the Deep Sea Drilling Project, holes were drilled at 193 sites, arranged in 19 traverses. The holes penetrated a wide variety of sea-floor materials, including clay, chalk, volcanic ash, and coarser debris that is interpreted as having come from icebergs, which rafted it out to sea and released it on melting. Many of the core-holes are deep enough to penetrate basalt below the layers of sediment. The most significant result of the program has been strong evidence favoring the concepts of sea-floor spreading and the drifting of continents.

Late in 1968, the United Nations declared the decade of the 1970s to be the "International Decade of Oceanic Exploration." It is part of a program that will foster international cooperation in marine research to further our understanding of the world's oceans.

### Elements and Atoms

Every substance is made up of one or more distinct varieties of matter, known as elements. An **element** is a substance that cannot be changed into simpler kinds of matter by ordinary chemical processes.[1] Some elements occur alone. Examples are metallic copper, gaseous nitrogen, and carbon (in two different forms, diamond and graphite). Much more commonly, two or more elements are combined to form a **compound.** An example is common salt, which consists of the elements sodium and chlorine.

The formation of compounds is governed by principles concerned with the size and properties of the extremely small units, or atoms, of which every element is composed. An **atom** is the smallest portion of an element that has the properties of that element; it may be thought of as the tiniest bit of an element that can exist individually or in combination with other atoms. Although atoms are far too small to be magnified to visibility, physicists have learned by indirect means a great deal about their internal structure.

2-4
## MATERIALS OF THE CRUST: ELEMENTS AND MINERALS

---

[1] Radioactive disintegration, and bombardment by cosmic rays, discussed in the next chapter, are not considered to be "ordinary" chemical processes.

Small as atoms are, they are made up of far smaller particles. Each atom contains at its center an assemblage of several kinds of particles, which together form the **nucleus.** The nuclei of all atoms contain particles of positive electrical charge, termed **protons,** and most atomic nuclei also contain electrically neutral particles, or **neutrons.** Because the protons and neutrons are densely packed, almost all the mass of an atom is in its nucleus. Each element is identified by the number of protons in the nucleus, and is given a name and a symbol. The number of protons ranges from 1 in the element hydrogen (H) to 92 in uranium (U). An **isotope** of an element is an alternative form of the element that has the same number of protons and generally similar chemical behavior but a different number of neutrons in the nucleus.

The remainder of the atom consists of particles called **electrons,** which have electrically negative charges. Electrons are visualized as moving very rapidly about the nucleus in spherical orbits, or "shells," of which there may be one or more. These shells take up most of the space occupied by the atom. As an illustration, if the sphere described by the single electron of the simplest atom, hydrogen, is enlarged so that its diameter is about that of a basketball, the diameter of the nucleus will still be far smaller than the period at the end of this sentence. The shells, when there are more than one, occur at different distances outward from the nucleus, the close-in ones being held most tightly and those farther out progressively less so. The outermost shell, which is of most concern to us, never contains more than eight electrons. The tendency is toward achieving that number whenever possible, and the most stable, least reactive elements are those with eight electrons in the outer shell.

An atom contains as many negative electrons as positive protons, and thus is electrically neutral. Most chemical reactions do not involve electrically neutral atoms, however, but involve **ions,** which are atoms that carry an electrical charge. The sodium atom has only one electron in its outermost shell; this electron is held rather weakly, and may be lost, whereupon the next inner shell, containing eight electrons, becomes the outside shell. Since the electron that is lost has a negative charge, its removal leaves the sodium atom positively charged. The resulting sodium ion is written $Na^+$, meaning sodium with one excess positive charge. The element chlorine, by contrast, contains seven electrons in its outer shell. If it gains one, to make the "preferred" arrangement of eight, it acquires an excess negative charge and becomes the chlorine ion, $Cl^-$. $Na^+$ and $Cl^-$ may combine by sharing electrons, to form a compound called sodium chloride, or common salt. In this compound, both ions hold eight electrons in their outer shells and thus are chemically stable. Since the ions are combined in a one-to-one ratio, the formula for salt is simply written NaCl, the value "1" being understood after each symbol.

The ionic charge assumed by each element is characteristic for that element. Some of the common ions in geological reactions are hydrogen, $H^+$; potassium, $K^+$; calcium, $Ca^{+2}$; iron, $Fe^{+2}$ (or $Fe^{+3}$, depending on the chemical environment); silicon, $Si^{+4}$; and oxygen, $O^{-2}$. Oxygen combines with other elements to form common groups of ions. Among the geologically important of these are carbonate, $CO_3^{-2}$, and sulfate, $SO_4^{-2}$. An element may combine with one of these groups of ions that has a balanced and opposite charge; for example, $Ca^{+2}$ with $CO_3^{-2}$, to form a stable compound, calcium carbonate, $CaCO_3$.

## The Nature of Minerals

A **mineral** is a naturally occurring solid inorganic substance that has an orderly internal structure and characteristic chemical composition, crystal form, and physical properties. "Naturally occurring" simply means that manufactured or artificial products are not minerals. If a mineral must have an orderly internal structure, it can only be a solid; air, water, and other fluids are excluded from this classification. Also excluded is material that is or has been alive, such as wood, although the solid matter of shells and bones qualifies under the definition.

The "orderly internal structure" clearly refers to the atomic framework formed when elements combine. Sodium chloride, NaCl, occurs in nature as a mineral called halite. In halite, the countless $Na^+$ and $Cl^-$ ions are precisely arranged in a three-dimensional framework, so that the mineral invariably crystallizes in the shape of a cube. Furthermore, when a halite crystal is shattered it breaks along three mutually perpendicular plane surfaces: it has, in other words, the physical property known as cubic **cleavage.** In quartz, another common mineral, silicon and oxygen are combined, to give the formula $SiO_2$. Quartz crystallizes in six-sided, or hexagonal, prisms and related forms, but in quartz the ionic network is strong in all directions, cleavage is not developed, and the mineral breaks instead along irregular surfaces called **fracture.** Also because of its tight ionic packing and strong internal structure, quartz exhibits to a high degree another physical property, **hardness,** which is the resistance that a mineral offers to scratching.

Halite and quartz thus serve to demonstrate that crystal

form and at least some of the physical properties of a mineral depend on the internal arrangement of the ions that compose it. Indeed, this orderly internal structure is at the heart of our definition of a mineral.

The geologist is interested in minerals for a good reason: they make up rocks, and rocks are the materials with which he is chiefly concerned. The physical properties of minerals, by which most of them can be readily identified, are given in laboratory manuals. Because minerals are of fundamental importance in geology, the following discussion should be studied thoroughly and should be supplemented by work with mineral specimens in the laboratory.

## Silicate Minerals

It is surprising to find that only two elements, oxygen and silicon, constitute nearly three-fourths of the earth's crust by weight, and that these two plus only six more make up 98.6 percent (Table 2-1). Since oxygen and silicon are by far the most abundant elements, we should expect to find that many minerals contain them; and indeed they do. In silicate minerals, the fundamental unit of internal structure is a four-sided, or tetrahedral, arrangement of ions in which one silicon ($Si^{+4}$) is surrounded by four equidistant oxygens ($O^{-2}$). In the **silica tetrahedron** (Figure 2-4), the ratio of silicon to oxygen is one to four; and since the oxygens carry eight negative charges and the silicon carries only four positive ones, the unit has four negative charges, thus: $SiO_4^{-4}$. Since $SiO_4^{-4}$ is not electrically neutral, it does not occur by itself, but only in combination with ions that carry a positive charge. These ions are mostly of metallic elements. For example, $SiO_4^{-4}$ combines with $Mg^{+2}$ and/or $Fe^{+2}$, to form a compound with the formula $(Mg, Fe)_2SiO_4$. This stable substance is the mineral olivine.

More commonly than combining with metallic ions as separate $SiO_4^{-4}$ units, the silica tetrahedron occurs in repetitive structures, such as single chains, double chains, and sheets (Figure 2-4). In these structures, adjacent tetrahedra share oxygen ions; in the layer or sheet type, for example, each tetrahedron shares three oxygens with its neighbors. In the silicate mineral **quartz,** which has a three dimensional structure of complex nature, the proportion of silicon to oxygen is 1:2, because each oxygen ion serves as the corner for two tetrahedra.

In the important rock-forming silicate minerals other than quartz, silica is present as chains or sheets of tetrahedra, and the one or more common metals present, generally in combination with oxygen, are in the form of ions.

Table 2-1 The Eight Most Abundant Elements in the Earth's Crust

| Name | Symbol | Percent by Weight |
|---|---|---|
| Oxygen | O | 46.60 |
| Silicon | Si | 27.72 |
| Aluminum | Al | 8.13 |
| Iron | Fe (from Latin *ferrum*) | 5.00 |
| Calcium | Ca | 3.63 |
| Sodium | Na (from Latin *natrium*) | 2.83 |
| Potassium | K (from Latin *kalium*) | 2.59 |
| Magnesium | Mg | 2.09 |
| | | 98.59 |

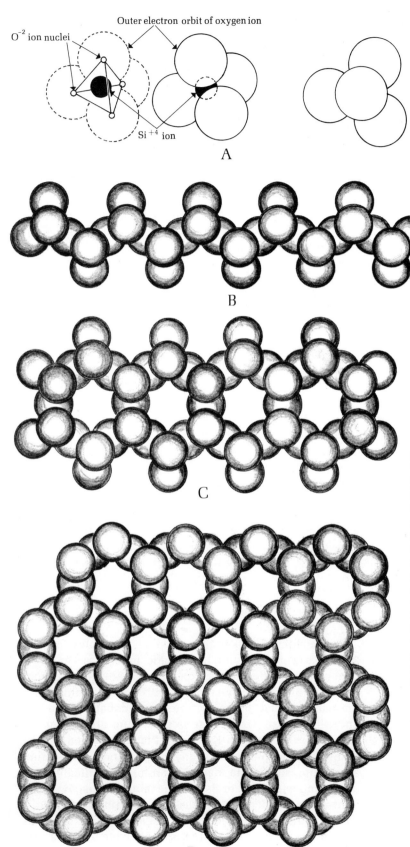

A

B

C

**Figure 2-4**
The silica tetrahedron ($SiO_4^{-4}$) and some of the structures that it forms. Cutaway view at left in A, and side view next to it, show the tetrahedral form taken by the four oxygen ions, with the single silicon ion at the center. At the right in A is a top view, in which the silicon ion is hidden. B shows a single chain, formed by adjacent tetrahedrons sharing a single oxygen ion. This is the basic structural pattern of augite. C shows a double chain, formed when tetrahedrons share alternately two and three oxygens. Hornblende has this structure. D is a sheet structure, in which each tetrahedron shares three oxygens with its neighbors. This is the structure of the micas and the clay minerals.

D

Outer electron orbit of oxygen ion

$O^{-2}$ ion nuclei

$Si^{+4}$ ion

The six most abundant metals are indicated at the top of Table 2-2. The table also shows that a few silicate minerals contain a hydrogen-oxygen ion, OH⁻, termed hydroxyl. A variety of ionic structures results, giving an assortment of crystal forms and physical properties. The internal structure of the mineral augite, for example, is based on the single chain of silica tetrahedra; that of the mineral hornblende, on the double chain. Perhaps the most easily visualized is the "sandwich" structure of the minerals collectively termed **mica.** In the dark mica, biotite (Table 2-2), for example, sheets of silica tetrahedra alternate with sheets of closely packed oxygens and hydroxyls in which are embedded ions of aluminum, magnesium, and iron. The cleavage of the mica minerals, which allows them to be split into thin plates or even films, is clearly a reflection of this sandwich structure.

*Table 2-2* Constituents of the Common Silicate Minerals

| Mineral | Silica | K | Na | Al | Ca | Mg | Fe | Other | OH |
|---|---|---|---|---|---|---|---|---|---|
| Quartz | X | | | | | | | | |
| Feldspars | | | | | | | | | |
|   Orthoclase | X | X | | X | | | | | |
|   Plagioclase | X | | X | X | X | | | | |
| Micas | | | | | | | | | |
|   Muscovite | X | X | | X | | | | | X |
|   Biotite* | X | X | | X | | X | X | | X |
| Hornblende* | X | | | X | X | X | X | X | X |
| Augite* | X | | | X | X | X | X | | |
| Olivine* | X | | | | | X | X | | |
| Garnet | X | | | X | X | X | X | X | |
| Chlorite | X | | | X | | X | X | | X |
| Serpentine | X | | | | | X | | | X |
| Talc | X | | | | | X | | | X |
| A clay mineral | | | | | | | | | |
|   Kaolinite | X | | | X | | | | | X |

*Ferromagnesian minerals.

Of the chief silicate minerals, which are shown in Table 2-2, quartz is common and widespread, as might be expected considering its constituents and simple composition, SiO₂. The most abundant of all the minerals in the earth's crust, however, are the **feldspars,** orthoclase and plagioclase. **Orthoclase** is potassium aluminum silicate, with the formula $KAlSi_3O_8$. **Plagioclase** is a general name for feldspar ranging in composition from sodium aluminum silicate to calcium aluminum silicate. Two other important minerals are the micas. **Muscovite,** the light-colored mica, is a potassium aluminum silicate; it contains hydroxyl, and hence is said to be **hydrous.** **Biotite,** the black mica, is a hydrous silicate of potassium, aluminum, iron, and magnesium. Because of its content of iron (*ferrum*) and magnesium, biotite is termed a **ferromagnesian** mineral. Other ferromagnesian silicate minerals include **hornblende, augite,** and **olivine. Garnet** is a complex silicate of iron, magnesium, aluminum, and other elements. **Chlorite** is a hydrous silicate of iron, magnesium, and aluminum. **Serpentine** and **talc** are hydrous silicates of magnesium. A group of extremely fine-grained minerals are collectively termed the clay minerals; a common one, **kaolinite,** is a hydrous aluminum silicate.

## Nonsilicate Minerals

The silicate minerals just enumerated make up the vast bulk of the rocks of the earth's crust. But there are, in addition, several other common minerals of significance. These are listed in Table 2-3, grouped by composition. Of the two oxides of iron, **hematite** is a true mineral but **limonite** is a general term for several hydrous iron oxides ("$nH_2O$" in its formula means "with a variable amount of combined water"). An oxide mineral familiar to everyone is **ice.**

A pair of related minerals that contain calcium, magnesium, and the element carbon are the carbonates **calcite,** $CaCO_3$, and **dolomite,** $CaMg(CO_3)_2$. Calcium, oxygen, and the element sulfur combine to form the sulfate mineral **anhydrite,** $CaSO_4$; with water of combination, this becomes **gypsum,** $CaSO_4 \cdot 2H_2O$. A chloride, the mineral **halite,** $NaCl$, makes up common rock salt.

The minerals of Tables 2-2 and 2-3 have been described as important because they constitute most of the rocks in the earth's crust. We may now remark that more than half of those minerals, where they occur in exceptional concentrations, are of interest because they are of value to man. Hematite, for example, is the world's leading ore mineral of metallic iron. In addition to the minerals previously discussed, there are several whose main significance is their economic value. Another source of metallic iron is an oxide, **magnetite,** $FeO \cdot Fe_2O_3$. Then, several of the metallic elements combine with sulfur to form sulfides. A common one is **pyrite** (fool's gold), $FeS_2$. **Galena,** $PbS$, and **sphalerite,** $ZnS$, are ore minerals of lead and zinc, respectively; **chalcopyrite,** $CuFeS_2$, is an ore mineral of copper. (The symbol Pb comes from the Latin *plumbum*, lead; Cu, from *cuprum*, copper.) Finally, some elements occur in nature all by themselves, that is, as **native elements.** Examples are **sulfur,** S, **copper,** Cu, and **gold,** Au (Latin, *aurum*). The element carbon occurs alone as two entirely different minerals, **diamond** and **graphite.**

About two thousand minerals are known. If you visit

Table 2-3 Composition of the Chief Nonsilicate Minerals

I. *Oxides:* element$^{+3}$ and $O^{-2}$
    Hematite, $Fe_2O_3$
    Limonite, $Fe_2O_3 \cdot nH_2O$
    *Oxide:* element$^+$ and $O^{-2}$
    Ice, $H_2O$

II. *Carbonates:* element$^{+2}$ and $CO_3^{-2}$
    Calcite, $CaCO_3$
    Dolomite, $CaMg(CO_3)_2$

III. *Sulfates:* element$^{+2}$ and $SO_4^{-2}$
    Anhydrite, $CaSO_4$
    Gypsum, $CaSO_4 \cdot 2H_2O$

IV. *Chloride:* element$^+$ and $Cl^-$
    Halite, $NaCl$

a geological museum, you will undoubtedly see specimens of others besides those we have named, which have been collected because of their beauty, rarity, or exceptionally fine crystal form. Yet the above group includes all those that you are likely to come across in your geological adventures. Let us now look briefly at the rocks that are made of these minerals.

## 2-5 MATERIALS OF THE CRUST: ROCKS

Most rocks are aggregates of minerals. In the identification and classification of rocks, therefore, their mineral content is highly significant. We need to know (1) what minerals are present (the rock's **composition**); (2) whether the individual mineral grains are large, medium, or small (the rock's **texture**); and (3) how the mineral grains are put together (the rock's **fabric**). If we know these factors, and add to them certain aspects of the rock mass as a whole, we can identify common rocks without difficulty.

Study of rocks from all parts of the earth's crust over a long period of time has shown that rocks may be divided into three principal classes on the basis of origin. Each class contains a number of specific rock types, yet the rocks of each class have certain distinctive aspects that differentiate them from those of the other two. In this chapter we confine our attention to the diagnostic features of the three main rock classes.

### Igneous Rocks

Most **igneous rocks** are made up of two or more of the following minerals: quartz, orthoclase, plagioclase, biotite, hornblende, augite, olivine. (Note that these are all silicates.) The texture of igneous rocks ranges from coarse-grained, in which individual minerals may be readily distinguished by the unaided eye, through medium-grained, to very fine-grained, in which the minerals can be distinguished only under the microscope. Careful study shows that the minerals of igneous rocks are in the form of grains that lie tightly against each other or even interpenetrate, forming a closely interlocking fabric (Figure 2-5). The typical igneous rock is **massive**, which means that its composition, texture, and fabric do not change much from one part of a laboratory specimen to another, or even within a large mass of igneous rock as seen in the field.

No doubt you have seen pictures of molten lava pouring down the slope of a volcano. When this fluid material cools and hardens, it forms a rock that has the characteristics outlined above, and hence is an igneous rock. (The word igneous is derived from the Latin *ignis*, mean-

ing fire.) Even though we find many masses of igneous
rock that have no present connection with volcanic activity, we are quite certain that they too were once molten.
Laboratory experiments show that the silicate minerals
of igneous rocks do not come into being at ordinary
temperatures, but at very high ones—several hundred
degrees Fahrenheit—and then only by crystallization from
a "silicate melt." Geological evidence, from the way in
which these rocks occur in the crust (that is, their **field
relations**), also leads directly to the deduction that igneous
rocks result from the *cooling and solidification of molten
fluid.*

## Sedimentary Rocks

Members of the second principal class, the **sedimentary
rocks,** differ substantially in mineral composition from
the igneous rocks. The one mineral that is abundant in
both classes is quartz. The only other silicates that are
common in sedimentary rocks are the clay minerals, such
as kaolinite. Other common minerals are the carbonates,
calcite and dolomite; the sulfates, anhydrite and gypsum;
the chloride, halite; and the iron oxides, hematite and
limonite.

The texture of sedimentary rocks ranges widely, from very coarse to extremely fine. Some rocks have a fabric of interlocking mineral grains, like that of igneous rocks (though involving quite different minerals). But many sedimentary rocks have a different kind of fabric, in which the mineral grains are stuck together by a cementing material. Thus sand, a sediment composed of loose grains of quartz, becomes a sedimentary rock, sandstone, when the grains are cemented together. Such rocks are said to have a granular fabric (Figure 2-6).

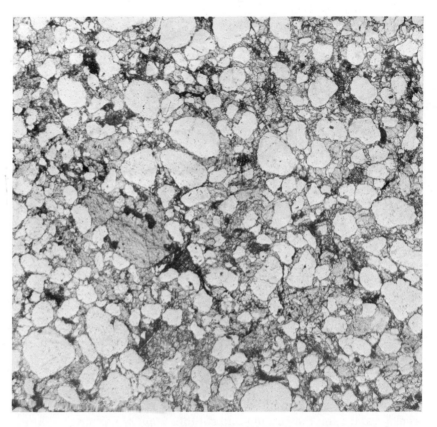

**Figure 2-6**
A sedimentary rock, sandstone, as seen through the microscope. Rounded grains of quartz are cemented firmly in calcite, giving a typically granular fabric. (Photo by C. D. Rinehart, U. S. Geological Survey.)

*Most sedimentary rocks result from the accumulation and later consolidation of mud, clay, silt, sand, or shell fragments at the bottom of a body of water; or from the chemical precipitation of dissolved salts when a water body dries up.* Evidence for these statements is abundant. Many sedimentary rocks contain shells, imprints, or other traces of organisms—that is, fossils. These rocks tend to be bedded, or stratified (Figure 2-7), showing clearly that different kinds of sediment accumulated in layers, one above the other, separated by surfaces called **bedding planes.** Many sedimentary rocks contain such features as ripple marks, exactly like those we can see forming in sediments at the bottom of present-day lakes or shallow

seas. Thus it is clear that the sedimentary rocks formed at ordinary temperatures, at or near the earth's surface.

**Figure 2-7**
Typically stratified sedimentary rocks. Hammer gives scale. (Photo by A. F. Shride, U. S. Geological Survey.)

## Metamorphic Rocks

Common minerals of the **metamorphic rocks** are mostly silicates. These include (1) all the minerals mentioned under igneous rocks except augite and olivine; (2) the light-colored mica, muscovite; and (3) a distinctive group of minerals that includes chlorite, talc, serpentine, and garnet (Table 2-4). The only nonsilicate minerals that occur in rock-making quantities are the carbonates, calcite and dolomite, which are also important in sedimentary rocks. In texture the metamorphic rocks range from coarse-grained to microscopically fine-grained. They have an interlocking crystalline fabric.

We find that the minerals of metamorphic rocks tend to be arranged in parallel sheets, bands, and streaks. This structural feature, termed **foliation,** is especially well developed in rocks made of platy or flaky silicate minerals, which have crystallized so that they lie parallel to each other (Figure 2-8). Some of the fine-grained metamorphic rocks are smoothly and evenly foliated, like a stack of papers; in coarser varieties the foliation is crude, and the

Table 2-4  The Common Rock-forming Minerals and the Rock
Classes in Which They Chiefly Occur

| Mineral | Igneous | Sedimentary | Metamorphic |
|---|---|---|---|
| *Silicates* | | | |
| Quartz | X | X | X |
| Orthoclase | X | | X |
| Plagioclase | X | | X |
| Muscovite | | X | X |
| Biotite | X | | X |
| Hornblende | X | | X |
| Augite | X | | |
| Olivine | X | | |
| Garnet | | | X |
| Chlorite | | | X |
| Serpentine | | | X |
| Talc | | | X |
| Kaolinite | | X | |
| *Oxides* | | | |
| Hematite | | X | X |
| Limonite | | X | |
| *Carbonates* | | | |
| Calcite | | X | X |
| Dolomite | | X | X |
| *Sulfates* | | | |
| Anhydrite | | X | |
| Gypsum | | X | |
| *Chloride* | | | |
| Halite | | X | |
| *Native Element* | | | |
| Graphite | | | X |

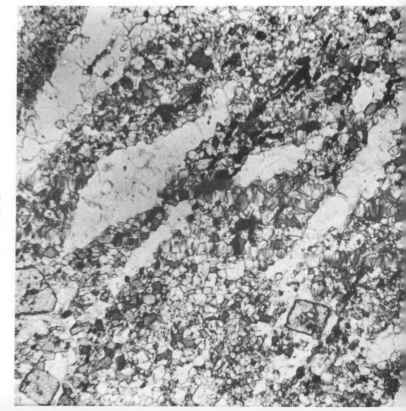

**Figure 2-8**
A metamorphic rock, schist, as seen
through the microscope. Mainly grains
of quartz (light gray) and biotite (dark
gray), in an interlocking fabric. The
rock shows marked foliation. (The large
diamond-shaped grains are an
aluminum silicate mineral; the small
ragged black areas are magnetite.)
(Photo by F. S. Simons, U. S.
Geological Survey.)

rocks appear as though they had been kneaded, like dough (Figure 2-9).

This foliation is not produced by the solidification of a molten liquid, or by the accumulation of sediment. Rather, it seems to result from the rearrangement, or recrystallization, of mineral matter in the solid state. There are a few metamorphic rocks that are without foliation, and in these the minerals show an exceptionally coarse texture or tightly interlocking fabric. Here too, recrystallization seems to have been at work. Finally, minerals such as chlorite and garnet, which are distinctive of metamorphic rocks, are found to result from the breakdown of earlier minerals and the rearrangement of their constituent elements.

Thus the metamorphic rocks are recognized as a distinct class because they show foliation, recrystallization, or the development of distinctive minerals. The inference is clear that in some way these rocks have been derived from rocks of the other two classes. A study of the rocks, their minerals, and their relations in the field leads us to conclude that *metamorphic rocks have been produced from preexisting rocks by heat, pressure, and the action*

**Figure 2-9**
The dark material is a metamorphic rock, which shows marked foliation. The light-colored material is igneous rock (granite), which invaded the other rock, cooled, and hardened. This took place under deep burial by younger rocks that have since been eroded away. (Photo courtesy of Ontario Department of Mines and Northern Affairs.)

*of hot solutions within the crust,* by which they have been changed in form and recrystallized (metamorphosed). (*Metamorphic* is from the Greek and means "formed after.") Thus every metamorphic rock is a derived rock. Some reveal quite clearly whether the parent rock was igneous or sedimentary; others have been so intensely altered that the rock from which they formed cannot be identified.

The most abundant rock-forming minerals are listed by composition in Table 2-4, together with the class or classes of rock in which each is significant.

## 2-6 SOURCES OF EARTH ENERGY

If we were to look at the earth from a vantage point some millions of miles away, we should perceive it as a tiny sphere, surrounded by cold and virtually empty space. Yet this sphere itself is neither cold nor empty. Life thrives on it. Clouds form, rains fall, streams flow; winds blow and waves crash against the shore. Volcanoes spew lava; and from time to time the solid earth itself quivers as shock waves pass through it.

All such activities involve energy—tremendous amounts, by human standards. Where does this energy come from? What makes the earth a dynamic body, rather than a static one?

### Gravity

A primary source of energy is the force of gravity, by which every particle of gaseous, liquid, and solid matter is attracted toward the center of the earth. It is gravity that holds the atmospheric envelope against the lithosphere, thus insulating the solid earth from the frigid near-vacuum through which it passes. And gravity is significant geologically, because under its influence streams and glaciers move downward, expending their energy of motion, or **kinetic energy,** in doing geologic work. Unstable masses of rock and soil slide downhill, and sediment settles to the bottom of lakes and seas. The general tendency of earth materials is always to move from higher to lower positions, under gravity.

### The Sun

Another source of energy is the sun. Solar energy is literally of vital importance, since it supports life. Heat from the sun stirs up the atmosphere, generating winds; and these in turn produce waves. The sun's heat also evaporates water from the oceans and lands. When the resulting atmospheric moisture is dropped on the land

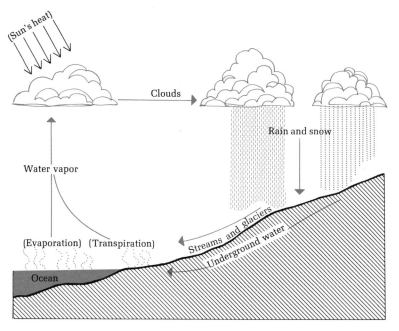

**Figure 2-10**
The hydrologic cycle.

as rain and snow, the result is streams, glaciers, lakes, and the waters underground. Together with gravity, the sun's energy powers the **hydrologic cycle** (Figure 2-10).

## The Lithosphere

Streams and glaciers move earth materials ever downward, toward the level of the sea; they wear down or **erode** the lands (the word means "gnaw away"). If just the processes of **erosion** had been operating during the earth's long history, the land areas would long ago have been reduced nearly to sea level, and the surface of the globe would be almost featureless. But this, as we know, is far from the situation. Whole continents stand well above sea level, and mountainous regions stand thousands of feet above it. Clearly, other processes have been at work, involving some source of energy other than gravity and the sun.

This source of energy is the lithosphere itself. Internal energy in the form of heat is spectacularly evident in volcanic regions, where we can see molten lava and superheated steam issuing from within the earth. Towering volcanoes attest that large amounts of rock move upward from within the solid earth and out onto its surface. Evidence from field observation also tells us that large quantities of molten rock have moved upward within the lithosphere, but not all the way to the surface. Note that the energy involved is greater than the pull of gravity, and the earth materials move outward from beneath the surface rather than inward toward the earth's center.

But many large regions that stand high do not consist of igneous rock and give no evidence of volcanic history. For example, the plateau into which the Colorado River has cut the Grand Canyon consists of flat beds of sedimentary rock, many of which were laid down beneath the sea; yet today the whole area has been bodily uplifted, and stands as much as 8,000 feet above sea level (Figure 2-11). Furthermore, in many mountainous regions we see rocks that have been not only lifted upward but also folded, broken, and otherwise deformed, even metamorphosed.

**Figure 2-11**
Sedimentary rocks in the walls of the Grand Canyon, Arizona. Most of these strata originated as layers of sediment in the sea. They now stand thousands of feet above sea level, because the whole region has been uniformly uplifted by internal earth forces. (Photo by E. M. Spieker.)

Both regional uplift and the deformation of rocks in mountains require enormous amounts of energy. The ultimate source of this internal energy is probably the spontaneous breakdown of certain elements into simpler ones. This process, known as radioactive disintegration (Section 3-4), emits great quantities of heat—so much, that geophysicists believe that the earth is not becoming cooler, as was long thought, but is at least maintaining its "body temperature." Some of this heat energy is transformed into kinetic energy. The release of kinetic energy within the lithosphere is not merely inferred from our study of ancient rocks. It is proved daily by the instruments that record earthquakes, and is emphasized from time to time by the loss of life and property. We meet here, then, another phase of internal energy that is strong enough to overcome the ever present force of gravity, strong enough to move rock masses outward from the earth's center.

Igneous processes, regional uplift, and mountain formation all tend to build the continents up rather than wear them down. Against these forces, from very early times, have been arrayed the forces of erosion. Today or at any time in the geologic past, the relations of land and sea, and the surface relief of the lithosphere, have depended on the status of the never ending conflict between these two sets of forces, one powered by the energy of the lithosphere and the other by gravity and the energy of the sun.

**2-7**
**SUMMARY**

Study of the records of earthquakes shows that the solid earth, or lithosphere, consists of a very dense core, a less dense mantle, and a thin outer rind, the crust, which ranges in relative density from 2.2 to 3.0. Thickness of the crust is 3 miles or so in the ocean basins where the crust is basaltic, and 20 miles or more in the continents where the rocks are dominantly granitic. The base of the crust, a discontinuity known as the Moho, is the approximate mirror image of the surface of the lithosphere. Maximum relief of the earth's surface is negligible compared with its diameter. Much evidence on crustal materials and history comes from exploration of the ocean basins.

Rocks, the constituents of the earth's crust, are made up of minerals. A mineral is an inorganic solid substance that is made of atoms (ions) of one or more elements. These ions are held together by electrical charges, and are arranged in certain precise patterns, which control the mineral's external form and physical properties. The bulk of crustal rocks consists of silicate minerals, combinations of elements with a fundamental structural unit known as the silica tetrahedron. Some eight or ten nonsilicate minerals are also quantitatively important. Rocks are classified in three groups on the basis of origin: igneous, produced from solidification of a silicate melt; sedimentary, formed mostly by settling of sediments in water; and metamorphic, formed by reconstitution of igneous or sedimentary rocks under conditions of stress in the crust. Each class of rocks has its characteristic assemblage of minerals.

Very large amounts of energy are involved in earth processes. Sources of this energy are gravity, the sun's heat, and radioactive disintegration within the lithosphere.

## SUGGESTED READINGS

Anderson, D. L. 1962. The plastic layer of the earth's mantle. *Scientific American,* vol. 207, no. 1, pp. 52–59. (Offprint No. 855. San Francisco: Freeman.)
Presents evidence supporting the concept of a plastic layer in the upper mantle at a depth between 37 and 155 miles. This layer has an important bearing on tectonic processes.

Bullen, K. E. 1955. The interior of the earth. *Scientific American,* vol. 193, no. 3, pp. 56–61. (Offprint No. 804. San Francisco: Freeman.)
Discusses the interior of the earth as revealed by study of earthquake waves. These studies indicate that matter in the earth's inner core is so highly compressed that it is twice as rigid as steel.

Cailleux, A. 1968. *Anatomy of the Earth.* New York: McGraw-Hill. 255 pp. (Paperback.)
An introduction to physical geology, which includes chapters on the external form of the earth, and its interior, crust, and composition.

Dana, E. S. 1963. *Minerals and How to Study Them,* 3rd ed., revised by C. S. Hurlbut, Jr. New York: Wiley. 323 pp. (Paperback.)
A good reference for both general study and field use. This book is devoted mainly to fairly detailed description of the mineral species, but also has sections dealing with crystallography and properties of minerals.

Keller, W. D. 1962. The earth a slave to energy. *Journal of Geological Education,* vol. 10, no. 1, pp. 1–8.
An interesting article that emphasizes the importance of energy, derived mainly from nuclear reactions, as the driving force responsible for geologic events and processes.

King-Hele, D. 1967. The shape of the earth. *Scientific American,* vol. 217, no. 4, pp. 67–76. (Offprint No. 873. San Francisco: Freeman.)
An account of speculation about the earth's shape and of measurements of the earth, as carried on from early times to the present with its information from artificial satellites. Satellite studies reveal that the earth departs from the spherical in many ways, including a slight tendency toward being pear-shaped.

Pearl, R. M. 1956. *Rocks and Minerals.* New York: Barnes & Noble. 275 pp. (Paperback.)
Written in popular language, this book provides a general treatment of rocks and minerals, including gems, crystals, meteorites, and artificial minerals, for the hobbyist or rock-hound.

# 3 SEQUENCE AND TIME

3-1 Sequence and Superposition

3-2 Determination of Sequence

3-3 Uniformity of Process

3-4 Geologic Time

3-5 The Geologic Time Scale

3-6 Relative versus Absolute Time

3-7 Time and Probability

3-8 Summary

Geologists long ago set themselves the challenging task of piecing together the history of the earth and of its inhabitants. A goal like this focuses the attention of geologists not only on the physical aspects of earth features, but also on their age and the order or sequence in which they formed. Consequently hills, valleys, mountain ranges, rocks, and fossils are examined geologically to determine, if possible, both the processes that formed them and the time that, in the orderly scheme of things, they came into being.

Before the middle of the seventeenth century, most ideas on the origin, causes, and order of formation of earth features were the results of rather loose speculation. They developed from limited observation of local features, and they depended on fanciful guesses about prehistoric processes. Although this early speculation contained some gems of insight, it lacked, as a body, the indispensable threads of continuity and universality. Modern geology began when inductive methods of science were applied in seeking universal answers to the physical and historical questions posed by all earth features.

Nicolaus Steno (Niels Stensen), an anatomist and churchman of mid-seventeenth-century Denmark, was apparently the first to use inductive methods in solving geologic problems. Steno's claim to distinction is based on a thin pamphlet, published in 1669, in which he showed that several outwardly dissimilar earth features (crystals, shells, and some rocks) are just different results of the same general process (deposition of solid matter from a fluid). Having established this to his satisfaction, Steno then pointed out that it should be possible to infer deductively the order (or sequence) in which the various parts of each solid substance came into being.

## 3-1 SEQUENCE AND SUPERPOSITION

Because of the limited nature of Steno's observations, many of his conclusions have not stood up under repeated testing. However, the concept of a general **law of sequence,** his principal contribution, is today the central, unifying generalization of geology. Although Steno provided us with no formal statement of the law of sequence, the following expression of it would probably have satisfied him: *The record of an event or sequence of events surrounds, overlies, or is impressed upon the record of earlier events or sequences of events.*

Use of this generalization makes it possible to determine deductively the relative order in which the parts of complex earth features came into being. For example, Steno observed that many sedimentary rock layers are deposited from water, and tend to accumulate in essentially horizontal beds. He further noted that at the time a rock layer was deposited, there was another, already deposited layer beneath it. Furthermore, at the time any rock layer formed, none of the beds that now lie above it existed. Thus, he concluded that in any undisturbed sequence of sedimentary rock beds, the beds on the bottom of the pile are the oldest, those above them are progressively younger, and the topmost beds are the youngest.

Steno cited other examples of sequence, involving the order of formation of mineral crystals, fossil sea shells, and even mountain ranges. But the situation described above has such great geologic utility, and such obvious simplicity, that the entire law of sequence is frequently described in the restricted terms of bed-on-bed superposition. However, the "law of superposition," as bed-on-bed superposition is generally termed, is merely one aspect of the general law of sequence and should be understood as such.

**3-2 DETERMINATION OF SEQUENCE**

The diagram in Figure 3-1 represents a cross section (a vertical slice through part of the crust), showing in a general way some features that might be seen on and beneath the land surface. Let us apply the law of sequence in determining the order in which the various features came into being.

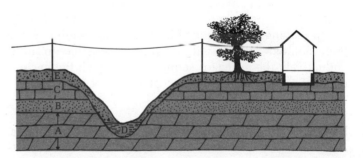

**Figure 3-1**
What is the sequence of events shown here?

In all such problems it is usually best to start with the question, What was there first? In the cross section, the layers of rock were clearly there first: the valley is cut into them, and the house stands on them. The oldest of such layers, we have learned, is the one on the bottom, in this case A; so the *formation of bed A* would be the first (oldest) event recorded in the figure. (Of course, bed A rests on something still older, but it is not shown.) In succession, then, would come the *formation of bed B* on top of A, and the *formation of bed C* on top of B. Bed C is the youngest bedrock shown. We now have to deal with loose materials lying upon it. Which was there first—the soil E, or the sediment in the valley D? Clearly the latter, because, where the two are in contact, E overlies D. And before D could have been deposited, there must have been a valley. Hence *cutting of the valley* is the next event; *deposition of valley sediment D* followed, and finally *formation of soil E*.

This completes the record of geologic events. But obviously man has been in the picture; let us utilize the law of sequence a little further. Which is the oldest—power line, tree, or house? As there would be no sense in building a power line to a nonexistent house, the house must be the older of these two; and since crews have trimmed the tree to allow passage of the line, the tree must have been there even earlier. So the order would be *growth of the tree, building of the house, construction of the power line*. The last two events might have been very close together in time.

A better subject for interpretation, because it is actual rather than imaginary and involves only geologic events, is pictured in Figure 3-2. Here we see the margin of a mountain range, in what is obviously arid country where vegetation is scanty and the geologic record lies open for inspection. One of the features in the view is a series of parallel bands, partly covered but still recognizable, that extends for some distance along the lower part of the mountain front. On checking these bands in the field,

**Figure 3-2**
In this view of the Slate Range, southeastern California, there is evidence of a vanished lake, a series of torrential streams, and an earthquake. The geological features and their story are discussed in the text. (Photo by G. I. Smith, U. S. Geological Survey.)

we find that they are old beaches, which record different levels of a lake that once occupied the low area in the foreground. This series of abandoned beaches is partly obscured by prominent deposits of gravel that slope outward from canyon mouths. As there is obviously no water in the area most of the time, we are led to assume that streams issue from the mountains only intermittently, probably following cloudbursts. Clearly the gravel deposits are younger than the lake beaches on which they lie. We then investigate the feature that shows as a thin dark band extending irregularly across the gravel deposits from side to side of the view. This turns out to mark the line of a fracture (fault), along which the gravel surface has been raised a few feet on the side toward the mountains. Such movements occur abruptly, and this one probably did so, in all likelihood causing an earthquake. As the fracture offsets the gravel deposits, it must be younger than they are. Or at least younger than most of them; in places, streams have washed over the fracture and covered it with younger gravels. To sum up, we can say with confidence that the features shown in the picture were formed in this order: mountains, lake beaches, older gravels, fracture, younger gravels.

In applying the law of sequence to Figures 3-1 and 3-2, we have been able to reconstruct, or deduce, a series of events from the arrangement of a group of features; in other words, to derive history from things. We stated with some assurance that Event A happened before Event B, B before C, and so on. Thus to make inanimate geologic features tell their story, so that each feature can be assigned an age in relation to the ages of associated features, that is, a **relative age,** is a large part of the geologist's job. To determine the **absolute age,** in years, of geologic features is another matter.

**3-3 UNIFORMITY OF PROCESS**

Although we can apply the law of sequence in determining the order in which geologic events occurred, this law tells us little about the actual events themselves, the "how" and "why" of geology. For some time before and after Steno's day, these questions were commonly answered by invoking odd or fanciful processes, for example the precipitation of all rocks from a "**primeval ocean**," or the formation of mountains and valleys by great "catastrophes."

Modern answers to questions on the origin of geologic features owe their substance to James Hutton (1726–1797), a Scottish lawyer, physician, farmer, and natural philosopher. Hutton concluded, from extensive observation of the rocks, that the same geologic processes operating today have been going on since the earth began. This conclusion, which has become known as the **principle of uniformity of process,** holds that an understanding of modern processes and their results is the key to an understanding of earth features produced in the past: that the present is the key to the past. Consequently, geologists today find it unnecessary to call on strange forces or processes to explain geologic features; on the contrary, the features are interpreted as the results of familiar processes (weathering, erosion, and the like), which operated in the past just as they do today.

To say that past processes were essentially the same as those of today is not the same as saying that they operated at the same *rate*. We believe, for example, that erosion by streams is going on now at a faster rate than at many times in the geologic past, because the continents are standing relatively high above sea level. Nor is the *scale* of modern processes necessarily the same as formerly. Along a few extremely arid present-day coastlines, seawater dries up and leaves a crust of precipitated salts; evidence from the rocks tells us that this same process took place repeatedly in the geologic past, but on a scale so vast as to produce thick beds of salt thousands of square miles in extent. Although geologic processes have remained constant in character, their rate and scale have varied greatly with the passage of time.

A feature of Hutton's principle that may have disturbed his contemporaries is the obvious fact that immense amounts of time must have been required if deep canyons were formed through erosion by seemingly puny rivers, or if great mountain ranges were built in response to forces that seem now to operate so slowly. Because biblical scholars confidently believed that the earth was just slightly more than 5,000 years old, it was felt by many that there had simply not been time enough since creation for modern processes to have produced the variety of known earth features. Gradually, however, Hutton's ideas were accepted by other keen observers, and this acceptance brought with it a growing notion of great earth age.

Decades later, after it had been established that streams carve the valleys in which they flow, other men began to think of the enormous time it must have taken for rivers like the Colorado to carve mile-deep gorges like the Grand Canyon. In addition, even greater antiquity was suggested by the fact that most of that canyon is cut in sedimentary rocks, which are the accumulated debris of many long cycles of weathering, erosion, transportation, and deposition. All these rocks had to accumulate in what is now

**3-4 GEOLOGIC TIME**

49

Arizona before the Colorado River began to cut its canyon.

Growth rings of giant sequoia trees in California show that some of them have been growing for more than 3,000 years, and a cypress tree in Mexico is said to be 6,000 years old. These trees are perhaps the oldest living things; yet they are rooted in soil that covers just the top layer of a thick pile of rocks. Certainly, then, the earth is more than 6,000 years old. But how much more?

## The Sedimentary Clock

About 450 B.C., the Greek historian Herodotus recognized that each annual flood of the Nile left behind a thin layer of mud over all the surface of its valley floor. He realized that the great thickness of sediments in the valley and at the mouth of the river must have grown through small annual additions of river-borne mud, and he concluded that deposition of all this material must have required many thousands of years.

Many attempts have been made to determine how long it takes for a foot of sediment to accumulate. For if an average rate of sediment accumulation can be determined, the total thickness of sedimentary rock in the crust can be multiplied by this figure and an estimate of earth age derived.

The rate at which sediments accumulate varies widely, of course, but studies in the Mississippi Valley some years ago suggested that on the average, it takes something like 900 years for a foot of sediment to be laid down. It is a somewhat more difficult matter to determine how much sediment has been deposited since the earth began, but the aggregate thickness of all sedimentary rocks is roughly 100 miles, or 528,000 feet. If it took 900 years for each foot of sediment to accumulate in the past, as seems to be the case in the Mississippi Valley today, the earth must be at least 900 × 528,000, or 475,200,000 years old. This is a tremendously greater life span than the 5,000 years assumed for earth age in Steno's and Hutton's times.

Actually, there are several reservations about the validity of estimates of earth age based on an assumed average rate of sediment accumulation. For one thing, the rate assumed (900 years per foot of sediment) is an average only for modern deposition in one valley system. Sediment may be accumulating more rapidly, or more slowly, today than in the past, for there is abundant evidence that continental outlines and heights have not always been the same. Besides, sedimentary rocks form just a thin veneer over the much thicker crust of igneous and metamorphic rocks. Because no one has yet devised a way to determine the rate of formation of such rocks, we can conclude from our analysis of sedimentary layers only that the earth is more than 475,200,000 years old. Even though this figure is probably highly inaccurate, and at best only a minimum, it provides the notion of great antiquity. Such a notion is an essential ingredient of all geologic thinking.

Since sediments accumulate on the continents or along their edges at different rates from time to time and from place to place, applying a single rate of sedimentation is a poor method of determining the actual age of continental parts of the crust. In deep ocean basins, on the other hand, which are far from land and constant as to physical conditions, the rate of sediment accumulation is relatively steady. This rate, which is very slow, is being measured and used today to give us information on the history of the ocean basins and thus of the earth as a whole.

## The Radioactive Clock

During the last seventy years or so, it has been found that the atoms of certain elements break down slowly into atoms of simpler, more stable elements. The process of breakdown is known as **radioactive disintegration,** and the elements that undergo it are said to be **radioactive.** Atomic variants, or isotopes (Section 2-4), spontaneously change into isotopes of other elements. Each isotope is designated by its **mass number,** which is the number of protons plus neutrons in the nucleus. Thus uranium 238 disintegrates, or decays, to lead 206, uranium 235 to lead 207, and thorium 232 to lead 208; potassium 40 decays to argon 40 and calcium 40; and rubidium 87 to strontium 87. Two aspects of this process are of key significance. First, the rate is extremely slow; for example, the time required to reduce the number of uranium 238 atoms in a given system by one-half is 4.5 billion years. Second, the rate of decay for any radioactive element is unaffected by pressure, temperature, the presence of fluids, or other external factors.

Even though the rates of radioactive disintegration are very slow, they are known, and the amount of "parent" and "daughter" isotopes can be measured by geochemical instruments and techniques. If we know the ratio between the amount of a parent isotope, say potassium 40, and the amount of a daughter isotope, argon 40, in a mineral crystal, we can calculate how long the process of decay has been going on since formation of the crystal, that is, its age in years. Age determinations by this technique are free of the variables and unknowns that hamper earlier methods of reckoning geologic time, and they provide geologists with an accurate clock by which the age of past events can be determined.

A limiting factor in using radioactivity is that only those minerals that contain radioactive elements can be dated. These turn out to occur mostly in igneous and metamorphic rocks, the radioactive element being contained in biotite, muscovite, hornblende, orthoclase, or less common silicate minerals. A second limiting factor is that the method can be relied on only where the containing rocks have constituted a "closed system" since the start of radioactive disintegration, where there has been no external gain or loss of parent or daughter.

The uranium-lead and rubidium-strontium methods may be used to date events that occurred at any time between the origin of the earth, about 4,600 million years ago, and a point in time about 10 million years ago. The potassium-argon method covers the same faraway end of the spectrum but can be used to date materials as young as 100,000 years.

A well-known method, useful only for dating materials less than 50,000 years old, uses **carbon 14,** or **radiocarbon.** This is a radioactive variety of carbon that is continually being formed in the upper atmosphere, as the result of bombardment of ordinary nitrogen by cosmic rays. The newly created carbon 14, along with stable atmospheric carbon, combines with oxygen to form carbon dioxide ($CO_2$), which is assimilated by plants and animals. As long as an organism is alive, the ratio of radioactive to stable carbon in its tissues and hard parts remains fixed. When the plant or animal dies, the C-14 content begins to decrease by decay of C-14 to nitrogen and is no longer replenished. As this reversion takes place at a known rate, the number of years from the time of death of the organism can be told from the observed radioactivity of the carbon 14, whether it is from wood, bone, shell, or other remains. Carbon 14 disintegrates fairly rapidly, so that after about 50,000 years there is not enough left to be accurately measured. The method has proved to be of great value in geology for dating recent events. It is also much used by archaeologists in determining accurately when certain peoples or cultures thrived.

An age of more than 4.0 billion years is reported for certain granitic rocks of South Africa; only slightly younger are rocks in southwestern Greenland, whose age is 3.95 billion years as determined by the rubidium-strontium method. These are the oldest earth materials so far known. Rocks ranging in age from 3.0 to 3.6 billion years have been dated from several other regions, including Minnesota, southern Africa, and northwestern Russia. Interestingly, radioactive dating of specimens from the

Apollo lunar missions shows that some of the rocks of the moon's crust are appreciably older (4.0 to 4.6 billion years).

Most of the earth's radioactively dated rocks are of igneous origin, and in their original molten form were injected into the rocks that now surround them. From the law of sequence, it is clear that the oldest rocks whose age is known are young parts of the sequences in which they occur. We have very little idea how much older the undated rocks in such sequences are. All we can say at present, from direct geologic evidence, is that the earth is certainly more than 4.0 billion years old. This immense figure would probably have been readily accepted by James Hutton, who as long ago as 1785 saw in the earth "no vestige of a beginning—no prospect of an end."

**Early Observations**

It would be convenient if we could fit events in earth history into an extension back into geologic time of our familiar calendar. Unfortunately, it is generally impossible to determine, with sufficient precision to make this possible, the time at which most past events took place. Indeed, before development of the radioactive dating process, it was not possible to fit most events into an extended modern calendar at all.

About the time Hutton was developing his theories of the earth, German and Italian geologists independently made the important observation that exposed parts of the crust in Europe could be divided into four main parts (sometimes termed **sequences**), which are stacked up on top of each other (Figure 3-3). The oldest division is made up of igneous and metamorphic rocks; these are overlain by a considerable thickness of dark, contorted sedimentary rocks, containing a few fossils. Rocks of this second division are overlain by flat, or only slightly deformed, sedimentary rocks, most of which contain many fossils. Finally, large parts of the surface are overspread by loose

## 3-5 THE GEOLOGIC TIME SCALE

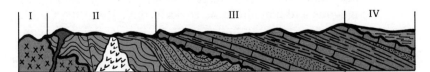

or weakly cemented sand, gravel, and clay, much of which contains a great profusion of beautifully preserved fossils.

From these observations, it was generally concluded by the end of the eighteenth century that the formation of the exposed crust (and hence, by inference, the earth

**Figure 3-3**
Generalized cross section of rocks from Wales on the left to the London Basin on the right. Roman numerals indicate the four main divisions into which early geologists divided European rocks.

as a whole) could be divided into four main stages, during each of which one of the rock divisions noted previously was formed.

Once it was recognized that exposed rocks, and, by implication, geologic time, could be divided into four main parts, an increasing number of geologists turned their attention to careful studies of the rocks in each division. These studies raised many questions, several of which threatened the very basis of the newly discovered divisions of geologic time. For example, fossil shells found to be typical of one division in one part of Europe perversely turned up in rocks of another division exposed hundreds of miles away. Also, it was found that there were considerable differences from place to place in the rocks that make up the four divisions.

## The Significance of Fossils

A solution of these problems, and a guide for most subsequent studies of earth history, appeared in 1799. William Smith, an English surveyor and geological hobbyist, noted, while working on a canal project, that rock layers in any local section could be distinguished readily from those above and below by the sorts of fossil shells and skeletons they contained. Furthermore, and probably more important, he observed that although a given fossil-bearing rock layer might change into a different rock type from one locality to another, its assemblage of fossil shells remained more or less the same. Thus Smith, and later others, came to identify and distinguish rock units over wide areas in Europe on the basis of the fossils they contained rather than their composition.

Smith's discoveries proved to be a significant turning point in the development of a classification of exposed crustal rocks and a geologic time scale. In the first four decades of the nineteenth century a great mass of information was assembled about the distribution of rocks and their fossils. By 1840, it was clear that the old, gnarled igneous and metamorphic rocks at the base of the visible European crust contain few or no fossils; the dark, contorted sediments next above are characterized by a few fossils of primitive creatures that seem mostly to have no close living relatives; the flat or only slightly deformed sediments above the latter contain many fossils of large, complex animals and plants, more modern in appearance than those below, but still strange and old-fashioned by comparison with fossils in the youngest division. These last are essentially like living animals and plants.

## Subdivisions of Rocks and Time

Recognition of the striking lateral persistence of fossils, and of the great vertical distinction between successive fossil assemblages, led in the 1840s to a suggestion that important divisions of rock, and also important units of time, might appropriately be defined on the basis of fossil animals and plants, rather than on the basis of rock types.

Thus, the oldest rocks of Europe (and elsewhere), since they lack abundant fossils, are believed to have formed in an interval of time when the earth was essentially devoid of life, or at least lacked the assortment of larger animals and plants that characterized later geologic time. This time interval has been variously termed the Azoic (without life), Agnotozoic (unknown life), or Cryptozoic (hidden life) era (or eon); but none of these terms has gained widespread usage. We believe that the early stages of earth history are more simply and adequately termed the **Precambrian Eras.**

Rocks of the second European division, just above the oldest unit, were termed **Paleozoic** (ancient life), and this name has remained in use. They are believed to have been deposited in an interval of time characterized by the existence of ancient forms of animal life. The time interval is known as the **Paleozoic Era.**

The third European rock division, which includes fossils of animals and plants intermediate in appearance between Paleozoic creatures and modern forms, defines an era of time termed the **Mesozoic** (intermediate life) **Era.** The uppermost division of fossiliferous rocks in Europe and elsewhere is the record of events in an interval of time termed the **Cenozoic** (recent life) **Era.**

The rock record of earth history varies greatly from one continent to another, and also from one part of a continent to another. Careful studies of the type and distribution of fossils, however, indicate that the decision of 1840 to subdivide the rock record, as well as the calendar, on the basis of the nature of the biosphere, was a wise one. That is, essentially the same vertical sequence of life forms has been recognized in each continent, even though the rocks containing the fossil record are quite different.

During the Paleozoic Era in North America, sedimentary rocks accumulated in various places to a total composite thickness of more than 100,000 feet. The nature of these rocks, and of the fossils they contain, changes considerably from bottom to top of this thick pile. It is fairly clear, then, that any history we might deduce from the Paleozoic division as a whole would be a very general one indeed. Primarily for this reason, the thick sequences of Paleozoic, Mesozoic, and Cenozoic rock are customarily divided into thinner intervals, or **systems,** of rock strata, representing shorter segments, or **periods,** of time.

Subdivision into systems and periods has been effected in essentially the same way as the larger subdivision into sequences and eras. That is, each of the several systems that make up the Paleozoic, Mesozoic, and Cenozoic sequences is recognized by the distinctive sorts of fossils it contains; in a parallel way, each of the periods that make up an era is characterized by the types of animals and plants that lived then.

The geologic calendar, or **time scale,** is shown in Table 3-1. Study this table carefully, bearing in mind the following points. First, the determination of absolute ages, indicated by the figures, allows us to show the duration of each period to scale. Thus the Cambrian was a long period and the Mississippian a relatively short one; but note that the Precambrian Eras were more than six times as long as all later time. Second, the names on the time scale serve a twofold purpose. They are applied both to intervals of time (Paleozoic Era, Ordovician Period) and to the bodies of rock deposited during those intervals of time (Paleozoic sequence, Ordovician System). Third, the table shows that Cenozoic time and rocks can be subdivided in either of two ways. Tertiary and Quaternary (that is, Third and Fourth) are holdovers from the old fourfold subdivision of European rocks. These terms are gradually being superseded; Paleogene and Neogene may be used instead (though not with exact equivalence).

As indicated in Table 3-2, the Quaternary Period is subdivided into two highly unequal parts, the Pleistocene and the Recent. The Pleistocene, which was an interval of extensive continental glaciation in the Northern Hemisphere, makes up most of the two-million-year duration of the Quaternary. Only the last 10,000 years or so, from the disappearance of the last ice sheet, constitutes the Recent.

The names on the geologic time scale are as essential in the study of geology as the names of the months in everyday life. They should be learned by heart.

### How Old Is Ancient?

When we reflect that all human history is compressed into Quaternary time, it becomes apparent that the term *ancient history* as conventionally used does not refer to times that are ancient by geologic standards. The point may be illustrated by a simple analogy. Suppose we let the entire 4.6 billion years of geologic time be represented by one year of 365 days. The origin of the earth, and

Table 3-1   The Geologic Time Scale

| Sequence of Rocks Era of Time | System of Rocks Period of Time | | Number of Years' Duration (Millions) | Number of Years Ago (Millions) |
|---|---|---|---|---|
| CENOZOIC | Quaternary (See Table 3-2) | Neogene | 65 | — 2 — |
| | Tertiary | Paleogene | | — 65 — |
| MESOZOIC | Cretaceous | | 70 | — 135 — |
| | Jurassic | | 55 | — 190 — |
| | Triassic | | 35 | — 225 — |
| PALEOZOIC | Permian | | 55 | — 280 — |
| | Pennsylvanian | | 40 | — 320 — |
| | Mississippian | | 25 | — 345 — |
| | Devonian | | 55 | — 400 — |
| | Silurian | | 40 | — 440 — |
| | Ordovician | | 60 | — 500 — |
| | Cambrian | | 70 | — 570 — |
| PRECAMBRIAN ERAS | (No widely recognized system-periods) | | 4030 | |

(40 inches of column omitted)

| | | | | — 4600 — |

Adapted from "The Phanerozoic Time-Scale, a Symposium." 1964. *Quart. J. Geol. Soc.* London, vol. 120S.

Table 3-2   Subdivision of the Quaternary Period into Pleistocene and Recent

Number of Years Ago (Thousands)

Recent = 0–10

200 —
400 —
600 —
800 —

*Pleistocene*

1000 —
1200 —
1400 —
1600 —
1800 —
2000 —

QUATERNARY

TERTIARY

thus the beginning of geologic time, comes with the start of the new year on January 1. Our scale, more than 500,000 years per hour, is such that by noon on January 1 more than six million years of earth history have already passed. The oldest rocks, and thus the beginning of the preserved geologic record, are radioactively dated at about March 1. Primitive organisms probably first appear in the oceans some time in May; but marine animals with shells that are preservable as fossils do not show up until far later, the middle of November. Thus ten and a half months go by before we leave the Precambrian and enter the Paleozoic.

By December 1, fishes have evolved and are numerous in the seas. For about three days beginning on December 6, great swamps exist in east central North America, in which extensive coal deposits of Pennsylvanian age are formed. Dinosaurs, the leading figures of the Mesozoic, appear on December 15 and are extinct by the day after Christmas. The Pleistocene glaciers form at 8 P.M. on New Year's Eve, and the ice age is over a bit less than four hours later. Some time in this four-hour period, man makes his appearance. Written records begin after 11:59 P.M. The United States becomes a nation one and a quarter seconds before midnight.

3-6
RELATIVE VERSUS
ABSOLUTE TIME

Until rather recently, we had little notion of how many years the various geologic periods and eras included. To be sure, many estimates had been made, based on thickness of sediments deposited, rate of vertical change in character of fossils, and numbers of significant events that took place. None of these estimates, however, could be taken very seriously. The geologic time scale, a relative scale that recognizes past events only in relation to their position among other events, has been perfectly satisfactory.

With the advent of the radioactive dating process, it became possible to determine rather closely the absolute length, in years, of most geologic periods. The information most recently available on the absolute length of geologic periods and eras is indicated along the right-hand side of Table 3-1.

Although more and more events in earth history have been assigned an absolute age, expressed in "years ago," the absolute scale is seldom used in everyday geology. For example, in referring to the time when the dinosaurs disappeared from the earthly scene, the geologist is much more likely to say "at the end of the Cretaceous Period" than to say "about sixty-five million years ago," although both expressions refer to the same point in time. (By the same token, you would refer to an event in the recent past by saying "last October fifteenth" rather than "one hundred seventy-four days ago.") The relative ages of rock strata and the events they record are immediately apparent in the field from their fossils and order of superposition, whereas absolute ages can be obtained only rarely, and then only after expensive laboratory work. For these reasons geologists usually speak of events that happened during the Ordovician or Cretaceous periods, say, or of fossils or rocks of Ordovician or Cretaceous age, rather than assign an age in years to the events or the fossils.

The determination of absolute age, calibrated in years, has the advantage of reminding us of the immensity of geologic time. Indeed, the geologist thinks in quite a different time framework from other scientists, quite often to the surprise and amusement of nongeologists. We may describe the extinction of the dinosaurs, for example, as being "rather abrupt" and as taking place at the "end" of the Cretaceous Period; yet the disappearance of these creatures may have required a million years or more. After all, one million years is only about two hundredths of 1 percent (0.022 percent) of the earth's total age, and in general, we are unable to divide geologic history into shorter segments than that. Of course, knowing the absolute age of an igneous event, a period of metamorphism, or an assemblage of fossils helps us to unravel geologic history with much more precision than we would be able to do otherwise.

A gambler knows that probability, or the odds, improves with repeated rolls of the dice. The chances of turning up all sixes on the first roll of eight dice are extremely slight, only about one in 1.5 million. With 200 throws, the odds improve to about one in 10,000, still highly improbable. But the longer the game is played, the greater the probability of rolling all sixes at least once; if the dice are thrown 5 million times, the odds become 95 in 100. Thus, if a gambler were sufficiently compulsive to roll dice for three straight years, an event that was extremely improbable at the outset would become extremely likely.

So it is with events in geologic time. Occurrences that are so rare as to be extremely improbable in the short span of a human lifetime become steadily more probable as we look back across the immense reaches of geologic time or forward into the future. In the earth's history, time is so long that any event that is physically possible becomes highly probable, and indeed the likelihood that it has happened at least once approaches certainty. Put

**3-7**
**TIME AND**
**PROBABILITY**

even more briefly: given enough time, what *can* happen *will* happen.

On August 19, 1969, an eight-hour storm dumped 31 inches of rain on Nelson County, Virginia, causing havoc from floods and landslides. Meteorologists estimate the chances of such an event at far less than once per thousand years. Yet it happened. What are the chances that the earth will encounter a meteorite large enough that the collision would excavate a great crater, fracture and fuse the rocks, generate earthquakes, or produce a huge tidal wave? Here the chances are undoubtedly far smaller than those of the rainstorm, but still the event is not impossible. When the geologist hears someone dismiss an unlikely event by saying, "Why, that wouldn't happen in a thousand years!" he is by no means certain that the matter is closed.

From the field of paleontology we take one further example of time and probability. To begin with, we note that there are about a thousand species of land snails in the Hawaiian Islands today. Most are unique to Hawaii, but they are thought to be the descendants of about twenty-five species that arrived there within the last 15 million years or so, as invaders from some continental area thousands of miles away. Now land snails are not equipped to swim or fly across oceans, and are mostly too large to be blown for great distances by the wind. Furthermore, any hypothesis that the colonists were transported from a distant source on the legs of a bird, or on a bit of driftwood, might well be met with skepticism, as the chances of one such transport being successful each year have been estimated at one in about 70,000. Those unaccustomed to time-influenced geologic thought would thus dismiss any such modes of snail transport as highly improbable, and look elsewhere for an explanation. But the paleontologist is aware that the matter should be considered further. As long as neither of the modes of transport hypothesized is impossible, the chances that one or both of them actually brought the snail pioneers to Hawaii would increase steadily with repeated tries. Improbable as either of these modes of transport may seem in the short run, they cannot be ignored as potential means of dispersal in the long run. As a biologist has written, "One has only to wait: Time itself performs the miracles."

Now, with all the talk in this section about catastrophic rainstorms and miracles, doesn't it seem that we are contradicting the principle of uniformity of process? This, you recall, states that the geologic processes of the past were like the familiar ones of today; and we specifically disallowed catastrophes as a general explanation for past geologic events. But the only happenings not allowed are those that violate physical laws and hence are physically impossible, no matter how much time is available. In this category would come the formation of all the earth's rocks through precipitation from an "ocean," or the explanation of some geologic feature as the product of water that flowed uphill. Uniformity of process is entirely acceptable, for it applies as well to the exceptional event as to the almost infinite monotony of ordinary everyday processes and events.

Events in earth history can be put in order by noting the sequence in which their records occur. By acceptance of Hutton's concept of uniformity of process, and first-hand acquaintance with modern processes and their results, the records themselves (whether rocks, minerals, fossils, valleys, or mountains) can be interpreted in the light of the process or processes that produced them. Earth history compiled in this way can be related to a relative time scale, based on vertical changes in character of fossil assemblages within rock masses. The absolute time scale, if understood in relation to the relative scale, provides a measure of rate in years of geologic processes, and emphasizes the antiquity of the earth. This antiquity is so great that ordinary geologic processes must have been punctuated repeatedly by unusual or exceptional events.

**3-8
SUMMARY**

SUGGESTED READINGS

Berry, W. B. N. 1968. *Growth of a Prehistoric Time Scale—Based on Organic Evolution*. San Francisco: Freeman. 158 pp. (Paperback.)
    Gives a fairly complete history of the development of the geologic time scale.

Deevey, E. S. 1952. Radiocarbon dating. *Scientific American,* vol. 186, no. 2. pp. 24–28. (Offprint No. 811. San Francisco: Freeman.)
    A report on radiocarbon dating written at the time when this method of measuring the age of organic materials was relatively new.

Eicher, D. L. 1968. *Geologic Time*. Englewood Cliffs, N. J.: Prentice-Hall. 150 pp. (Paperback.)
    Tells the story of how the concept of geologic time developed and how time is measured, and relates geologic time to rocks.

Harbaugh, J. W. 1968. *Stratigraphy and Geologic Time*. Dubuque, Iowa: W. C. Brown. 113 pp. (Paperback.)
    An introduction to stratigraphy and geologic time, in which methods of dating and correlating rocks are included, and also a chapter on the geologic time scale.

# 4 IGNEOUS ROCKS AND PROCESSES

4-1   Magma and Its Mobility

4-2   Extrusive Processes and Products

4-3   Intrusive Processes and Products

4-4   By-products of Igneous Processes

4-5   Summary

At many places in the modern world, volcanic eruptions pour out vast amounts of molten lava, which rapidly loses its heat and congeals into rock. Rocks identical with such volcanic rock are recognizable in regions where there are now no active volcanoes; in fact, rocks that are clearly volcanic in origin are found at nearly every level of the geologic column in some part of the world. It is clear that volcanic activity has long been important in shaping the lithosphere.

## 4-1 MAGMA AND ITS MOBILITY

Studies of modern and ancient volcanic features have led geologists to conclude that masses of molten rock, or **magma,** are formed from time to time within the lithosphere. In oceanic regions such as the Hawaiian Islands, evidence from earthquake activity shows that the magma is generated in the upper part of the mantle; in continental regions, much magma is apparently formed in the lower crust. The reasons for the occurrence of these pockets of liquefied rock are not fully understood. Arching or cracking of the crustal rocks may locally reduce the pressure enough to allow rock to melt, or radioactive heating may raise the temperature to the melting point. However formed, the bodies of magma are mobile, and because they are less dense than the surrounding rocks, they tend to work their way upward toward the surface of the lithosphere. Their eventual cooling and solidification give rise to the igneous rocks.

Note that magma and igneous rock are not synonymous. Magma is the parent molten material, as it exists before any of its gases or other components have been lost in passing through the crust on the way toward the surface. An igneous rock is the hardened remains of a magma. Magma may be compared with paint in the unopened container; igneous rock, with the same paint after it has dried on the wall.

One major group of igneous rocks we term **extrusive,** because they are formed from material that is poured out, or extruded, on the surface of the lithosphere. We can see this happening today in volcanic regions. A second group, the **intrusive** igneous rocks, are formed when the rising magma stops, cools, and solidifies within the crust before reaching the surface. No one can watch while this happens, and indeed we would never see the products of igneous intrusion if it were not for erosion, which gnaws away at the overlying rocks and at length lays bare the intrusives for our inspection.

## 4-2 EXTRUSIVE PROCESSES AND PRODUCTS

The processes here to be described involve the eruption or outpouring of igneous material on the surface of the crust. They make a good starting place for a study of

**Figure 4-1**
Ropy basalt flow of Quaternary age, Craters of the Moon National Monument, Idaho. Wrinkles formed when partly solidified surface lava was dragged along by still-fluid lava underneath. (Photo by H. T. Stearns, U. S. Geological Survey.)

physical geology, because they produce brand new rocks, along with new landforms such as volcanoes. Furthermore, in many places the processes may be observed in operation and their products may be readily examined.

## Quiet Eruptions

A major type of extrusive activity is characterized by the quiet outpouring of highly fluid lava. (**Lava** is not a rock name, but an informal term for extruded molten rock; it is applied to both the flowing liquid and the resulting solid rock.) On cooling, the lava forms a dark, heavy, very fine-grained rock called **basalt,** which is made up chiefly of the minerals plagioclase and augite (Table 4-1). The upper surface of most flows is smooth and ropy, plainly showing the taffylike consistency of the lava from which the rock formed (Figure 4-1). The basalt in the upper part of many flows is more or less **vesicular,** that is, shot through with gas-bubble holes **(vesicles)** where gas from the molten lava escaped. Highly cellular basalt, honeycombed with vesicles, is termed **scoria.**

The great fluidity of molten basalt (Figure 4-2) means that it moves rapidly; some active flows have been clocked at as much as 20 miles per hour. Where the lava erupts from a central pipe or vent, it spreads widely and builds a cone with very gentle slopes. Such a structure is known as a **shield volcano** (Figure 4-3), from its resemblance in profile to the shield carried by ancient warriors; a more apt comparison would be with an upside-down saucer. Mauna Loa and its sister volcanoes of the Hawaiian group are shield volcanoes. Though subdued in profile, these are immense structures, tens of miles wide at the base, and standing as much as five miles above the floor of the Pacific Ocean on which they are built. Basaltic lava may periodically overflow the top depression, or **crater,**

**Figure 4-2**
Lava flows, Grand Canyon, Arizona. (Photo by J. R. Balsley, U. S. Geological Survey.)

of a shield volcano, but more commonly it issues from cracks on the sides of the cone. Each time this happens, another stream of lava is added to the volcano.

Quiet eruptions of basaltic lava may also come from **fissures,** or cracks in the crust, many miles long. Colossal amounts of basalt have been extruded in this way, not all at once but as a series of flows over a long time, separated by intervals when soil and even trees formed on one flow before it was buried by the next. In the Neogene Period, an area of some 200,000 square miles of mountainous country in what is now Washington, Oregon, and southern Idaho was inundated by flow after flow of basalt. These flows continued until finally the rough terrain was completely buried by lava more than 5,000 feet thick, and the regional feature we call the Columbia River Plateau was built up. The edges of the superimposed **plateau basalts** can be seen in the walls of the Columbia and Snake River canyons, at Grand Coulee (Figure 4-4), and in many other places. There are other famous examples of plateau basalts, notably in India.

### Explosive Eruptions

In many volcanic regions the lava erupted is stiff and viscous. It has a lower temperature than basalt, and tends to cool and crust over more rapidly. When such lava

**Figure 4-3**
A shield volcano. Extinct Mt. Giluwe, New Guinea.
Several composite cones have been built on the main
structure. (Photo courtesy of Division of Land
Research, CSIRO, Canberra, Australia.)

**Figure 4-4**
Columbia River plateau basalts near
Grand Coulee, Washington. Note
columnar structure, especially in lowest
bed. (Photo by John S. Shelton.)

67

Table 4-1    Simplified Classification of the Igneous Rocks

| Texture | Composition | | |
|---|---|---|---|
| | Quartz<br>Orthoclase<br>Biotite | Na-rich Plagioclase<br>Hornblende<br>Biotite | Ca-rich Plagioclase<br>Augite<br>Olivine |
| Coarse<br>Even-grained<br>Porphyritic | Granite<br>Porphyritic granite | Diorite | Gabbro |
| Fine<br>Even-grained<br>Porphyritic | Felsite<br>Porphyritic felsite | | Basalt<br>Porphyritic basalt |
| Glassy,<br>Vesicular | Obsidian<br>Pumice | | Scoria |
| Pyroclastic | Tuff<br>Volcanic breccia | | |

**Figure 4-5**
Light-colored volcanic tuff of Quaternary age lying on Permian sandstone, Jemez Mountains, New Mexico. Stream erosion, exemplified by the canyon in the foreground, has removed much of these rocks.

congeals in a volcanic vent, the vent is plugged; the pressure of steam within the magma below may then build up until explosion results. When the pressure buildup is regularly relieved, the explosive activity is mild and continuous. When the pressure is allowed to accumulate for a long time, violent explosions may result.

**Felsite** is a general name for the lava so produced (Table 4-1). Like basalt, it is very fine-grained in texture. Felsite is lighter in color, however, and also richer in silica; it has quite a different composition from basalt, being made up chiefly of biotite, orthoclase, and quartz. Many felsites

contain relatively large grains, or **phenocrysts,** of quartz or feldspar, scattered throughout the rock. The resulting texture is said to be **porphyritic.** A porphyritic texture indicates two stages of cooling in the magma: an earlier slow one, in which the phenocrysts grew, and a later one, in which the remainder of the magma was more rapidly cooled. Though porphyritic texture is especially characteristic of felsite, it is found in other varieties of igneous rock, intrusive as well as extrusive.

Flows of felsite may be interlayered with thin flows of a rock called **obsidian,** which has the same bulk composition as felsite but consists of glass rather than mineral grains (Table 4-1). Occasionally, felsitic lava may be so highly charged with gas that it erupts as an effervescent foam, blobs of which are blown from the vent. Suddenly chilled on contact with the air, this forms the highly porous, cellular glass called **pumice.** (Note that obsidian and pumice, being made of glass, are exceptions to the rule that rocks are aggregates of minerals.) Gobs of less frothy lava, blown from the vent while still liquid but solidifying before they fall to the sides of the cone, are termed **bombs.**

Explosive eruptions also produce large amounts of shattered rock fragments, collectively called **pyroclastics** (literally, "broken by fire"). In order of increasing particle size, these are termed **ash, cinders,** and **blocks.** When consolidated, volcanic ash forms an extremely fine-grained rock known as **tuff** (Figure 4-5). The coarser pyroclastics produce a lithified mixture of angular fragments called **volcanic breccia** (Figure 4-6).

**Figure 4-6**
Volcanic breccia, consisting mostly of jumbled angular blocks of light-colored pumice and shiny black obsidian. Thomas Range, Utah. (Photo by M. H. Staatz, U. S. Geological Survey.)

69

The cone that builds up around an explosive vent is thus a chaotic assemblage of lava flows and layers of pyroclastics. Such cones are **composite volcanoes** (Figures 4-7, 4-8, 4-9). The viscous nature of felsitic lava, and the heaping up of pyroclastics, give composite volcanoes much steeper slopes than shield volcanoes. The typical composite cone has a graceful profile that rises from a wide base to a tall peak. Most of the world's spectacular volcanic mountains are composite cones. Mayon in the Philippines, Fujiyama in Japan, Vesuvius in Italy, and Shasta and Rainier in this country are famous examples.

At some vents, the proportion of pyroclastics greatly exceeds that of lava, and a **cinder cone** is built up (Figure 4-7). Cinder cones are smaller than the other types mentioned, and have steep sides and a relatively large crater. Lava flows, if present, break through the sides of the cone. Cinder cones may build up with great rapidity. Parícutin in Mexico, which started "from scratch" in a cornfield in 1943, was 550 feet high within a month and more than 1,000 feet high at the end of its first year.

**Figure 4-7**
Cerro Negro volcano, Nicaragua, in eruption in 1968. A composite cone, the volcano was going through an explosive phase when the picture was taken and was building a cinder cone that reached a height of 800 feet. Ash clouds were shot as much as 10,000 feet above the cone. Earlier lava flows in the foreground appear highly fluid, but actually are of viscous blocky lava that moved only about one foot per hour. The active phase shown here lasted about 6 weeks. Cerro Negro erupted again early in 1971. (Photo by Mark Hurd Aerial Surveys, Inc.)

**Figure 4-8**
Mount Popocatepetl, a composite
volcano rising nearly 18,000 feet above
sea level some 60 miles east of Mexico
City. This volcano has been inactive
since 1922 and is undergoing erosion.
(Photo by S. E. White.)

**Figure 4-9**
Wall of South Inyo Crater, a small
extinct composite cone in eastern
California. Above loose sliderock can
be distinguished a thick flow of felsitic
lava (A), volcanic ash (B), and a bedded
unit consisting of mixed rock types (C).
A widespread pumice layer (arrow),
obscured by slopewash, lies at the
contact between B and C. (Photo by
C. D. Rinehart, U. S. Geological
Survey).

## Fiery Clouds and Calderas

A composite cone, long dormant or thought to be extinct, may suddenly come to life with extreme violence. An example is Mont Pelée, on the island of Martinique in the West Indies. After lying quiet for more than fifty years, in 1902 it became explosively active. Eventually it emitted a series of **fiery clouds** composed of burning gases and glowing bits of incandescent lava, immensely turbulent, extending to heights of 10,000 feet or more. The clouds were ejected horizontally or nearly so, traveling downslope at speeds as great as two miles per minute and obliterating all in their path. One fiery cloud destroyed the town of St. Pierre with its 28,000 inhabitants; the heat was intense enough to fuse a keg of nails and melt glass. Similar eruptions have been recorded at Katmai in Alaska in 1912, at a volcano in the Philippine Islands in 1951, and at Mt. Agung, on the Indonesian island of Bali, in 1963. Evidences of prehistoric fiery clouds are observable at a number of places, including the vicinity of Crater Lake, Oregon. The chief requirement for the production of a fiery cloud seems to be an exceptional buildup of gas pressure in the depths of the volcanic cone.

Explosive eruptions may be so severe that, in effect, a volcanic cone is disemboweled. Support for the top of the cone is removed, and the whole structure collapses inward. The result is a **caldera,** or huge pit, as much as several miles across, that marks the site of the former composite cone (Figures 4-10, 4-11). Crater Lake in Oregon is a famous example. Despite its name, the lake is not in a crater, but in a caldera. Some 6,500 years ago a composite volcano that stood on this spot discharged about 10 cubic miles of pumice in a series of fiery-cloud outbursts, and then fell in upon itself, leaving a pit six miles wide and 4,000 feet deep. The volcano Vesuvius

**Figure 4-10**
Crater of Great Sitkin volcano, Alaska, a composite cone. More than one-half mile in diameter, the crater was produced by an explosive eruption that blanketed the surrounding area (an island in the Aleutians) with pumice and other pyroclastics as much as 20 feet thick. A domelike new cone was growing when the picture was taken; heat has melted snow to form irregular contact. (Photo by F. S. Simons, U. S. Geological Survey.)

has been built in a caldera that was formed within historic times, when an earlier cone blew up and collapsed. It was ash from this explosion that buried the town of Pompeii. Many other imposing calderas are known.

The chief features of volcanism are summarized in Table 4-2.

| Table 4-2   Summary of Extrusive Processes and Products | | | |
|---|---|---|---|
| Type of Activity | Materials Ejected | Rock Types | Features Produced |
| Quiet | Gases; fluid lava | Mainly basalt | Shield volcanoes; plateau basalts |
| Explosive | Gases; viscous lava; bombs, ash, cinders, blocks | Felsite, obsidian; pumice, tuff; volcanic breccia | Composite volcanoes, cinder cones; calderas |

## Distribution of Volcanoes

We find that volcanoes tend to occur in the same regions as earthquakes and young mountain ranges. The region of most pronounced volcanic and seismic activity on the

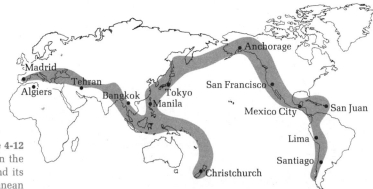

**Figure 4-12**
Most of the earth's volcanoes are in the shaded "circum-Pacific belt" and its western extension to the Mediterranean region. These belts are also the sites of more than 90 percent of the world's earthquakes (Section 14-7). Some of the reasons for this concentration of volcanic and seismic activity are indicated in Chapters 16 and 17.

earth's surface is a mountainous belt that encircles the Pacific Ocean (Figure 4-12). This belt extends from the southern tip of South America along the Andes, western Mexico, the coastal ranges of North America, the Aleutian Islands, Japan, the East Indies, and New Zealand, to recently discovered volcanic mountains trending across Antarctica toward South America. Another belt trends eastward from the Mediterranean to southern Asia. The conclusion seems obvious that these belts mark zones of weakness or unrest in the earth's crustal rocks.

## 4-3
## INTRUSIVE PROCESSES
## AND PRODUCTS

In the processes of igneous intrusion, magma moves upward into the outer crust but stops and solidifies short of the surface. The resulting igneous rocks are exposed to our view only if the cover of overlying rocks is removed by erosion.

### Observation and Deduction

Thus in dealing with intrusives we can see only the rocks, not the processes that formed them. Our efforts at understanding intrusive processes must be largely deductive, directed toward explaining events from the records they have left behind. We can start by making some simple observations, from which fairly obvious conclusions can be drawn.

1. Intrusive rocks are made of the same group of silicate minerals as the extrusives: quartz, the feldspars, biotite, hornblende, augite, olivine. Origin from a silicate melt analogous to lava is clearly indicated.
2. The texture of the rock in small intrusive masses, and in those that were injected near the surface, is fine-grained, whereas the texture of the rock in large masses that congealed at depth is medium- to coarse-grained. Since small shallow intrusions must have cooled much more rapidly than large deep ones, the conclusion is

74

plain that texture is a function of the rate of cooling: the slower the cooling, the coarser the texture. (Laboratory experiments substantiate this conclusion.) A change in the rate of cooling of a given magma from slow to rapid produces porphyritic texture.

3. Clearly the intrusives were once very hot, as they have baked and otherwise altered the rocks that they invaded (termed the **country rock**).

4. Since intrusives penetrate the country rock and cut across its stratification and other structures, by the law of sequence the intrusives must be younger than these rocks and structures.

## The Reaction Series

A fair idea as to what actually goes on in the making of an intrusive rock has been gained in the laboratory. An appropriate mixture of silicates is melted, so as to produce, in effect, a magma. Then the temperature is gradually lowered and the process of solidification is watched. One fact is soon apparent: the minerals do not all start to crystallize at once. The first solid to form is the silica tetrahedron (Section 2-4); thereafter, metallic ions join with the tetrahedron in diverse patterns, to form the igneous silicate minerals. Each mineral crystallizes in a specific temperature range. Once formed, the mineral crystals are not inert, but react with the remaining melt to form different minerals as cooling proceeds. For this reason, the order in which minerals form is called the **reaction series.** First of the ferromagnesians to appear is olivine; as the temperature declines, this mineral reacts with the remaining melt to form augite. With further cooling, hornblende and biotite appear. First of the feldspars to form is the calcium-rich variety of plagioclase; as the temperature declines, this grades into sodium-rich plagioclase, which is succeeded by the potassium-rich feldspar, orthoclase. The last mineral to form is quartz. The results of these studies show that a single magma may form a variety of igneous rocks. The processes of crystallization may be interrupted, for example, if early-formed crystals settle out of the melt and thus cease to react with it. Or a part of the melt at some stage in cooling may be drained away, or there may be an abrupt change in the temperature.

## Smaller Intrusive Masses

In the discussion of volcanoes, we referred several times to the pipe or conduit through which the magmatic material reached the surface. When a volcano has gone

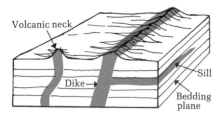

**Figure 4-13**
The smaller intrusive masses.

through its life cycle and becomes extinct, the forces of erosion are free to wear it away, and we find numerous places where this has happened. All that is then left is a spire or cylindrical tower of igneous rock, which represents the feeder pipe of the now removed volcano. Such features are termed **volcanic necks** (Figure 4-13). They generally stand high because the rock of which they are made is more resistant to erosion than the surrounding rocks. Shiprock, a well-known volcanic neck, stands 1,300 feet above the arid plains of northwestern New Mexico (Figure 4-14).

**Figure 4-14**
Shiprock, a volcanic neck in northwestern New Mexico. Note the radiating dikes. (Photo by John S. Shelton.)

Intrusive rocks often take a tabular (sheetlike) form. A **sill** is a sheet of intrusive rock that lies parallel with the structure of the enclosing rocks (Figures 4-13, 4-15); a **dike** is a similar body that cuts across this structure (Figures 1-13, 4-16, 4-17). Sills are commonly injected along the bedding planes of sedimentary rocks, or the foliation of metamorphic rocks (Section 2-5). Dikes, however, tend to follow fractures that transect these features. Occasionally a single intrusive sheet may take the form of both sill and dike, as shown in Figure 4-18. Sills and dikes range widely in size, from mere films a few inches or feet in length to masses hundreds of feet thick and miles in extent.

76

**Figure 4-15**
The dark band crossing the center of the view is a sill, which has been intruded into limestone. The bleached and baked zones above and below the sill attest its heat during emplacement. Mount Gould, Glacier National Park, Montana. (Photo by Eugene Stebinger, U. S. Geological Survey.)

**Figure 4-16**
Dikes of basalt intruding Precambrian rocks, northwestern Adirondack Mountains, New York. (Photo by E. L. Shay.)

**Figure 4-17**
Pegmatite dike in Precambrian gneiss, near Denver, Colorado. (Photo by J. R. Stacy, U. S. Geological Survey.)

**Figure 4-18**
The basaltic dike that cuts across the shales and sandstones passes into a sill in the upper right corner of the view. Note the columnar jointing in the dike. Grand Canyon; Colorado River in the foreground. (Photo by W. K. Hamblin.)

The type of rock in a sill or dike depends on composition of the parent magma and on the rate of cooling. Magma rich in iron and magnesium, and low in silica, produces basalt (Section 4-2) in small dikes or sills that cooled rapidly; in larger ones that lost their heat more slowly we may find coarser-grained rocks such as **diorite** (chiefly plagioclase, hornblende, augite, biotite) and **gabbro** (plagioclase, augite) (Table 4-1). Magmas low in ferromagnesian minerals and high in silica produce felsite (Section 4-2) in small intrusives, and in larger ones such coarse-grained equivalents of felsite as **granite** (mainly orthoclase, quartz, biotite [or hornblende]) and related rocks.

Dikes are especially abundant in areas of former volcanism, and indeed some volcanoes are fed by dikes as well as by central pipes. Three prominent dikes radiate from Shiprock (Figure 4-14). Dikes may also be associated with large intrusions.

### Sill or Buried Flow?

Suppose we find a place where the edges of some flat-lying stratified rocks are exposed, as shown in Figure 4-19. All the rocks are sedimentary except one layer, which is dark-colored igneous rock. How can we tell whether the igneous layer is a buried lava flow, which covered bed A, cooled and solidified, and then in turn was buried by bed B; or a sill, which was intruded between beds A and B after both had been formed?

**Figure 4-19**
Layer of dark igneous rock between beds of sedimentary rock.

First we should examine the dark igneous rock itself. Assume that we find this to be very fine-grained, and hence classify it as a basalt. Further, we find that the basalt is somewhat vesicular toward the top. These facts suggest rapid cooling and the evolution of gas; in other words, an extrusive origin for the rock. If we then discover that the upper surface of bed X is somewhat uneven, with traces of a buried soil, and that the bottom of bed B shows no effects of heat, we have to conclude that rock X was extruded as a flow. Its upper surface was exposed to weathering for some time before being blanketed by bed B. Under these circumstances, the sequence of formation of the three units would be A, X, B.

If, on the other hand, bed X is medium-grained in texture, without vesicles, we suspect slow cooling under cover. If the lower surface of bed B shows baking or other effects of heat, and if perhaps small dikes of rock X extend upward into B, then there is no doubt that X has been intruded as a sill. Under these circumstances, the three rock units formed in the order A, B, X.

This is an example of the way in which the geologist uses the evidence presented by the rocks in order to interpret their history.

**Batholiths**

Volcanic necks, sills, and dikes are dwarfed in comparison with the great masses of intrusive rock termed **batholiths.** These have several distinctive features, one of which is their size. The Sierra Nevada batholith in California, to name one that is far from the largest, is 400 miles long, 30 to 50 miles across, and more than 30,000 cubic miles in volume. A second characteristic of batholiths is that they consist of granite, or of granitelike rocks rich in quartz and feldspar. A third is that batholiths have no known floor: deep erosion in batholithic regions has revealed no underlying rock that is different from the granite of the intrusion itself. A fourth is their place of occurrence, which is in crustal belts that have been deformed by earth movements. Massive batholithic injections seem to be a part of the cycle of mountain building in such belts.

Most of our information about the history of batholiths comes from their marginal zones, where granite is in contact with country rock. These zones present many complexities. In some of them the granite abuts sharply against wall rock (Figure 2-9), but in others the contact is vague and gradational. Some contact zones cut indiscriminately across the structures of the country rock; others are more or less parallel to its structural trend, or "grain." Blocks of country rock are found in granite (Figure 4-20); they were apparently pried off the walls of the magma chamber and frozen in the granite when it solidified.

**Figure 4-20**
A fragment of metamorphic rock embedded in light-colored granite, Mt. Airy, North Carolina. The dark fragment is a piece of the country rock, carried to its present position by the granitic magma and frozen there when the magma cooled.

Commonly a wide zone adjacent to a batholith is much altered in composition and texture, by heat and emanations from the magma; these alterations constitute a form of metamorphism (Section 15-2). As might be expected, granite also penetrates the country rock in the form of dikes and sills. Around some batholiths we find dikes made of a rock called **pegmatite.** Most pegmatite has the composition of granite, but it is unusual in having an exceptionally coarse grain size. The interlocking grains of feldspar, quartz, and mica may average an inch, several inches, or a foot or more in diameter (Figure 4-21). This extremely coarse texture seems to be in conflict with what we said about small intrusive bodies being fine-grained because of rapid cooling; but there is an obvious explanation. Pegmatite dikes, being offshoots of large batholiths, are not the result of magma's being injected into cold rock, but into rock strongly heated by the adjacent batholith. Hence they do not cool much faster than the batholith does. The growth of large mineral grains is furthered by hot gases and solutions emanating from the parent intrusion.

Pegmatite dikes are notable mineral-hunting grounds. The large display specimens of feldspar, quartz, and mica that you see in laboratories and museums undoubtedly

Figure 4-21
Pegmatite, Black Hills, South Dakota. The hammer rests against a single crystal of a lithium-bearing mineral called spodumene; an even larger spodumene crystal is about a foot above the hammer. Remainder of the pegmatite is mainly massive quartz. (Photo by J. J. Norton, U. S. Geological Survey.)

came from pegmatites. In a few regions, these rocks also yield many rare and unusual minerals.

Batholithic magmas move into the country rock in part by pushing it aside and in part by loosening blocks of wall rock and incorporating them into the magma. Yet this does not seem to be the whole story. In some regions where batholiths and their host rocks have been deeply eroded, there is evidence that the host rocks have been "granitized," or altered to granite through partial melting in place, "soaking" by granitic fluids, or some sort of ionic diffusion from a magmatic source. This conclusion is unavoidable wherever sedimentary or metamorphic rocks gradually merge into the main granite of a batholith, and where traces of stratification or foliation can be detected in rock that is now clearly granite. Thus we must conclude that some batholiths are produced by massive intrusions of magma, and others by a change of preexisting rocks to granite in the solid state.

The question of how it is that large intrusive masses make room for themselves is only one of several fundamental problems. What, for example, is the source of magmatic heat? Precisely how does "granitization" take place? Why are some intrusives (and extrusives too) basaltic in composition, whereas others are granitic? In the present state of geological knowledge, the answers to these questions are largely speculative.

## Hot Springs and Geysers

A spring is a flow of underground water that emerges at the surface. If, in its subsurface migration, water passes through rocks heated by magma, it will naturally emerge at temperatures well above those normal for the surface. The resulting **hot springs** are common in areas of recent or current volcanism.

A **geyser** is a boiling spring that intermittently erupts water and steam. Instead of being allowed to circulate freely and lose its heat progressively, like the water in a teakettle or an ordinary hot spring, the water in a geyser is constricted by a highly irregular conduit or channel. The weight of the water that seeps into the conduit and fills it up produces enough pressure on the lower part of the column to keep it temporarily from boiling. Eventually, however, the water at the bottom of the conduit becomes superheated, a few steam bubbles throw some water out at the top, the pressure is slightly decreased, and the whole water column flashes into steam and erupts. The period of time between eruptions depends on how long it takes the conduit to fill again and the process to repeat itself.

Famous examples of both hot springs and geysers are those in Yellowstone National Park (Figure 4-22). The park is situated on a volcanic plateau built up of pyroclastics (the "yellow stone" exposed in the canyon walls) and flows of lava and obsidian. Although the region is no longer volcanically active, enough heat remains in the underlying rocks to affect the underground waters and generate abundant steam.

## Geothermal Energy

In some areas of one-time volcanic activity, wells can be drilled to yield steam, which is used in turbines to generate electricity. Although **geothermal energy** has been used in Italy since 1904, it was not until 1958 that the second commercial development took place, in New Zealand. Interest has developed rapidly since then, primarily because this source of energy is low in cost and relatively nonpolluting. Other countries now using geothermal energy include Iceland, Japan, Mexico, Russia, and the United States; at least six more countries are planning development.

4-4
**BY-PRODUCTS OF IGNEOUS PROCESSES**

Figure 4-22
Old Faithful Geyser, Yellowstone
National Park. The low mound consists
of mineral matter precipitated
from the hot water.

The only installation in the United States is at "The Geysers," about 90 miles north of San Francisco, where a power plant (Figure 4-23) generates about 400,000 kilowatts per year (compared with nearly 10 million kilowatts from the company's conventional plants). The "Geothermal Steam Act," which became law in December 1970, is expected to stimulate exploration for additional sources in the western United States, where many areas of volcanic hot springs are known. (The geysers and hot springs of Yellowstone are off limits for commercial development because of the area's national-park status.) Among the requirements that must be met before a large-scale power plant can be built are (1) a temperature in the underground steam source, or reservoir, of at least 200° C (392° F); (2) a low content of dissolved mineral matter in the water and steam; and (3) a heat source sufficient to maintain reservoir temperatures for 20 to 00 years

### Economic Products

The two most abundant kinds of igneous rock, namely basalt and granite, are locally of much economic importance. Large tonnages of basalt are quarried, crushed,

**Figure 4-23**
Wells drilled for geothermal steam at
The Geysers, Sonoma County,
California. The wells are being tested
for temperature and pressure before
being put into service. The steam
comes out of the ground at about 450° F
and 150 pounds per square inch; it
then enters the turbines at
approximately 350° F and 100 psi.
Power plant beyond wells. (Photo
courtesy of Pacific Gas and Electric
Company.)

screened, and mixed with cement to form concrete for
highways, foundations, and other structures. Certain
masses of granite that have uniform texture and pleasing
appearance (Figure 4-24) are cut and polished for architectural stone and for monuments and memorials.

Many metallic elements of great industrial value—copper, lead, zinc, chromium, uranium, and numerous others—
are widely distributed in the rocks of the earth's crust,
but are generally in extremely small amounts. Only
exceptionally are such elements concentrated into **ore
deposits,** or bodies of rock from which metals can be
obtained commercially. Igneous processes, especially
those involved in batholithic intrusion, have been instrumental in forming ore deposits of various types.

**Figure 4-24**
Wall of a granite quarry, northern
Vermont. The rock is a fine-grained
gray granite that is uniform throughout.
Small dikes are present at center
background and at right. (Photo
courtesy of Rock of Ages Corporation.)

Concentration of heavy metallic minerals within a cooling magma is the process that is believed to have produced Europe's largest deposit of iron ore (the magnetite of northern Sweden), the great nickel deposit at Sudbury, Ontario, and the chromite ores of South Africa. The process is analogous to that which takes place in a smelter, where the molten metal sinks and the lighter silicate fraction (the slag) rises.

We learned earlier that the country rock adjacent to batholiths is likely to be much altered by heat and by gaseous fluids that emanated from the magma. A feature of this alteration is often the formation of ore deposits, particularly at those places where the country rock is limestone. This rock, which consists chiefly of the mineral calcite, is especially susceptible to **replacement,** or the simultaneous removal of one type of material and the substitution of another, on a volume-for-volume basis. (Replacement is not confined to zones marginal to batholiths; we will meet it again in other connections.) Replacement ore deposits, then, occur chiefly in limestones adjacent to intrusive masses. The common metallic minerals produced are simple sulfides. Ore deposits of copper, lead, zinc, and iron have been formed in this way. So have certain nonmetallic minerals of value, such as talc and garnet.

The famous deposits of gold in the Mother Lode of California, of copper at Butte, Montana, and of uranium at Great Bear Lake, Ontario, all occur in still a different type of deposit, namely in **veins.** A vein forms when an open crack or fissure becomes partly or completely filled with mineral matter, generally precipitated from hot watery solutions of magmatic origin. Though veins are commonest in the country rock near batholiths, in places they are found extending well downward into the batholithic rock itself, showing that, at the time of vein formation, the magma was consolidated in its outer parts. Veins are by no means made up entirely of metallic minerals; the most common vein mineral is valueless quartz. Some veins yield minerals containing nonmetallic elements of value, for example fluorine.

It should not be concluded that all batholiths are characterized by ore deposits; far from it. Whether such deposits are formed at all depends largely on the composition of the magma and on the nature of the wall rock. Since vein deposits in particular seem to form in and above the upper parts of intrusive masses, the depth to which a given batholith has been eroded has a direct bearing on the occurrence of such deposits. Finally, whether we are able to find all the ore deposits that exist depends on how intelligently we can solve the numerous geologic puzzles connected with their occurrence.

Bodies of molten rock, or magma, which form in the upper mantle or lower crust, give rise to fine-grained extrusive and coarser-grained intrusive rocks. Extrusive rocks include basalt, felsite, obsidian, and pumice, together with tuff and other pyroclastics. Features produced are plateau basalts, shield and composite volcanoes, cinder cones, and calderas. A mountainous belt characterized by volcanoes and earthquakes borders the Pacific Ocean.

Intrusive rocks form more slowly, through progressive crystallization of minerals with falling temperature, in an order known as the reaction series. Small intrusive bodies include dikes, sills, and volcanic necks; the largest intrusives are great masses of granite, called batholiths, with dimensions in the scores or hundreds of miles. Batholithic intrusions bring with them so much heat that they generally metamorphose the country rock. They may also send out dikes and other offshoots of a very coarse-grained granitic rock called pegmatite. The margins of some batholiths are sharp, but those of others are gradational and seem to reflect the conversion of country rock to granite.

By-products of igneous activity include hot springs and geysers, geothermal steam, and several types of ore deposit. The latter include deposits produced by the settling-out of heavy metallic minerals within the magma; by replacement of the country rock, especially limestone; and by the filling of veins.

**4-5 SUMMARY**

## SUGGESTED READINGS

Ernst, W. G. 1969. *Earth Materials,* chap. 5, pp. 92–109. Englewood Cliffs, N. J.: Prentice-Hall. (Paperback.)
A good presentation on the formation of igneous rocks and the features produced by igneous activity.

Keller, W. D. 1969. *Chemistry in Introductory Geology,* 4th ed., pp. 1–38. Columbia, Mo.: Lucas. (Paperback.)
A simplified but sound approach to the chemistry of materials that make up the crust of the earth, including magma and its products.

Spock, L. E. 1962. *Guide to the Study of Rocks,* 2nd ed., chap. 4, pp. 47–87. New York: Harper & Row.
An authoritative and relatively detailed coverage of the many different types of igneous rocks.

Tuttle, O. F. 1955. The origin of granite. *Scientific American,* vol. 192, no. 4, pp. 77–82. (Offprint No. 819. San Francisco: Freeman.)
Discussion of the problem of whether granite is formed by "granitization" or crystallization from a melt.

Williams, H. 1951. Volcanoes. *Scientific American,* vol. 185, no. 5, pp. 45–53. (Offprint No. 822. San Francisco: Freeman.)
A general discussion of volcanoes including their effect on man and the geologic activity indicated by the mountains they build.

# 5 WEATHERING

5-1   New Minerals from Old

5-2   The Agents of Weathering

5-3   Kinds of Weathering

5-4   Differential Weathering

5-5   Regolith and Soil

5-6   Summary

A freshly congealed flow of basalt is made up of plagioclase and augite that formed at a temperature of several hundred degrees as the flow gradually cooled. If this temperature were maintained, these minerals would persist indefinitely without change. But cooling continues, to stop only when ordinary surface temperatures are reached. Thereafter, dew condenses on the rock, moist air penetrates its vesicles and crevices, and rain water percolates through it. Neither plagioclase nor augite can long withstand this drastic change from the conditions under which they were produced. Both minerals, where directly exposed to air and moisture, start to alter into different compounds (minerals) that are chemically stable under the new circumstances.

On a basalt flow a few decades old, we find pockets of yellow-brown clay in the low places, where bunch grass or hardy shrubs have become rooted. Much older flows are completely blanketed with clayey soil that supports a continuous cover of vegetation. In each instance the clay was clearly derived from alteration of basaltic rock; yet analysis of the clay shows no trace of either plagioclase or augite. Some of the elements of these complex silicates have been rearranged into entirely different minerals. Other elements are present only in greatly diminished amounts. Some of the quantity present in the original basalt minerals must have been partly removed during the process of alteration.

Similarly with granite. The quartz, orthoclase, and biotite of this rock, formed at high temperatures under a thick rock cover, remain unchanged so long as their environment is undisturbed. But sooner or later erosion removes the protective cover of overlying rock, the granite is gradually brought into the zone of weathering, and its minerals alter to minerals that are at home under conditions at or near the surface. In areas where granite has been deeply weathered, the soil grades downward into crumbly, decayed rock, and this in turn into fresh unaltered granite at a depth of many feet.

Logically enough, the resistance that the igneous minerals present to the surface agents of alteration is the reverse of the order of formation of these minerals in the reaction series (Section 4-3). The minerals formed at the highest temperatures, such as olivine, augite, and calcium-rich plagioclase, are least resistant to surface attack, and alter rapidly; those produced at lower temperatures are less prone to attack and more resistant. Quartz, the igneous mineral that forms at the lowest temperature in the reaction series, is highly resistant to change under surface conditions.

**Weathering** is the alteration that rocks undergo by exposure to air, water, and organic matter. The examples cited above are igneous rocks, but of course sedimentary

**Figure 5-1**
Effects of weathering on an old gravestone, Middletown, Connecticut. Stone was set up in 1802. The rock is a sandstone of quartz and feldspar grains, cemented by calcite and hematite. Disintegration has resulted mainly from frost wedging and the mechanical effects of chemical weathering, especially alteration of the feldspar. (Photo by G. F. Matthias.)

and metamorphic rocks are also susceptible to weathering. Some varieties of rock are easily weathered and others are highly resistant, but none is immune. Everyday evidences of weathering are the blurred lettering on old gravestones (Figure 5-1), the roughened or etched surface of cut stone exposed to the elements on the outside of buildings, and the crumbling of foundation blocks near the ground where moisture soaks in and remains for periods of time.

## 5-2
## THE AGENTS
## OF WEATHERING

▪ *Oxygen.* The atmospheric gas oxygen is chemically active and combines readily with a number of elements. Iron, freed during the weathering of ferromagnesian silicates such as augite and biotite, is especially likely to be oxidized. Two of the common minerals produced are the oxides hematite and limonite.

▪ *Carbonic acid.* Small amounts of carbon dioxide gas ($CO_2$) in the air unite with rain water in the following simple reaction:

$$CO_2 + H_2O \longrightarrow H_2CO_3 \text{ (carbonic acid)}.$$

Carbonic acid is a very weak acid, which over long periods is effective as a solvent on certain minerals. It tends to unite with calcium, magnesium, sodium, and potassium, but instead of forming new minerals, such reactions generally yield soluble compounds that are removed in percolating waters.

▪ *Water.* Besides entering into the formation of carbonic acid, water combines directly with some compounds to produce hydrates. Thus in the zone of weathering the mineral anhydrite $CaSO_4$ may combine with water to form a more stable mineral, gypsum, $CaSO_4 \cdot 2H_2O$. More commonly, water dissociates into the positively charged hydrogen ion, $H^+$, and the negatively charged hydroxyl ion, $OH^-$. In these forms it enters into many weathering reactions, especially those that alter the feldspars.

▪ *Ice.* Water has still another role to play. Unlike most substances it expands when it changes from liquid to solid. The change to ice involves a volume increase of about 9 percent, and the force of this expansion is tremendous. The formation of ice is a locally important agent of weathering. Its effects are purely physical, not chemical like those of the other agents mentioned.

▪ *Plants and animals.* Rocks may be split apart by the roots of trees and bushes, or etched and roughened by the growth of lichens on the surface. Burrowing animals may bring bits of partly altered rock to the surface and thus expose them to additional weathering. The work of plants and animals is difficult to measure, but it is locally impressive and in the aggregate is probably large.

Two kinds of weathering may be distinguished: **disintegration,** or mechanical weathering, and **decomposition,** or chemical weathering. For convenience we discuss them separately, but we should recall that such neat divisions are man-made rather than natural. In nature, the two kinds of weathering almost always go on simultaneously.

5-3
KINDS OF
WEATHERING

### Disintegration

Any natural agent at or near the surface that mechanically breaks up rock is a part of disintegration. A significant kind of rock breakage is the process that we call **frost wedging.** When an irregular crevice or crack in rock becomes filled with water and the temperature drops below freezing, the upper part of the water column freezes first and may seal the crevice tightly. If the rest of the water then freezes, expansive stresses are produced that can be relieved only by pushing the walls of the crevice apart, thus further splitting the rock. Frost wedging is especially important on high mountains, above the timberline, where bare rock is continually exposed and where warm days may alternate with freezing nights (Figure 6-5). Cliffs, pinnacles, and canyon walls in high country are commonly the sites of intense frost wedging.

When the outer part of a rock is repeatedly soaked with rain water, some of its minerals may alter to minerals containing $H_2O$ or $OH^-$. Such hydrous minerals invariably take up more space than the unweathered minerals from which they are produced. The resulting expansion sets up stresses in the near-surface part of the rock that becomes wet. Although these stresses are chemical in origin, they are purely mechanical in effect; the resulting disintegration may be termed the **mechanical effects of chemical weathering.** Thin flakes and scales are loosened from the surface, typically of massive rocks such as granite. Where water attacks such rocks along intersecting sets of fractures, the mechanical effects of hydration tend to be concentrated on the edges and corners of the fracture blocks. In time, angular blocks of fresh rock may be converted into rounded masses consisting of concentric shells of decayed rock, arranged somewhat like the layers of an onion (Figure 5-2). The splitting-off of thin curved shells of weathered rock is termed **exfoliation.** It is one of the mechanical effects of chemical weathering.

Flat or gently curved sheets of rock, superficially similar to those produced in exfoliation but very much larger,

Figure 5-2
Exfoliation in a dark igneous rock of basaltic composition, Deep River Triassic basin, North Carolina. (Photo by Stephen Tysinger.)

characterize some outcrops of massive rocks such as granite. **Sheet structure,** as this feature is known, involves tabular masses of rock that range in thickness from a foot or two to more than 25 feet. In quarries, the sheets can be seen to become thicker with depth; the partings between rock sheets tend to parallel the bedrock surface. Most geologists conclude that sheet structure is caused by "unloading," the relief of confining pressure owing to removal of overlying rock by erosion. The uncovered rock expands toward the nearest exposed surface, hence the parallelism between sheet structure and surface of the bedrock.

The various processes of disintegration, like the convict on the rock pile, are simply engaged in "making little ones out of big ones." Their geological significance lies chiefly in the fact that rock material broken into pieces has a larger surface area than rock in large masses, and therefore is much more readily attacked by chemical processes. Thus we may regard the processes of disintegration as mainly preparatory for those of decomposition.

### Decomposition: Granite

Since the processes of decomposition are chemical in nature, they must be considered as they affect individual minerals (chemical compounds). The decomposition of a rock is really the total effect of the decomposition of its constituent minerals. As an example, let us consider the changes that affect a granite being decomposed in the zone of weathering. The main minerals of granite are quartz, orthoclase, and biotite.

Quartz ($SiO_2$) is one of the most chemically stable of all minerals. Thus the quartz grains in the granite are little affected by decomposition. As the other minerals become altered and the rock decomposes, the quartz simply remains as loose grains (sand).

Orthoclase, the main constituent of granite, is a silicate of potassium and aluminum. It is not especially stable under weathering conditions. Its decomposition will go about as indicated in the following equation:

$$2KAlSi_3O_8 + H_2CO_3 + H_2O \longrightarrow Al_2Si_2O_5(OH)_4 + K_2CO_3 + 4SiO_2.$$

orthoclase          a clay mineral    potassium    silica
                                      carbonate

Note that only three substances are involved in the reaction to the left of the arrow: the orthoclase itself, carbonic acid, and water. Of the three products to the right of the arrow, only one, the clay mineral, remains in place and accumulates. Potassium carbonate and silica are removed in solution, in water that percolates into the ground or trickles away on the surface. (The $SiO_2$ in dissolved form, produced in the reaction, is not quartz. Quartz is in no way involved in the reaction.)

Biotite is a more complex silicate, containing iron and magnesium as well as potassium and aluminum. It decomposes in the same general way as orthoclase, but with

**Figure 5-3**
Deeply weathered granite. The rounded forms have been produced by exfoliation. Grains of quartz and partly decomposed feldspar form the granular material in the foreground and surrounding the granite masses. Joshua Tree National Monument, California. (Photo by Mary R. Hill.)

one difference. Since biotite contains iron, and this element readily oxidizes, we find that the decomposition of biotite produces, in addition to clay and soluble compounds, some iron oxide, let us say in the form of limonite, $Fe_2O_3 \cdot nH_2O$. Like the clay, this mineral is insoluble and accumulates. Since even a little limonite is a strong pigment, its effect is to color the clay yellow-brown.

In sum, then, the thorough decomposition of a granite yields two groups of products. The first is an insoluble residue, which remains at the site of weathering (Figure 5-3). This residue includes the unaltered grains of quartz, the clay minerals from alteration of orthoclase and biotite, and the limonite from alteration of biotite. It is, in short, a yellowish brown sandy clay. The second group consists of compounds in solution, which are removed from the scene in water and eventually find their way into streams and the sea. The chief soluble compounds are silica, potassium carbonate, and magnesium bicarbonate.

## Decomposition: Limestone

As another example we may take a rock that is simpler in composition, namely a slightly impure limestone. The main mineral of limestone is calcite, $CaCO_3$; let us say that this limestone consists of 90 percent calcite. The remaining 10 percent is clay, distributed throughout the rock.

Calcite combines with carbonic acid in a simple reaction, as follows:

$$CaCO_3 + H_2CO_3 \longrightarrow Ca(HCO_3)_2.$$
$$\text{calcite} \qquad\qquad \text{calcium bicarbonate}$$

The product of this reaction, calcium bicarbonate, is readily soluble in surface waters. Therefore 90 percent of the original rock is dissolved and taken away from the site of weathering in the waters that drain it. Clay minerals, we have seen above, are themselves the products of chemical weathering. Thus under surface conditions the clay is stable; it simply accumulates as an insoluble residue. In many places we find such residual clay as a surface layer or blanket lying on still undissolved limestone (Figure 5-4).

## End Products of Decomposition

Like granite and limestone, most rocks on decomposition yield two types of end products: soluble compounds that are removed, and insoluble minerals that accumulate. To understand the general direction that decomposition of a given rock will take, we must consider it mineral by mineral, noting the reaction of each mineral with water,

oxygen, and carbonic acid, the chief agents of chemical weathering. We have already indicated some general tendencies of common minerals in the zone of weathering:

1. Quartz tends to persist unaltered.
2. Orthoclase (and its sister feldspar, plagioclase) decompose to clay and soluble compounds.
3. Biotite, and other ferromagnesian minerals, yield clay, iron oxides, and soluble compounds.
4. Calcite dissolves, especially if abundant carbon dioxide is present to provide carbonic acid.
5. Clay, once formed, tends to persist.

The great significance of water is apparent from the preceding paragraphs. Another factor that promotes chemical processes, in nature as in the laboratory, is high temperature. Hence rock decomposition proceeds fastest in the tropics. Under the steady soaking of warm tropical rains, rocks are commonly decayed to depths of many feet; here even clay may be partly dissolved, leaving simpler minerals of maximum stability. Decomposition is steady, though less rapid, in humid climates of the temperate zones, as in the eastern United States. In arid and semiarid regions, though temperatures may be high, decomposition is very slow because of lack of moisture. Limestone, for example, in humid regions forms lowlands or gentle slopes mantled with residual clay, soil, and vegetation, whereas in desert regions limestone is highly resistant to weathering and stands out as cliffs or ridges of bare rock. In places that are either very dry or very cold, decomposition is at a minimum; it may be less significant than mechanical disintegration.

**Figure 5-4**
Limestone quarry, south central Indiana. The bedrock is blanketed by red clayey soil, the residue of layers of rock that have been removed in solution. This residual clay also fills deep channels in the bedrock, three of which are shown in the quarry wall. The rock, the Bedford limestone of Mississippian age, is widely used as a building stone. (Photo courtesy of Indiana Geological Survey.)

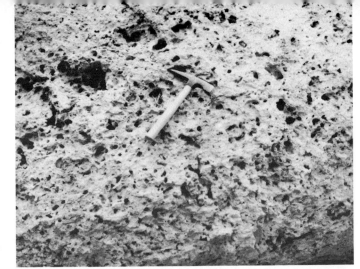

**Figure 5-5**
Solution-pitted limestone.

**Figure 5-6**
Differential weathering in
Cretaceous sandstone,
west of Steamboat
Springs, Colorado.
Retaining wall at lower
left gives scale. (Photo by
J. H. Rathbone.)

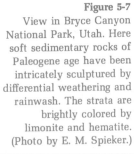

**Figure 5-7**
View in Bryce Canyon
National Park, Utah. Here
soft sedimentary rocks of
Paleogene age have been
intricately sculptured by
differential weathering and
rainwash. The strata are
brightly colored by
limonite and hematite.
(Photo by E. M. Spieker.)

Frost, carbonic acid, and the other agents of disintegration and decomposition unfailingly attack the least resistant parts of an exposed rock. Shaly or weakly cemented sandstone beds, fractured zones in igneous rock, the most soluble parts of limestones—all are selectively attacked. The resulting rock surfaces are typically etched, pitted, and uneven (Figures 5-5, 5-6). This type of surface is said to be produced by **differential weathering.**

The impact of raindrops, and the washing action of the water as it trickles over the surface immediately after falling, are termed **rainwash.** Coupled with rainwash, differential weathering produces striking effects in places where soft, easily sculptured rock is exposed to the atmosphere (Figure 5-7).

## 5-4 DIFFERENTIAL WEATHERING

Loose materials that rest on solid rock are inclusively termed **regolith** (*rhegos* is a Greek word meaning blanket). Clay, sand, and other materials derived from rock decomposition in place may be termed **residual regolith** (Figures 5-8, 5-9). The nature and amount of residual regolith depend on the nature of the rock from which it was produced, the amount of moisture available, and the time during which decomposition has gone on. **Transported regolith** (Figure 5-10), as the name indicates, is loose material brought into an area from elsewhere by natural agents. Stream gravels and deposits left by glaciers are examples. Transported regolith is not a product of weathering of the local bedrock, and commonly has no genetic relation to it. Many types of transported regolith have undergone chemical weathering since they were laid down; the agents, processes, and products are much the same as those involved in the decomposition of solid rock.

**Soil** is that part of the regolith that supports plant life. As such, it is man's most valuable natural resource. The systematic study of soils forms the subject matter of soil science, or **pedology.**

The main constituents of soil are clay, sand, and other mineral matter derived from the weathering of rock or regolith, plus decomposed organic matter, or **humus.** Such solid material makes up about half of a good soil; the other half consists of pore space, which contains air and moisture. The water that percolates downward through the soil is a weak solution of carbonic acid, and to it are added acids from decaying vegetation. Thus soil water is a relatively strong solvent. It tends to remove the more soluble compounds (calcium carbonate, for example) from the uppermost soil, and redeposit some of them a few inches to a few feet in depth below the surface. At the same time, the downward-moving water tends to wash

## 5-5 REGOLITH AND SOIL

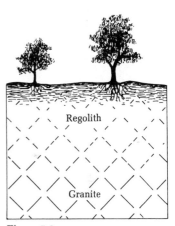

**Figure 5-8**
Residual regolith developed by decomposition of granite. The boundary between regolith and rock is not sharp. (Compare Figure 5-3.)

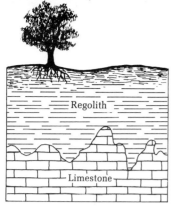

**Figure 5-9**

Residual regolith resulting from decomposition of limestone, lying on a solution-pitted, rough bedrock surface. (Compare Figure 5-4.)

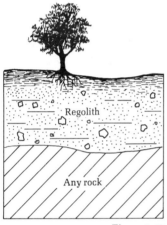

**Figure 5-10**

Transported regolith, unrelated to the underlying rock. Its soil has developed by weathering of minerals in the regolith since it was deposited.

some of the finest clay particles from the surface zone and deposit them with the precipitated mineral matter below.

Typically, then, soil is a zoned material; soil scientists refer to the zones as **horizons.** The assemblage of horizons, from the surface down to the unaltered parent material, is a **soil profile.** At the very surface is the O horizon, consisting of organic debris, more or less decomposed or matted. Then comes the A horizon, a light-colored zone that is relatively sandy because of removal of the finer clay particles. Next below is the B horizon, a darker zone that tends to contain compounds dissolved from above, together with washed-down clay. Hematite and limonite, the oxides of iron, are common in the B horizon. The lowest, or C, horizon is the parent material, the rock or regolith from which the soil developed. Subdivisions of the A, B, and C horizons are commonly recognized.

Several factors are significant in the formation of soil. The *parent material* provides the mineral matter with which the process begins. The *climate,* especially moisture and temperature, affects the type and vigor of chemical reactions. The *activity of organisms* produces decomposed organic matter and the acids derived from it. *Topography,* or character of the land surface, has a strong bearing on how much rainfall runs off and how much soaks in. As there is much variation in all these factors on the earth's land surface, it is not surprising that a wide spectrum of soil types is recognized.

**5-6 SUMMARY**

Weathering is the alteration that rocks undergo as a result of exposure at the earth's surface. The agents of weathering are oxygen, carbonic acid, water, ice, and organisms. Mechanical weathering, or disintegration, includes frost wedging, a result of alternate freezing and thawing of water in rock crevices; and exfoliation, a mechanical result of chemical weathering, which rounds the corners and edges of masses of fractured rock. Thick curved sheets of rock may superficially resemble the products of exfoliation, but are believed to form as a result of rock expansion on unloading by erosion.

Chemical weathering, or decomposition, acts on a rock's constituent minerals. Among the common minerals of igneous and metamorphic rocks, quartz is stable and resistant; feldspar is altered to clay and soluble compounds; biotite and the other ferromagnesian minerals are altered to clay, soluble compounds, and iron oxides.

Calcite, the chief mineral of limestone, may be wholly removed in solution, leaving only a residue of clay or sand that was in the unweathered rock. In weathering, the amount of water available and the prevailing temperatures are highly important. Limestone, for example, though readily dissolved in humid climates, forms resistant cliffs and ridges in desert climates. The etching and pitting of rocks of uneven resistance is termed differential weathering.

Regolith, the loose material that rests on solid rock, may be of two types: residual, formed in place by weathering of the bedrock; or transported, brought in from elsewhere and thus unrelated to the rock on which it lies. Soil, that part of the regolith that supports plant life, consists of mineral matter and partly decayed organic material, or humus. Soils typically show a soil profile, a vertical succession of zones of weathering, called horizons. Parent material, climate, organic activity, and topography are all important factors in the formation of soil from rock or regolith.

## SUGGESTED READINGS

Basile, R. M. 1971. *A Geography of Soils.* Dubuque, Iowa: W. C. Brown. 152 pp. (Paperback.)
A good ready reference for soils that includes discussion of soil formation and classification, world distribution of soils, and a glossary of soil science terms.

Bloom, A. L. 1969. *The Surface of the Earth,* chap. 2, pp. 16–39. Englewood Cliffs, N.J.: Prentice-Hall. (Paperback.)
Good coverage of mechanical and chemical weathering, the role of climate in weathering processes, and the development of soil.

Keller, W. D. 1969. *Chemistry in Introductory Geology,* 4th ed., pp. 39–65. Columbia, Mo.: Lucas. (Paperback.)
An authoritative and fairly complete treatment of the chemical weathering of rocks.

Kellogg, C. E. 1950. Soil. *Scientific American,* vol. 183, no. 1, pp. 30–39. (Offprint No. 821. San Francisco: Freeman.)
Discusses the development and evolution of soil materials and the tremendous influence of soil on the life of man.

McNeil, M. 1964. Lateritic soils. *Scientific American,* vol. 211, no. 5, pp. 96–102. (Offprint No. 870. San Francisco: Freeman.)
Lateritic soils, which are common in the tropics, present an obstacle to increasing food production in these areas because they are deficient in organic matter and soil nutrients.

Tuttle, S. D. 1970. *Landforms and Landscapes,* chap. 2, pp. 9–24, Dubuque, Iowa: W. C. Brown. (Paperback.)
A concise presentation of weathering, with emphasis on the response of earth materials to changes in environment and the importance of weathering to human existence.

# 6 DOWNSLOPE MOVEMENTS

6-1  Slopes and Gravity

6-2  Types of Downslope Movement

6-3  Geological Significance of Downslope Movements

6-4  Downslope Movements and the Works of Man

6-5  Summary

An absolutely horizontal land surface is a rarity in nature; almost everywhere the ground surface slopes. Angles of slope are very gentle in plains regions, more pronounced in hilly terrain, and steep to vertical in the mountains. Since the force of gravity is universally present, and many earth materials beneath slopes are not inherently very strong, it follows that these materials tend constantly to move downward under their own weight. In this chapter we inquire into the resulting **downslope movements.**

The character of the material, and the angle of slope at which it lies, are obviously significant factors. Solid rock, if not cut by fractures or other planes of weakness, is stable even on vertical faces. Rock that is fractured, thinly stratified, or inherently weak yields much more readily to the influence of gravity. Regolith is ordinarily weaker than rock and may move down slopes that are almost imperceptible to the eye.

A highly important factor in promoting downslope movements is water. Here we are not referring to streams, but to water that soaks into rock and regolith and saturates them. The importance of water is twofold: it increases the weight of the mass, and it separates particles from one another, reducing cohesion and allowing easy sliding or rotation. A mass of clayey soil that is completely dry is compact and fairly strong, but let it be saturated with water and it becomes a heavy, plastic mass whose equilibrium on a sloping surface can only be precarious. Downslope movements are especially common after prolonged rains or spring thaws.

## 6-1 SLOPES AND GRAVITY

The many possible combinations of slope angle, earth material, degree of water saturation, and type and rapidity of motion produce a wide variety of downslope movements under gravity. For our purposes we will consider all such movements under these four headings:

## 6-2 TYPES OF DOWNSLOPE MOVEMENT

1. Sudden movement of rock masses: rockslides, rockfalls.
2. Slower but perceptible movements, mainly of regolith: mudflows, debris flows, slump.
3. Imperceptibly slow movements of regolith: creep, solifluction.
4. Formation of talus masses.

The general term **landslide** may be applied to any perceptible downslope movement of earth materials.

### Sudden Movements of Rock Masses

A few minutes before midnight on August 17, 1959, an eight-state area in the western United States was shaken by one of the strongest earthquakes ever recorded in this country. The shocks were most intense in southwestern Montana, especially along the Madison River west of Yellowstone National Park. The cause of the shock waves was the abrupt displacement of bedrock along two faults (Sections 14-5, 14-7). The earth buckled and cracked; buildings collapsed, and considerable stretches of highway were damaged beyond repair. The waters of Hebgen Lake, behind a dam in the canyon, sloshed back and forth as though in a pan being jostled and tilted. The dam was cracked and weakened, but did not give way. Had the earthquake occurred in a densely populated area, the destruction and loss of life would have been catastrophic.

**Figure 6-1**
Madison Canyon rockslide of 1959. The slide started at the left edge of the view, 1,300 feet above the valley floor. About 90 acres of land, to a depth of as much as 150 feet, moved down and across the canyon and some 400 feet up the opposite side. The slide dammed the Madison River and created the lake in the foreground. When this picture was taken, engineers were cutting a channel (white strip) through the slide material to allow the lake to drain. (Photo by J. R. Stacy, U. S. Geological Survey.)

At first the only downslope movements were those of boulders, which were jarred free and crashed down into the canyon. Shortly after the main shock, however, some 40 million cubic yards of rock, weighing at least 80 million tons, broke loose from high on the south wall of Madison River Canyon and slid abruptly down to the valley floor. Here it came to rest as a mass of shattered debris 200 to 400 feet thick, which buried a campground, killing 28 persons, and formed a dam across the river (Figure 6-1).

Later investigations have shown that the bulk of the slide is composed of metamorphic rocks rich in mica. These rocks, which formed the upper part of the slope before the slide occurred, were deeply weathered and very soft. They were held in place by a bed of dolostone that extended along the lower part of the slope and acted as a natural retaining wall. The earthquake shocks caused the dolostone to fracture, allowing the softer rock above it to slide into the canyon. Blocks of dolostone broke free and were carried down the slope, across the canyon floor, and many feet up the opposite side, where they can now be seen protruding from the chaotic mass of broken metamorphic rock that makes up most of the slide mass.

An occurrence of this type is a **rockslide.** The one at Madison Canyon is by no means unique. In 1925, there was a large rockslide in the valley of the Gros Ventre River east of Grand Teton National Park, Wyoming. Here a thick bed of sandstone slid downslope on an underlying bed of water-saturated clayey shale. The slide mass was somewhat larger than that at Madison Canyon. It descended about 2,000 feet, crossed the valley, and moved about 350 feet up the opposite side. Many other slides, large and small, are known.

The instantaneous descent of a rock mass from a steep or vertical face is a **rockfall.** There has been a long series of famous and destructive rockfalls. A spectacular example occurred on October 9, 1963, when some 300 million tons of rock fell from a mountainside into the reservoir behind the Vaiont Dam in northern Italy. The dam did not fail, but the slide material displaced about one-third of the water, which swept over the dam and down the valley for miles, destroying everything its path. Over 3,000 lives were lost. The reservoir is now filled with rock material for more than a mile back from the dam, to a height of as much as 575 feet above reservoir level. The rockfall lasted no more than 15 to 30 seconds.

Hundreds of unreported big slides and falls have undoubtedly taken place in historic times in more remote mountain regions, and since both are characteristic of rough terrain, it is to be expected that more will occur in the future. Their cumulative effect, like that of all other downslope movements, is to reduce the relief of the lands, widen valleys, and bring shattered rock material down to places where streams or other agents of erosion can remove it.

## Slower but Perceptible Movements, Mainly of Regolith

After heavy rains in hilly or mountainous areas, regolith may flow downslope as a water-saturated mass. The movement is a **mudflow** if the moving material is entirely mud (Section 7-4), and a **debris flow** if it consists of broken-up rock and regolith of all particle sizes. There are all gradations from stiff, viscous masses that just move under their own weight to highly fluid masses that may move down valleys with great rapidity. In the explosive eruption of 1963 at Mt. Agung, on the island of Bali east of Java, ejection of fiery clouds of ash was followed by torrential rains. These mixed with the ash to produce mudflows of great violence and speed. One partly buried a village and overwhelmed 200 people.

On steep slopes, masses of regolith and disintegrated bedrock may break free along their upper edges and slide downhill, in a process known as **slump.** Slope failure of this kind was dramatically demonstrated on March 27, 1964, when a major earthquake rocked Alaska and gave a severe shaking to the city of Anchorage. A residential district, situated on a gravel terrace underlain by moist clay, underwent destructive slumping, as indicated in Figures 6-2 and 6-3. Slumping involves masses that are bounded on the upslope side by a scar or break, and on the downslope side by a bulge where the sliding material pushes forward. Slumping of a given mass may take only a few minutes, as at Anchorage, or it may go on slowly for weeks or months. As usual, the presence of water is a significant factor.

**Figure 6-2**
Air view of destructive landslide at Turnagain Heights, Anchorage, Alaska, that accompanied the Good Friday earthquake of March 27, 1964. Houses were on a terrace underlain by gravel and clay, which failed by slumping as shown in Figure 6-3. (Photo by George Plafker, U. S. Geological Survey.)

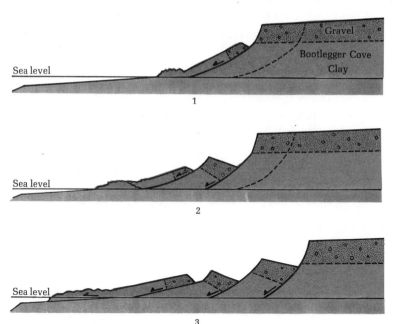

1

2

3

**Figure 6-3**
Development of the Turnagain Heights landslide by progressive slumping. Note the sharp break in the slope above each slump block, the backward rotation of the blocks, and the bulge on the low ground in front (left). Compare with Figure 6-2. (U. S. Geological Survey.)

## Imperceptibly Slow Movements of the Regolith

Most of us have seen retaining walls pushed away from the vertical, fence posts tilted downhill, or old gravestones all leaning in the same direction. Such features are evidences of an insensibly slow movement of the regolith under gravity, called **creep.** All slopes developed on regolith, even those with a good cover of grass or other vegetation, are susceptible to this type of very gradual downward movement. Although creep is scarcely noticeable, in humid regions it is instrumental in moving large quantities of material to lower levels.

In arctic regions, the ground freezes to great depths. In the short warm season, the upper few inches or feet thaw out, but the deeper material stays frozen and water cannot drain downward. The surface layer thus becomes water-soaked and unstable, and tends to flow down even very gentle slopes. **Solifluction** is the imperceptibly slow downslope movement of water-saturated regolith. In cold climates, it is responsible for the transport of much debris from higher to lower elevations.

## Formation of Talus Masses

On a quiet day in a mountain valley, it is not uncommon to hear an occasional rock fragment rattle down from the cliffs above, or even to see little cascades of particles trickle down vertical rock walls. Such rock fragments, pried from cliffs and peaks chiefly by frost wedging,

In figure labels: Gravel, Bootlegger Cove Clay, Sea level

accumulate as a **talus,** or heap of loose blocks that slopes outward from the cliff that supplies it (Figure 6-4). The material of a talus is termed **sliderock.** Extensive talus masses, their surface inclined at the **angle of repose** (maximum stable slope) of the particles of sliderock, commonly extend along the foot of steep canyon walls (Figures 4-4, 6-5).

**Figure 6-4**
Small talus cone formed by particles of soil and regolith accumulating at the foot of a steep slope.

**Figure 6-5**
Talus below limestone cliffs, western Wyoming. The sliderock fragments are produced by frost wedging.

Although the processes we have described are fundamentally simple, as they are merely the various ways in which unstable masses of earth material move downward under gravity, they are of great importance in helping streams, glaciers, and waves to sculpture the land surface.

In the discussion of geologic time in Chapter 3, it was stated, "Streams carve the valleys in which they flow." This can scarcely be contradicted, yet it is by no means

## 6-3 GEOLOGICAL SIGNIFICANCE OF DOWNSLOPE MOVEMENTS

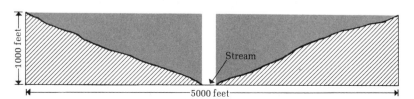

the whole story. Swift sediment-laden streams do indeed wear away the solid rock; but obviously this can go on only in the stream channel itself, where sediment and water are in direct contact with the rock over which they move. In the average V-shaped stream valley, however, the stream itself occupies but a very small part of the total area (Figure 6-6). The bulk of the rock that is removed to form such a valley is fed to the stream by slump, creep, talus accumulation, and the other movements that bring material down slopes (Figure 6-7). To be accurate, then, we should say that valleys are carved by stream erosion plus downslope movements.

By eroding its channel, a stream deepens its valley and increases the slope of the valley walls; but the stream

**Figure 6-6**
Cross section of an imaginary stream valley. That part of the valley that has been cut by the stream lies between the two areas in color; it amounts to less than 10 percent of the total rock removed to make the valley. All the rest has been contributed to the stream by downslope movements.

**Figure 6-7**
Slumping of the soft Cretaceous shale furnishes debris to the stream and widens the valley. Bad River, South Dakota. (Photo by D. R. Crandell, U. S. Geological Survey.)

uses some of its energy in acting as a conveyor belt to carry away material fed to it by downslope movements. A rough balance between these two functions is generally maintained. Occasionally, however, this balance may be drastically upset. For example, the rockslide that occurred in the Gros Ventre River valley, Wyoming (Section 6-2), provided the stream with material at a far faster rate than the stream could remove it; the river was dammed and a lake backed up behind the slide mass. When the waters finally overtopped this dam, they cut a channel through it very rapidly, releasing a flood that caused loss of life and property down the valley. By no means all the slide material was removed at that time; the Gros Ventre River, together with more normal and gradual downslope movements of the slide material itself, is still at work clearing its valley of the mass poured into it in 1925. As we have seen, a similar event occurred in Madison Canyon, Montana, in 1959 (Figure 6-1). To avoid damage from suddenly released waters, engineers dug a channel through the slide material, thus aiding the Madison River in clearing the valley bottom.

Downslope movements also contribute to erosion by valley glaciers (Section 12-3). Rock fragments dislodged by frost wedging from high peaks and crags may come to rest on a glacier occupying the valley far below, rather than accumulate in a talus. **Avalanches** of mixed snow and rock fragments also hurtle downward from time to time. Thus gravity delivers much rock material directly to the ice that can remove it from the area, and valley glaciers have a conveyor-belt function like that of streams.

Slumping and sliding of rock masses are notable along many steep coasts (Section 7-8), where waves beat against the base of a seaward-facing cliff and undermine it. The effect is to help reduce the land toward sea level, by delivering rock materials into the zone of wave action where they are ground up and moved seaward or along the shore.

## 6-4 DOWNSLOPE MOVEMENTS AND THE WORKS OF MAN

We have mentioned in passing a few of the ways in which man and his works are affected by downslope movements. Obviously man's constant efforts to remold the surface of the lithosphere more nearly to his liking rank among the prominent causes of such movements.

The catastrophic rockfall at Vaiont Dam in Italy (Section 6-2) apparently took place because the water in the reservoir behind the dam saturated the rocks in the adjacent walls of the gorge, making them unstable beneath the weight of the rock on the mountainside above. Heavy rains added to the problem. Engineers were aware of the situation and were reducing the reservoir level at the time of the fall. It is highly probable that there would have been no rockfall had there been no reservoir.

Another example of rock failure owing to saturation by the waters of a reservoir is shown in Figure 6-8. Here the landscape is more subdued than at Vaiont Dam, the rocks are relatively soft and weak, and failure took place in a series of slump movements over a period of several days rather than in a single fall in a matter of seconds. Again, if man had not constructed Grand Coulee Dam, backing lake waters many miles into what was formerly a stream valley, the cliff in Figure 6-8 would undoubtedly have remained stable, eventually being worn back by weathering, rainwash, and the less dramatic aspects of downslope movement.

Failure of the type shown in Figure 6-8 may occur in the building of major works. Great rockslides and earth-flows delayed completion of the Panama Canal for months, and increased the overall cost by millions of dollars. The possibility of becoming involved with such slides must affect all planning and cost estimates for the second trans-isthmus canal that is periodically proposed. During construction of the Grand Coulee Dam on the Columbia River, the excavation for one end of the con- crete structure was threatened by an unstable mass of water-saturated sand and silt. This was finally controlled by an unusual, if not unique, procedure: pipes were embedded in the soft material, and a refrigerant was circulated through them. Thus the mass was frozen in place until the necessary construction was completed.

Storm waves may erode rock material from the base

**Figure 6-8**
Multiple landslide in soft Pleistocene deposits, Washington. In foreground is Franklin D. Roosevelt Lake, the reservoir behind Grand Coulee Dam on the Columbia River. Saturation of the terrace sediments as the lake rose weakened them and caused failure. The slide developed in a few days. (Photo by F. O. Jones, U. S. Geological Survey.)

of cliffs along shorelines, removing the rocks' support and allowing them to slump seaward. This has happened repeatedly along the steep California coast. Such landslides not only damage or destroy houses built at the top of the cliff, but also bury highways or other man-made features along its base. Cliff-top property has also been damaged in this way along the south shore of Lake Erie, and elsewhere on the Great Lakes.

In recent decades, the pressure of population growth has made it necessary to construct housing developments in areas that geological prudence would suggest are unsuitable. In the Los Angeles region, a notable example, the flat ground has long since been utilized, and new houses must be built in canyons and on mountainsides. In large areas of new housing, steep slopes are terraced, each property being separated by retaining walls from the ones above and below. Such intensive development would be questionable under even the best of geological conditions, and it is even worse in southern California, where much of the underlying bedrock is relatively soft and weak and the region is periodically shaken by earth tremors. The cover of vegetation is scant and rainfall is concentrated in short seasons. For all these reasons, downslope movements are a recurring geologic hazard.

Even in geologically stable regions, such as the east central United States, downslope movements may be caused by man's activities. Certain soft, clayey shales, for example, commonly make trouble for highway departments, as the clays readily slump after heavy rains, even in road cuts with very gentle slopes. More than one handsome new interstate highway has been temporarily blocked by large slumped masses of rock and regolith.

Arctic soils that demonstrate solifluction (Section 6-2) are so water-laden in the warm season that they will not support the weight of a man, let alone that of a truck or a piece of heavy machinery. For this reason, ground transportation in Arctic areas (for example, on the north slope of Alaska, where large oil deposits have been found) is done in the winter, when the ground is frozen. Much of the controversy about the Alaska pipeline has arisen from the fact that it would cross ground that is frozen rock-hard in the winter and has the consistency of jelly in the summer. An important cold-weather problem would be to isolate the pipeline so that its heat would not cause it to melt the frozen ground and sink in, and summer problems would be physical support for the pipe and access to the line in case of need.

**6-5**
**SUMMARY**

The downward movement of rock and regolith under the influence of gravity takes a variety of forms. Masses weighing millions of tons may slide abruptly down slopes, or fall instantaneously from mountainsides or cliffs. Flows of water-saturated mud, or of mud and broken-up rock debris, generally move more slowly. A mass of regolith and disintegrated rock is said to slump if it moves as a unit, breaking free on the upslope side and producing a bulge where it pushes down the slope. Imperceptibly slow movement of soil and regolith is known as creep. In cold regions, slow movement called solifluction takes place in the warm season, when the upper few feet of soil become saturated with water that cannot drain away because the underlying ground remains frozen. Rock fragments that fall from a cliff, which are termed slide-rock, may accumulate along the cliff base as a talus, or heap of loose blocks.

Water is highly important in promoting downslope movements, as it adds weight and decreases cohesion within the mass of rock, regolith, or soil.

Downslope movements collectively tend to reduce the relief of the land surface and thus are a very important leveling process. They feed loose material to streams, glaciers, and the sea, all of which are agents of grinding-up and removal.

Man's work in modifying the land surface by building such structures as dams, reservoirs, canals, highways, pipelines, and large housing developments often produces downward movements of earth materials under gravity, or is drastically affected by them.

## SUGGESTED READINGS

Bloom, A. L. 1969. *The Surface of the Earth,* chap. 3, pp. 40–52. Englewood Cliffs, N.J.: Prentice-Hall. (Paperback.)
General coverage of downslope movement of rock materials, including some case histories of rapid movements.

Crawford, C. B., and Eden, W. J. 1963. Nicolet landslide of November 1955, Quebec, Canada. *Engineering Geology Case Histories,* no. 4, pp. 45–50. Geological Society of America.
A description of the Nicolet slide, together with discussion of the geologic factors that contribute to slides common in the marine clays of the valleys of the St. Lawrence and Ottawa rivers.

Grantz, A., and others. 1964. Alaska's Good Friday earthquake, March 27, 1964. *U. S. Geological Survey Circular 491.* 35 pp.
A report on the Alaskan earthquake based on data gathered during the two weeks following the quake, which includes photographs and information on landslides resulting from the earthquake.

Kiersch, G. A. 1965. Vaiont Reservoir disaster. *Geotimes,* vol. 9, no. 9, May–June, pp. 9–12.
Case history of a tremendous rockfall that destroyed a reservoir and caused a disastrous flood. Includes "before" and "after" pictures.

# 7 EROSION AND DEPOSITION BY STREAMS, WAVES, AND WIND

7-1   The Significance of Streams

7-2   Energy Factors in Streams

7-3   Erosion: How Streams Acquire Their Load

7-4   Transportation: How Streams Carry Their Load

7-5   Deposition: Loss of Mechanical Load

7-6   Stream Deposits

7-7   A Man-modified River: The Nile

7-8   Waves and Currents

7-9   Wind

7-10  Summary

A volcanic neck (Section 4-3), standing several hundred feet above its surroundings, is evidence of two opposing kinds of geologic activity: first, the building of a volcano on the site; second, the removal of this volcano through erosion. While the cone was being built, lava and ash accumulated more rapidly than they could be worn away by the elements; but as soon as the fires cooled, the extinct volcano became prey to weathering and erosion. Perhaps glaciers came into being on the summit slopes, while below them the waters from rains and melting ice gathered into streams that furrowed the mountainside with gullies and later with deep canyons. Finally, all that was left was the congealed lava in the pipe that once fed the volcano.

Such a series of events is but a minor skirmish in the endless conflict between the forces that build up the land surface and those that wear it down. The earth's outer crust contains innumerable records of other past battles in this conflict. Deep batholiths dome the crust upward; the domed rocks are eventually worn away and batholithic granite is exposed to the sky. Rocks that were complexly deformed in the depths of once mountainous regions are now at the surface, because the mountains have been removed by erosion. The sedimentary rocks of old sea floors are lifted bodily thousands of feet above sea level, only to be cut away by streams flowing in valleys such as the Grand Canyon (Figure 2-11).

Downslope movements, ice, wind, waves, and streams are all allies on the downwearing side of the conflict. The first of these is important everywhere. Though ice, wind, and waves are each a significant agent of erosion at certain places on the earth, of much greater importance are streams. Running water, flowing across the land as streams in valleys, picks up and carries away vast amounts of rock debris: in humid regions and arid ones, in mountains and on plateaus and plains. In so doing, streams act as the most important agents of land reduction.

As a land area is gradually worn away by streams, what happens to the material removed? Though there are commonly many interruptions on the way, its ultimate resting place is the sea floor. Streams act as agents of transportation as well as erosion, and big rivers carry tremendous quantities of debris to the sea, as such features as the Mississippi Delta testify. The work that streams do in transferring rock material from high places to low ones is discussed in this chapter. Later on, we will turn our attention to the various kinds of landforms that streams make as they sculpture the earth's surface.

**7-2
ENERGY FACTORS
IN STREAMS**

Streams are part of the hydrologic cycle (Figure 2-10). In this cycle, the sun's heat evaporates moisture from the earth's surface, mostly from the oceans; the resulting water vapor condenses, and some of it falls on the land as rain and snow; and that part of the precipitation that does not evaporate immediately, or sink into the ground, flows downward off the land surface, returning once more to the sea.

Any mass that moves from a high position to a lower one under gravity acquires kinetic energy (Section 2-6), the capacity to do work. (The energy of falling water drives the turbines at a hydroelectric plant.) Hence we may best consider streams as *energy systems*. They are analogous to machines, their power being derived from the force of gravity.

**Turbulence**

Every machine, the automobile engine as one example, expends a large part of its energy merely in overcoming its own friction and in dissipating the heat that it generates. A car moves by that part of the total energy left over after the engine's internal friction has been surmounted. So it is with streams. Studies of water flow in the laboratory, using small jets of dye, have shown that the water of a stream moves down its channel as a complex maze of eddies and swirling crosscurrents. In this internal milling about, or **turbulence,** kinetic energy is changed into heat energy and is dissipated. The largest part of a stream's total energy is used up in this way, internally, in friction of the particles of water against each other. The energy that remains is expended in friction against channel sides and bottom, and against particles of rock debris carried by the stream—in other words, in geologic work.

**Figure 7-1**
Stream channel at low-water stage. Discharge is so small (probably less than 1 cubic foot per second) that available energy is at a minimum and little geologic work is being done. Bedrock is sandstone of Mississippian age. West Branch of Rocky River at Olmsted Falls, Ohio; midsummer. Compare with Figure 7-2.

## Discharge

The volume of water that moves past a point in a unit of time is a stream's **discharge,** expressed in cubic feet per second. This is a highly variable factor. Long dry seasons reduce stream discharge to a minimum; heavy rainfall or the sudden melting of snow increase it, sometimes to flood stage. Since a small stream (or machine) has less energy than a large one, the available energy of a stream varies tremendously with time, from nearly zero at the low-water stage to very high at the high-water stage (Figures 7-1, 7-2).

Every big stream receives smaller streams (tributaries) that flow into it, each adding its discharge to that of the main stream. Hence discharge generally increases downstream. The discharge of the Mississippi River at its source in central Minnesota is so small that one can walk across the stream on stepping-stones, but this method of crossing soon becomes impossible as the discharge increases down the valley.

## Gradient

**Gradient** is the slope of a stream's channel, expressed in feet of vertical fall per mile of length. Streams in high mountains may have gradients of several hundred feet per mile; at the opposite extreme, meandering rivers like the Mississippi in its lower course drop only one foot per mile or even less. Gradient is obviously one of the factors that affect the rapidity with which a stream of water flows, and hence the stream's energy.

If a typical stream is surveyed and its gradient plotted on a **long profile** (a profile of the stream channel from source to mouth), it is found that the gradient decreases

**Figure 7-2**
Stream channel at same point as in Figure 7-1, in early spring. Discharge is about 3,000 cubic feet per second, and turbulence is high. Much sediment is being transported and the channel is being vigorously scoured.

**Figure 7-3**
Generalized long profile of a typical stream, from headwaters to mouth.

progressively downstream. The long profile of most streams has a form like that shown in Figure 7-3. Yet keep in mind that, on a profile like the one in the figure, several hundred miles of stream are shown by a line a few inches long. This gives the impression that stream profiles are invariably smooth curves, which is quite incorrect. The gradient of a stream may change a number of times along its course, owing to the varied nature of the rocky crust that it flows over. The Yellowstone River, for example, after leaving the lake at its source, pursues a leisurely course for some 12 miles in a wide grassy valley (which attracts moose and fishermen), at a gradient of less than 10 feet per mile; but after thundering over two falls formed by dikes, the river enters soft pyroclastic rocks in which it flows for many miles in a deep canyon at a gradient of some 56 feet per mile (Figure 7-4). Such irregularities are smoothed out on a long profile.

### Velocity

A stream's **velocity** is its rate of flow, measured in feet per second or miles per hour. Velocities range from less than one to more than 15 miles per hour; an average velocity for most streams is about five miles per hour.

**Figure 7-4**
Canyon of the Yellowstone River, Yellowstone National Park. The river's discharge, gradient, and velocity are all high. As the canyon is deepened by stream erosion, it is widened by downslope movements. (Compare Figure 6-6.)

All the factors we have mentioned affect velocity. Turbulence, or the reduction of kinetic energy in internal friction, acts as a brake on a stream's forward motion and thus decreases velocity. The effect of gradient is obvious: a stream of given size will flow faster down a steep slope than down a gentle one. As for discharge, we can easily see at first hand the close relationship of this factor to velocity. Observing the local brook or creek after a long dry spell, we see that discharge is small and also that velocity is sluggish (Figure 7-1). But on looking at the same stream after heavy spring rains, we find that the increase in discharge is accompanied by a marked increase in velocity. The slow-moving trickle has become a swirling torrent (Figure 7-2).

Indeed, very high discharge may produce relatively high velocity even in spite of low gradient. Measurements of stream velocity show, surprisingly, that rivers such as the Missouri actually flow as fast as their higher-gradient tributaries in the far-off headwater areas, or even somewhat faster. The reason lies in the enormously greater discharge and channel depth of the large rivers. It is apparent that with streams, as with machines, sheer size is the most important factor in governing the output of energy.

## Hydraulic Action

Slump, creep, and other kinds of downslope movements are steady suppliers of rock debris and regolith to valley bottoms (Figure 6-7). This kind of loose material is removed chiefly through direct impact by the moving water, or **hydraulic action.** At low water, a stream may have only enough energy to pick up some of the finest particles, whereas at flood stages, with high turbulence, the same stream may wash away sand, pebbles, and even boulders.

## Abrasion

**Abrasion,** or mechanical wear of rock against rock, is especially significant in sediment-laden streams that flow over solid rock. In such streams, the mineral grains and rock particles act as a rasp or file, abrading the bottom and sides of the rocky channel as they are dragged along by the water currents (Figure 7-5). Particles dislodged from the channel are added to the sediment load of the stream. As the rock and mineral grains are swept along in the turbulent crosscurrents of the stream, considerable abrasion also takes place among the particles themselves, breaking some into smaller bits and rounding and polishing others.

**7-3
EROSION:
HOW STREAMS
ACQUIRE THEIR
LOAD**

**Figure 7-5**
Stream channel at low water. Pebbles act as tools of abrasion at high-water stages. Light-colored rock is easily eroded limestone; darker, more resistant rock is chert (Section 8-4). (Photo by E. L. Shay.)

Interesting features are produced by stream abrasion in rocky channels. For example, part of the current may be deflected by a boulder or other obstruction so that a little whirlpool develops; if sand grains and pebbles are caught in this eddy and swirled round and round, they may at length abrade the rock beneath them and

**Figure 7-6**
Large pothole cut in Devonian shale. Pool is about 8 feet across. Watkins Glen, New York. (Photo by E. L. Shay.)

produce a cylindrical depression called a **pothole** (Figures 7-6, 7-7). Many potholes look as though they had been drilled into the rock, as in effect they have, the eddying current providing the energy and the sand and pebbles providing the "teeth" of the drill. Immediately below a waterfall, where energy from the falling stream is at a maximum, hydraulic action combines with abrasion to erode a deep **plunge pool** in the channel (Figure 7-8).

Note that abrasion takes place only in streams that carry mineral grains and rock particles. Many a clear stream tumbles over ledges and down a rocky channel without doing any appreciable abrasion, even enough to remove a film of green algae from its place of growth on the submerged rocks. Though such streams have much energy, they lack the tools with which to wear away the rock.

**Figure 7-7**
Potholes at low water. Granitic rocks, Sierra Nevada, California. (Photo by Mary R. Hill.)

**Figure 7-8**
The rock below a waterfall is scoured out by abrasion and hydraulic action to form a plunge pool. The coarser debris comes to rest in a gravel bar just downstream.

Waterfall

Rock

Plunge pool

Gravel bar

119

## Solution

Most streams carry a measurable amount of chemical compounds dissolved in the water. Though some of these compounds may be acquired by solvent action of the water on the rocks over which it flows, most are the result of earlier chemical weathering and are contributed by underground waters that seep into the stream along its course.

**7-4
TRANSPORTATION:
HOW STREAMS
CARRY THEIR
LOAD**

## Mechanical Load

The mineral and rock fragments transported by streams range in size from extremely fine to very coarse. The finest is **clay,** which consists of particles less than 1/256 mm (about 1/100,000 inch) in diameter. **Silt** grains are 1/256 to 1/16 mm in diameter, and **sand** grains 1/16 to 2 mm. **Mud** is an everyday term for a mixture of water, clay, and silt. The material coarser than sand, collectively termed **gravel,** includes the size range from small pebbles up to large boulders.

Particles of clay, silt, and fine sand may be held in the stream current as a result of turbulence, and carried as **suspended load;** coarser grains may be pushed, dragged, and shoved along the bottom, forming what is termed **bed load.** There is no sharp distinction between them. A given particle may be carried as suspended load for a distance and then dropped to become part of the bed load until picked up again. Indeed, most grains pursue a sort of hop-skip-and-jump course down a stream channel.

The maximum size of particle that a stream can hold in suspension depends on the available energy, a function chiefly of discharge and velocity. Hence at low water a stream may be able to move only clay-size particles in suspension and fine sand on the channel bottom, whereas at maximum flow it may scour up and carry coarse sand, and move pebbles and even boulders along the channel as bed load. The shape and specific gravity of the various particles are other factors that help determine what is carried and what is pushed or rolled.

## Solution Load

Mineral matter dissolved in the water forms part of a stream's load. Since this material is carried chemically rather than mechanically, it is in effect a part of the water itself. Thus solution load is unaffected by velocity, discharge, or other aspects of stream flow.

Significant products of rock decomposition are chemical compounds that have been removed from the site of weathering in solution. Among the most abundant of these are silica and the carbonates of potassium, sodium, calcium, and magnesium. Waters containing these weathering products ordinarily sink into the ground, where their dissolved load may be increased, decreased, or altered in composition. Eventually, most of these waters find their way into streams and thereby add their dissolved salts to the stream's solution load.

Small streams in high mountains, where temperatures are low and chemical weathering is at a minimum, contain very small amounts of dissolved salts. At the other extreme are the world's great rivers, which are veritable sluiceways not only of mechanically held debris but also of chemically transported mineral matter.

The process of solution transport is invisible; furthermore, most of the dissolved material delivered to the ocean is not deposited, but is simply added to the great oceanic storehouse of dissolved salts. For these reasons we are likely to underestimate the significance of solution load. A recent international program of sampling river waters has shown that about one-third of all the material carried to the ocean by big rivers is carried as solution load. An increasing proportion of this dissolved matter consists of chemicals manufactured by man, used in the home, on the farm, or in industry, and eventually washed into the sea.

Suspended load and bed load are dropped when a stream loses the energy necessary to keep them moving. This obviously happens when a stream enters standing water— a lake or the sea—and loses all forward motion. Deposition may also take place at any place along the stream's channel where velocity is checked, downstream from a boulder or other obstacle, for example; or where discharge is reduced, as when flood waters subside.

**7-5
DEPOSITION:
LOSS OF
MECHANICAL LOAD**

### In and along the Channel

A vigorously flowing stream with a moderately high gradient is a high-energy system, as we have used the concept. If we find such features as potholes, waterfalls, and rapids along its course, we may deduce that such a stream tends to erode rather than deposit. Any deposits that form at times of low discharge (low energy) are likely to be swept away with the next high water.

Not that deposits are totally lacking in the channels of such streams. For example, there is commonly a gravel bar downstream from a waterfall, wherein accumulates

**7-6
STREAM DEPOSITS**

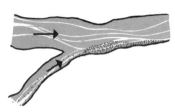

**Figure 7-9**
Sand bar below a stream junction. The tributary brings more sediment into the channel of the main stream than the latter can remove.

the coarser rock debris removed from the plunge pool by abrasion and hydraulic action (Figure 7-8). A briskly flowing tributary may contribute more load to its main stream than the latter can carry, causing a sand or gravel deposit to form downstream from the junction (Figure 7-9). Exceptionally, a stream may be fed with so much sand and gravel that it takes on what is called a **braided** pattern, its channel becoming a maze of bars among which the water flows in many subchannels (Figures 7-10, 12-13).

**Figure 7-10**
A braided stream: the Platte River, Nebraska. In this infrared aerial photo, water appears black and fields white. (Photo by R. L. Handy.)

It is in the valleys of what we may call **alluvial streams**, however, that most deposits are formed. Such streams flow on broad valley floors, across unconsolidated stream-deposited sediment (**alluvium**). Most alluvial streams are large rivers, such as the Missouri and the Mississippi.

Alluvial rivers characteristically swing back and forth across their valley floors in wide, looping curves known as **meanders.** The profile across the channel where the stream is relatively straight, between bends, tends to be roughly symmetrical, whereas the profile at the bends shows a gently sloping bank on the inside of the bend and a steep bank on the outside (Figure 7-11). Studies have shown that velocity and turbulence are at a maxi-

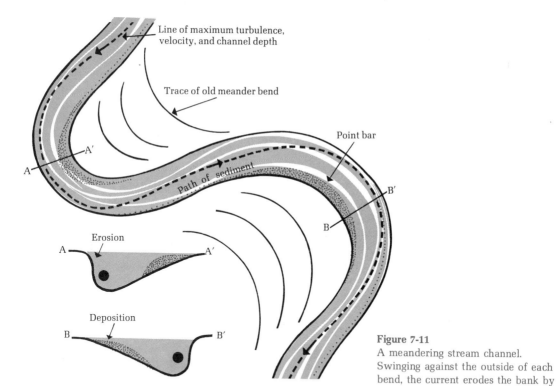

Line of maximum turbulence, velocity, and channel depth

Trace of old meander bend

Point bar

A'

A

Path of sediment

B'

B

Erosion

A — A'

Deposition

B — B'

**Figure 7-11**
A meandering stream channel. Swinging against the outside of each bend, the current erodes the bank by hydraulic action. Much of the sediment that falls into the water by bank caving is deposited in a point bar on the inside of the next meander bend downstream. Black dots in channel cross sections indicate zones of maximum turbulence.

mum on the outside of meander bends. Here the river washes against the steep bank and erodes the soft alluvium by hydraulic action, until it is undermined and caves into the water. The alluvium thus contributed to the stream is moved down the channel, and much of it is deposited on the inside of the next meander bend, where the water is relatively slow-moving. Deposits termed **point bars** are thus formed (Figures 7-11, 7-12). They are crescent-shaped bars built on the inside of meander bends, and their material comes mostly from bank caving on the outside of the next bend upstream. The tendency of alluvial rivers to erode on the outside of bends and deposit on the inside means that the meandering channel gradually migrates laterally across the valley floor.

**Figure 7-12**
Point bars (white) along the Iowa River. Note also the cutoff meanders. (Photo courtesy of U. S. Department of Agriculture.)

When an alluvial river rises in flood and eventually overtops its banks, it drops much of its load immediately, because velocity decreases abruptly as soon as the water leaves the confining channel. As the overflow moves slowly away from the channel, willows and other vegetation help to slow its motion and decrease its energy. The result is that a ridge of fine sediment is built up right along each side of the channel. Such ridges are termed **natural levees** (Figure 7-13). A big river like the lower Mississippi is bordered by natural levees 12 to 20 feet high; a single flood may add from 6 inches to 2 feet of fine sand and silt. Natural levees are present only along principal rivers that are heavily loaded and flood relatively often. Along smaller alluvial streams, even though they meander on a flat valley floor, natural levees are small or nonexistent.

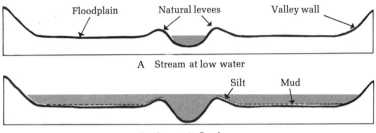

A   Stream at low water

B   Stream in flood

**Figure 7-13**

Profiles across the floor of an alluvial valley. Sediment is added to the natural levees and the floodplain only in times of flood. Note that the stream channel is deepened during the flood.

Between the levees and the walls of the valley lies a wide flat **floodplain,** which is covered by water at times of flood (Figures 7-14, 7-15). Floodplains receive a layer of fine sediment with each inundation. Such deposits are a misfortune if the floodwaters invade communities, but they may be a benefit to croplands. The annual layer of silt formerly laid down by the Nile replenished the fertility of its floodplain for centuries.

**At Stream Mouths**

When a stream enters standing water, it loses kinetic energy and hence deposits its load. The resulting deposit is a delta, so called from its resemblance in plan view to the Greek letter Δ. Deltas range in size from tiny ones made by rills entering a roadside ditch to immense features with long and complex histories, like the deltas of the Amazon, the Nile, and the Mississippi (Figures 7-16, 7-17). Experimental building of little deltas in glass-sided tanks in the laboratory, and the records from holes drilled into natural deltas, show that the sediments of which deltas are made are deposited in layers that slope gently toward the open water. In addition to their layered structure, deltaic sediments commonly have a consider-

**Figure 7-14**
Ohio River in flood at Rockport, Indiana, in 1937. View is upstream. Channel is at the right; note submerged natural levees in foreground and upper right. (Photo courtesy of U. S. Geological Survey.)

**Figure 7-15**
City Park at Iowa City, Iowa, subject to flooding because it is on the floodplain of the Iowa River.

able degree of **sorting.** The coarser, heavier materials— coarse sand, let us say—tend to be dropped first, and fine sand, silt, and clay to be deposited successively farther out from the stream mouth and down the delta front.

In deserts it is common for streams to flow out of

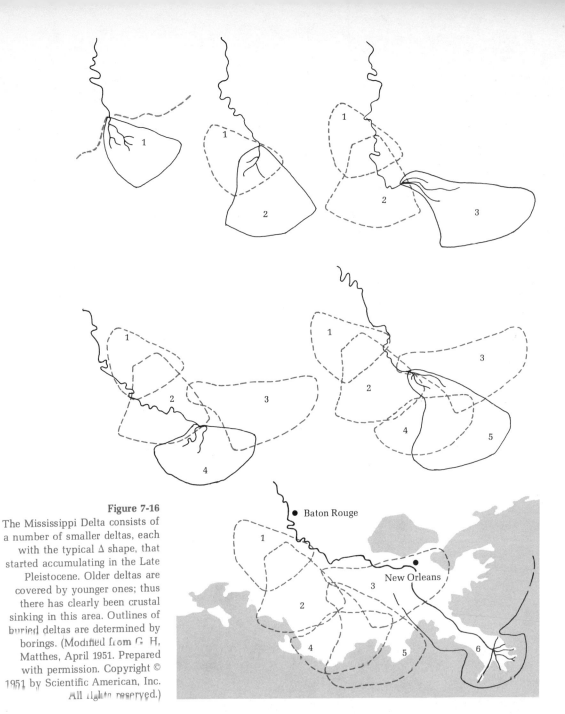

**Figure 7-16**
The Mississippi Delta consists of a number of smaller deltas, each with the typical Δ shape, that started accumulating in the Late Pleistocene. Older deltas are covered by younger ones; thus there has clearly been crustal sinking in this area. Outlines of buried deltas are determined by borings. (Modified from G. H. Matthes, April 1951. Prepared with permission. Copyright © 1951 by Scientific American, Inc.

mountain valleys and come to an end in parched and arid desert basins. In much of the American Southwest, for example, low rainfall and high evaporation prevent permanent lakes from forming. Low places on the land surface, closed depressions, are barren desert. The occasional rains that do occur generally fall in the surrounding mountains, and are likely to be of the cloudburst or sudden-downpour variety. The water from such rains, falling on bare slopes largely devoid of vegetation, picks

126

up heavy loads of rock debris as it pours into the adjacent canyons. In a few moments, a dry sunbaked canyon may be occupied by a roaring torrent capable of moving all sizes of debris up to that of large boulders. When such a stream, which may only last a few minutes or hours, debouches onto the desert floor at the foot of the mountains, the water, no longer confined by valley walls, spreads out and loses velocity. Discharge is also lost, as the water sinks rapidly into the loose gravelly deposits left from previous "flash floods." Consequently, the

**Figure 7-17**
Point where the Mississippi River (lower right, flowing away from observer) divides into distributaries, or "passes," on the present delta. This point, the "Head of Passes," is immediately to the right of the numeral 6 on the lower map of Figure 7-16. (Photo by S. R. Sutton, U. S. Army Corps of Engineers.)

**Figure 7-18**
Alluvial fans along the east front of the Panamint Mountains, California. Death Valley in the foreground. (Photo by John H. Maxson.)

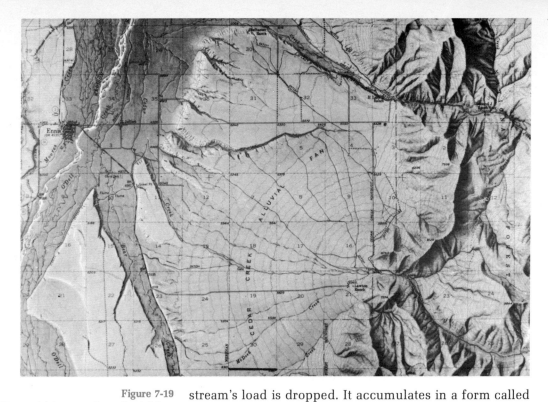

Topographic map of an alluvial fan, which has been built out from the mountains on the right into the valley of the Madison River. The apex of the fan is at the mouth of the canyon that supplied the sediment. Distance shown by the black line is one mile. (Part of the Ennis quadrangle, Montana, shaded-relief edition. U. S. Geological Survey.)

stream's load is dropped. It accumulates in a form called an **alluvial fan,** which is a deposit of sand, pebbles, and boulders, all mixed together (that is, poorly sorted), whose upper surface slopes outward from the canyon mouth at the apex of the fan. Mountain ranges in arid regions are commonly flanked by alluvial fans that have been built out into the adjacent desert basins (Figures 3-2, 7-18, 18-8).

Although alluvial fans are most spectacularly developed in deserts, they are not uncommon in humid regions. The fine specimen shown in Figure 7-19, for example, lies in the northern Rocky Mountains. An alluvial fan is to be expected wherever a heavily loaded stream emerges from the confines of a valley and deposits most or all of its load on a land surface.

## Temporary Nature of Stream Deposits

All the deposits that occur along stream channels, including bars of various kinds, natural levees, and floodplain deposits, as well as deltas and alluvial fans, are only temporary features when considered from the viewpoint of geologic time. They merely represent places where debris from the land is halted, for a shorter or longer period, on its journey "down the geologic gutter" to the sea. Sooner or later, the material lodged in such deposits is removed, and eventually finds its way to a resting place below sea level. It may come to rest in a large delta, or it may be seized upon by waves and currents and spread out on the shallow sea floor as layers of sediment.

128

The Nile, one of the world's great rivers, rises in eastern Africa and flows for some 4,000 miles northward to the Mediterranean (Figure 7-20). In order to impose some control on flooding, a dam was built at the city of Aswan, 600 miles above Cairo, in 1902. This dam was twice strengthened and raised. Then, in the decade 1960-70, an immense new structure, the High Dam, was built some 4 miles upstream. With a volume 16 times that of the Great Pyramid, the High Dam is designed "to fully control and harness" the waters of the great river.

**7-7**
# A MAN-MODIFIED RIVER: THE NILE

**Figure 7-20**
The Nile River. Construction of the Aswan High Dam has been of industrial benefit, but the silt from upstream now settles in Lake Nasser, no longer replenishing the fertility of the densely settled floodplain and the famous delta as it has done for centuries.

The High Dam adds 1.2 million acres of land to the overall farming area, and allows conversion of an additional 70,000 acres to year-round irrigation. Its turbines generate electricity for an aluminum plant, a phosphate plant, and an iron and steel complex near Cairo. Lake Nasser behind the dam, 312 miles long and an average of 7.7 miles wide, is expected to provide an important fishing industry. Egyptian pride in the Aswan High Dam is quite understandable.

But such dams create problems as well as solve them. Something like 100 million tons of silt per year, which was formerly carried downstream to the fertile floodplain and the famous Nile Delta, now settles and accumulates near the head of Lake Nasser some 900 miles upstream. The river below the dam, which is clear because the lake acts as a settling basin, now has excess energy, and has eroded deeply along the river channel; four check dams have been built and more are planned. Far to the north, the outer edge of the delta is being wave-eroded, because silt no longer builds it outward; it may be necessary to construct a series of dikes in the sea, at great expense, to protect the delta edge.

The Nile is by no means the only important stream to have been artificially modified. Many of the great rivers of North America, including the Columbia, the Colorado, and the Mississippi and its major tributaries, are the sites of dams, locks, deepened channels, and other features imposed for man's convenience. Invariably it is found that the processes we have discussed in this chapter are in delicate balance in the natural state, and drastic interference can only mean pronounced changes in this balance.

**7-8**
**WAVES AND**
**CURRENTS**

Along the world's shores there is a high concentration of energy where the restless waters of the ocean come into contact with the solid lithosphere. Features of much geologic and scenic interest are produced. Yet shorelines are linear zones that are extremely narrow when considered from a global point of view. Waves and currents, which do the geologic work in these zones, are not comparable to streams as large-scale sculptors of the earth's surface, but they act as distributors of solid debris along the shore and on the adjacent shallow sea bottom.

### Agents

Winds that blow across the immense oceans of the earth exert friction against the water and produce **waves.** Waves in the open sea are oscillatory forms, which pass through the water much like waves that cross a wheat field on a windy day. After such a wave has passed, each water particle (or stalk of wheat) returns to the same position it had before.

Large waves commonly travel for long distances, often far beyond the stormy area in which they are produced, and sooner or later they encounter the edges of islands or continents. Shoaling of the sea floor then causes the waves to "drag bottom," lose their oscillatory nature, and advance shoreward in a turbulent mass of surf. These waves differ from those of open water in that the water itself, as well as the wave form, moves forward. During periods of calm weather the action of waves is at a minimum; however, during storms, each wave that breaks may hurl thousands of tons of water against the land.

Waves may approach the land directly onshore, that is, with the wave front parallel to the trend of the shoreline. In this case most of the activity in the shore zone consists of forward wave motion and seaward back-wash. More commonly, waves encounter the land diagonally. Under these circumstances some water in each advancing wave is deflected along the shoreline, and a slow drifting of the water, or **longshore current,** is set up. These wave-induced currents are very active agents of transportation in the shore zone.

## Processes

### Hydraulic Action

Through direct impact, waves are relentless attackers of weak or fractured rock. Storm waves may exert pressures of a ton or more per square foot, and their pounding is locally earth-shaking. Blocks of rock weighing many tons may be torn loose and shifted about. In places, shattering of the rock is aided when the air in crevices is violently compressed by crashing waves.

Along more gently sloping sandy shores, hydraulic action is commonly more than sufficient to move sand grains and pebbles. Longshore currents pick up and move sand and finer materials.

### Abrasion

The wear of rock against rock is perhaps more intense in shore zones than in any other environment. Along the "stern and rockbound coast" of Acadia National Park, Maine, for example, one can see that angular pieces and blocks of fresh pink granite have been converted into rounded boulders by constant abrasion in the zone of surf. Along such shores, the surf zone acts not only as a mill, in which the rock debris acquired by hydraulic action is ground up, but also as a horizontal saw, cutting laterally into the land, using this same rock debris as tools of abrasion.

Pebbles on sandy beaches are known for their smoothly worn surfaces, and pieces of brick and bits of glass found on such beaches almost always show more or less rounding as a result of being washed back and forth and abraded by the sand.

### Transportation and Deposition

Every advancing wave on a beach moves sand grains in one direction, and each returning rill shifts them in another. To the casual observer these infinitely repeated actions may appear quite aimless. Actually, however, they are governed by two fairly obvious factors. First, we have seen that waves generally approach the shore at some diagonal rather than head-on; such waves wash sand grains obliquely up on the beach. The backwash, however, moves them directly seaward, down the beach slope. Thus individual grains or pebbles are drifted gradually along the beach in a sawtooth, or zigzag, path, in a process aptly termed **longshore drift.** Secondly, because the beach slopes toward the water, each grain tends to move a little farther seaward than landward; the result is that lighter and finer grains are eventually shifted seaward till they come to rest in water too deep to be disturbed by waves. Coarser and heavier grains remain in the surf zone, where they are abraded to finer sizes before being moved outward. Waves are excellent sorting agents, capable of separating pebbles and boulders high up on a beach from sugary clean sand along the main part of the beach, and these in turn from finer sediments transported offshore beneath the water surface. Longshore drift may intercept the fine sediments in their gradual seaward path, and distribute them laterally.

## Products

### Types of Shoreline

The geologic features produced by waves and currents depend largely on the character of the interface between land and sea, that is, on the type of shoreline. Of these there is a considerable variety. One type is at the outer edge of great deltas, such as those of the Mississippi and the Nile; another is formed where lava moves down the sides of volcanoes into the sea, as in Hawaii and Iceland. Shorelines may be low and swampy, like that of the Everglades in Florida, or fringed by coral reefs, as exemplified by the famous Great Barrier Reef along the northeastern coast of Australia. A recent classification of shorelines includes two primary classes, six subclasses, and sixteen varieties.

In the paragraphs that follow, some of the chief features produced by waves and currents along two common and extensive types of shoreline are described.

### Features of Mountainous Shorelines

Mountainous shorelines are characterized in cross section by rather abrupt descent into deep water (Figure 7-21), and in plain view by marked irregularity, with prominent headlands separated by coves or deep embayments (Figure 7-22). Such shorelines may be produced in several ways. In New England and eastern Canada, a region

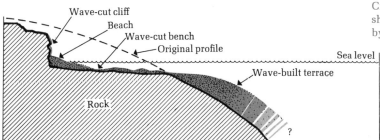

**Figure 7-21**
Cross section of a mountainous shoreline, showing features produced by wave attack.

**Figure 7-22**
Point Reyes, on the California coast north of San Francisco. A complex of resistant rocks in the foreground displays a prominent wave-cut cliff. Beyond, the curving shoreline of Drakes Bay (right of center) has been simplified and smoothed by spits and bay barriers. The San Andreas fault zone crosses the view in the middle distance.

Sir Francis Drake is believed to have anchored his ships here in the summer of 1579. (Photo by Aero Photographers.)

**Figure 7-23**
Air view of Destruction Island, coast of Washington. As shown by its flat top, the island is a remnant of an uplifted wave-cut bench. Ribs of resistant rock extend out onto present wave-cut bench. Note sand spit on near end of island. (Photo by A. E. Weissenborn, U. S. Geological Survey.)

deeply eroded by ice sheets (Section 12-4) was partially submerged, so that high hills became islands, valleys became deep embayments, and divides between valleys now extend seaward as promontories or headlands. The steep shorelines along our West Coast were formed as a result of mountain building that has taken place along the continental margin.

A geological rule might be informally stated, "The more prominent the feature, the more vigorous the attack by erosion." This holds especially true along irregular shorelines: a promontory that extends seaward receives the fullest force of wave attack. The chief erosional features produced are a **wave-cut cliff** (Figure 7-22) and a **wave-cut bench** (Figure 7-23). Some of the rock and mineral fragments produced by hydraulic action and abrasion remain along the water's edge as a **beach.** Seaward from the wave-cut bench there may be a **wave-built terrace,** although the precise nature and outer limits of this feature are imperfectly understood.

If longshore drift is active, much sediment is moved parallel to the shore, rather than outward from it. Tests with colored particles have shown that the sand in a typical beach is in slow transit along the shoreline. Indeed, beaches have been referred to as rivers of sand. A sand bar built from the end of a beach into the mouth of an adjacent bay is termed a **spit** (Figure 7-24). If the bar completely crosses the bay, so as to seal it off from the open sea, the term **bay barrier** is applied. A **tombolo** is a sand bar that extends between two islands or connects an island to the mainland.

Promontories are thus progressively blunted, as cliffs retreat and wave-cut benches widen. The mouths of bays are partly filled by drifted sand, and their landward ends

tend to be filled by deltas built by the streams that flow into them. The net effect of all this activity is the straightening and simplification of the shoreline. How rapidly this effect is achieved depends on the severity of wave attack and on the resistance of the rocks in the shore zone. Weak rocks may be readily undermined along the wave-cut cliff, with resultant slumping and rapid removal by hydraulic action. Tough, massive rocks yield very slowly (Figure 7-22). Sooner or later, however, a stage of approximate equilibrium is reached.

That this series of events may be interrupted is clearly shown along certain coasts by the presence of wave-cut benches and cliffs that now stand many feet above sea level (Figures 7-23, 7-25). Repeated uplift is recorded along stretches of the southern California coast by as many as 13 of these benches, the highest of which is well over 1,000 feet above the sea. Along coasts with a history of downsinking, on the other hand, there are undoubtedly submerged benches and cliffs, which were "drowned" when the land sank or sea level rose. Abandoned beaches and wave-cut features along the slopes of inland mountains mark the various levels at which lakes once stood (Figure 3-2).

*Features of Low-Plains Shorelines*

Low-plains shorelines slope very gently seaward from a land surface that is nearly flat (Figure 18-11). Typically

**Figure 7-24**
Sodus Bay, Lake Ontario. A spit extends into the view from the left; its end has been cut off in building an entrance channel to the bay. To the right of the channel, a small island is connected to the mainland by a narrow tombolo. Note how wave erosion of the hilly terrain (upper right) has combined with deposition across the bay mouth to straighten and simplify the shoreline. Elongate hills and islands are drumlins (Section 12-4). (Photo by Wahl's Photographic Service, Inc.)

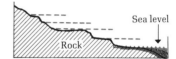

**Figure 7-25**
Cross section of a steep coast, showing a series of raised wave-cut cliffs and benches. Each dashed line marks the position of sea level at an earlier time.

they are produced by moderate regional uplift of a broad continental shelf (Figure 2-3; Section 18-1), bringing the edge of the sea into contact with what was formerly the shallow sea floor. Low-plains shorelines extend from New Jersey southwestward along the Atlantic and Gulf coasts.

The extremely gradual seaward slope on low-plains shorelines means that advancing waves "drag bottom" far offshore, and it is here that their turbulent energy is concentrated in a line of breakers. The chief features produced are low ridges of sand that extend parallel to the shore and some distance out from it. These are termed **longshore bars** if submerged, and **barrier islands** if built above sea level. A remarkable series of barrier islands extends along much of the Atlantic Coast; Atlantic City and the city of Miami Beach stand on such islands. The long narrow bodies of shallow water that separate barrier islands from the mainland are called **lagoons** (Figure 7-26).

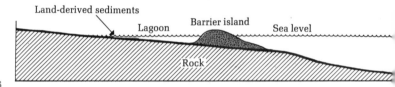

**Figure 7-26**
Cross section through lagoon and barrier island, as developed along a low-plains coast. Vertical scale is exaggerated; seaward slope of the land is much gentler than shown.

Barrier islands are easy to describe but difficult to explain. Some students of the subject believe they originated as spits; others think they were produced when turbulent waters in the line of breakers scoured sand from the bottom and heaped it up into a deposit that was eventually built above sea level.

Low-plains shorelines are modified by waves and currents so as to present a simple, relatively smooth front to the sea. This is exceptionally well shown along the North Carolina coast, where highly irregular estuaries of the Pamlico River and other streams are separated from the open ocean by the long narrow barrier islands termed the Outer Banks.

## Man and Shorelines

Waves may undercut cliffs along the shore and cause landsliding, to the dismay of the owners of clifftop homes. Unfortunately, wave attack from the open sea against weak rock is very difficult to arrest, as storm waves are likely to tear up protective walls or other structures if these are anchored in weak bedrock. On irregular coastlines, where nature has already provided a measure of shelter from incoming waves, bays may be protected, or spits extended, by **breakwaters.** These are essentially large-scale stone walls, held together with concrete and veneered with large loose blocks of stone each weighing many tons. Built with a slope to seaward, they cause advancing waves to dissipate much of their energy among the massive blocks. Breakwaters in the Great Lakes are commonly made by driving two parallel rows of wooden posts, or **piles,** into the lake floor and filling the space between with blocks of stone.

Although the sand in most beaches is slowly moving along the shoreline, the beach is maintained so long as the supply of sand is continuous; if the supply fails, waves will erode the beach, leaving coarser material. If the sandy beach happens to be a resort area, its erosion will have serious economic results. Sand may be dredged at the up-drift end of the beach and supplied when needed; or low retaining walls, called **groins,** may be built at right angles to the beach in order to arrest the migrating sand.

Longshore drift may deliver unwanted sand to a bay or harbor, making it necessary to operate a dredge to keep the channel deep enough for navigation. The dredged-up sand may be added to a preexisting spit or bar; used to construct marinas or man-made islands; or supplied to the beach farther along the shoreline from the harbor.

The work of waves and currents along the world's inhabited shorelines clearly presents a variety of problems. Equally clearly, solution of these problems is so costly that it ordinarily can be borne only by the municipal, state, or federal government. Wasted money and effort can be avoided if the geologic processes involved are thoroughly understood.

## 7-9 WIND

We consider that man lives at the bottom of a sea of air. One of the facts of life in this environment is that this "sea" is not stagnant. Its constant agitation is reflected in the moving currents of air that we call **wind.**

Like currents of water, wind is capable of picking up loose soil and regolith and moving it about. Wind work is important in deserts, but it is by no means restricted to them. All that is needed is dry ground and a supply of material that is fine-grained enough to be moved. The sand dunes to be found on many beaches and barrier islands testify to the energy of strong winds in a setting that is far from arid. Great dust storms have originated in regions that are not deserts, but plains that have been plowed and have then become dust-dry owing to an unusual series of rainless seasons. Wind work is negligible in regions where moisture and vegetation tie down soil and regolith.

Wind action, unlike that of streams or waves, is not concentrated against a particular part of the lithosphere, but is diffused over great areas. Further, the density of air is so low that wind is incapable of moving coarse sediment. For these reasons, wind is a relatively minor agent of modifying the lithosphere. Its ability to transport appreciable volumes of fine-grained sediment is demonstrated at times, particularly after volcanic explosions (Section 4-2) that pour clouds of pyroclastic dust into the atmosphere.

## Erosion

### Deflation

The term **deflation,** which means blowing away, is appropriately applied to one of the processes of wind erosion. We can see it in action on a windy day in a bare field or along a dusty street. Wind, like running water, is highly turbulent. Dust is carried in a maze of updrafts, eddies, and crosscurrents; coarser grains advance by jumping along the surface, much like the bed load at the bottom of a stream (Section 7-4).

Wind is a highly effective sorting agent. Fragments the size of coarse sand are too heavy to pick up. Medium-grained sand (grains about 0.1 to 0.3 mm across) is blown along within a few inches of the ground, exceptionally a foot or so above it. This sand is almost completely separated from finer sand, and this in turn from silt and dust, which are lifted high into the air and may be transported for scores or hundreds of miles before settling out of suspension.

In areas where the regolith is dry and consists of mixed fine and coarse material, for example on the surface of a debris flow (Section 6-2) or an alluvial fan (Section 7-6), silt- and clay-size particles are removed by deflation, but the coarser fragments remain behind. The surface is lowered by the amount of fine material blown away, and eventually becomes "paved" with a residue of fragments too coarse to be moved. Such a mosaic of rock fragments,

**Figure 7-27**
Dunes, Great Sand Dunes National Monument, Colorado. Wind blows from left to right. Stony desert pavement in foreground. (Photo by J. H. Rathbone.)

termed **desert pavement,** prevents further erosion by the wind. An example of desert pavement is shown in the foreground of Figure 7-27. Vast areas of the world's deserts consist of this sort of stone-veneered surface.

### Abrasion

Wind-driven grains of quartz sand are quite active tools of abrasion. Since these grains are seldom lifted more than a foot above the ground, however, rock carving by abrasion is sharply limited in the vertical dimension. An often cited evidence of the effectiveness of abrasion by windblown sand is the fact that telephone poles in sandy areas have been cut through at their bases.

Interesting products of abrasion are wind-cut stones, or **ventifacts.** A partly buried rock fragment may have its exposed surface planed, or **faceted,** by sand abrasion. If the fragment is later tilted into another position, a second facet develops. Along with faceting of ventifacts goes the development of a high polish.

## Deposits

### Dunes

When wind that is blowing sand along the ground encounters an obstruction such as a bush or boulder, its forward energy is decreased and some of the sand comes to rest. The resulting heap of sand enlarges the obstruction, and a **dune** begins to grow. The typical dune has a long slope on the side toward the wind, and a short, steeper slope on the lee side. The wind blows sand grains up the windward side and sweeps them over the crest into the "wind shadow" on the lee side (Figure 7-28). Here they fall to rest, accumulating at the angle of repose for loose sand. This leeward face of a dune is termed the **slip face.**

If sand in dunes accumulates in the manner just outlined, then the dune as a whole should show stratification parallel to the slip face; and this we find to be characteristic of dunes. The situation shown in Figure 7-28, in which all the layers are inclined in the same direction, would be produced by a wind blowing constantly from directly left to directly right. Since wind directions are notoriously variable, however, sand is commonly swept over the crest of a dune first in one direction for a time, then in another. The result is a feature termed **cross-strat- ification,** in which the layers, or strata, are inclined at an angle to the top and bottom of the dune or other unit

**Figure 7-28**
Cross section through a migrating dune. Sand grains are blown from the windward side and settle on the leeward side, or slip face. Thus the dune is stratified parallel to the slip face.

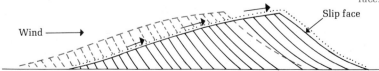

Wind ⟶                                   Slip face

in which they occur. (Sand and gravel deposited by running water are also typically cross-stratified.) The presence in ancient sandstones of cross-stratification of the kind seen in dunes today is clear indication that the sandstone is a "fossil dune" deposit (Figure 8-11).

A second consequence of the processes outlined above is dune movement, or migration. In Figure 7-28, it is clear that subtraction of sand grains from the windward side and their addition to the slip face produces a gradual downwind shift of the whole dune. Desert dunes may migrate hundreds of feet during a year; coastal dunes, in the presence of moisture and perhaps a little vegetation, move more slowly. Dune migration can be reduced or stopped by planting suitable grasses or other cover to hold the sand.

The majority of dunes form "waves of sand" that lie at right angles to the direction of prevailing winds (Figure 7-29). A modification of this transverse alignment is a common form of dune called the **barchan.** Viewed from above, each barchan has the form of a crescent, with the horns pointing downwind. Barchans tend to form where winds are strong and the supply of sand is limited.

The smallest dunes are only a few feet high. At the opposite extreme are hills of sand like those in the Great Sand Dunes National Monument in south central Colorado, which reach heights of 600 feet or more (Figure 7-27). At a few times in the geologic past, wide areas of dune sand have been covered with younger sediment and consolidated into rock, thus becoming a part of the geologic record (Figure 8-11).

**Figure 7-29**
Dunes, Death Valley area, California. Wind direction is from upper left to lower right. Scale is indicated by clumps of bushes. (Photo by Sarah Davis.)

### Loess

**Loess** (pronounced approximately *luss*) is wind-deposited silt, uniformly buff in color and without stratification.

Although soft enough to be readily dug with a shovel, loess does not slump but will stand in steep or even vertical faces, for example in road cuts (Figure 7-30). The reason lies in the cohesiveness and the sharp, angular character of the constituent silt grains, which tend to resist slumping or settling; and also in the large volume of void space in the loess, which allows water to percolate into it instead of concentrating on the surface and washing it away. That loess is a wind-laid deposit is clearly shown by the fact that it blankets hills and valleys indiscriminately, even in regions where these features have a relief of hundreds of feet.

The thickest deposits of loess are found on the downwind, or leeward, side of great deserts. The best known of such occurrences is in northern China, where hundreds of square miles are underlain by loess 100 feet thick or more. This material is the product of deflation by strong, long-continued winds that blew eastward from the dry reaches of central Asia, particularly from the Gobi Desert.

A second kind of loess occurrence is in regions that were marginal to Pleistocene ice sheets (Section 12-5). Rock debris, ground down to silt size by glacier abrasion, was left behind in large amounts when the ice sheet melted away. It was then picked up by winds and deposited over thousands of square miles beyond the former ice front. Loess deposits in northern Europe and in North America originated in this way. Loess is widespread in Illinois, Iowa, Nebraska, and adjacent states, and along the east side of the Mississippi River valley in Tennessee and Mississippi. Loess characteristically weathers to produce fertile soil.

Streams, the most important of the several agents of erosion, are considered as energy systems. They possess internal energy in the form of turbulence. Stream discharge, or volume, generally increases from source to

**Figure 7-30**
Loess, though unconsolidated, typically stands in vertical faces like this. Pleistocene; North Omaha, Nebraska. (Photo by R. D. Miller, U. S. Geological Survey.)

**7-10
SUMMARY**

mouth, whereas the gradient, or slope of channel, tends to decrease. Velocity, or rate of flow, is dependent on turbulence, discharge, and gradient, and hence is highly variable. Streams acquire mechanical load through hydraulic action and abrasion, which produce such stream-channel features as waterfalls, plunge pools, and potholes. Solution load is acquired through inflow of underground water, and is eventually added to the salts in the sea. Clay, silt, sand, and coarser fractions of the mechanical load are carried either in suspension or as bed load along the stream bottom. Suspended and bed loads are dropped when a stream loses energy, as may happen at places along the channel, especially in large alluvial rivers. Point bars, natural levees, and floodplains are among the deposits formed. Material that accumulates at a stream's mouth forms a delta if laid down in water, and an alluvial fan if deposited on dry land. Stream deposits are temporary features from the standpoint of geologic time. Stream processes are in delicate balance in nature, and man upsets the balance only at his peril.

The work of waves and currents is important in a narrow zone along shorelines and seaward from them. Waves erode by hydraulic action and abrasion; they are also sorting agents, separating sand from finer and coarser materials. Currents tend to shift sand along the shore and to move the finest sediments seaward. Features produced by waves and currents depend in large part on the type of shoreline. Mountainous shorelines tend to be simplified and straightened as headlands are worn back with the development of wave-cut cliffs and benches, and as deposition takes place across the mouths of bays forming spits and bay barriers. Cliffs and beaches that stand many feet above sea level testify to recent local uplift of the land. Low-plains shorelines are characterized by barrier islands, the general effect of which is to straighten and simplify the coast. Barrier islands are separated from the mainland by a lagoon.

The work of wind is widely diffused over large areas. Silt and dust may be removed from a land surface by deflation, leaving coarser fragments behind to form desert pavement. Sand is moved about locally. Wind-driven sand is a potent agent of abrasion, as indicated by the faceted and polished stones called ventifacts. Dunes are heaps of sand, typically cross-stratified, which migrate when sand grains are swept from the low-slope windward side over the crest and onto the steeper leeward side, or slip face. Wind-deposited silt is termed loess. Large areas are blanketed with loess, derived either from desert regions or from areas of pulverized rock debris that were exposed when the Pleistocene ice sheets melted. Loess blankets hills and valleys alike, and weathers to fertile soil.

## SUGGESTED READINGS

Bascom, W. 1960. Beaches. *Scientific American,* vol. 203, no. 2, pp. 80–94. (Offprint No. 845. San Francisco: Freeman.)
Deals with the formation and preservation of sand beaches, and points out what happens to the longshore flow of sand when man interferes by constructing groins and break-waters.

Bloom, A. L. 1969. *The Surface of the Earth,* chap. 4, pp. 53–80. Englewood Cliffs, N.J.: Prentice-Hall. (Paperback.)
A treatment of streams and channels that includes the quantitative measurement of running water in addition to a relatively complete coverage of geologic features formed by stream erosion and deposition.

Dunbar, C. O., and Rodgers, John. 1957. *Principles of Stratigraphy,* chap. 1, pp. 2–27. New York: Wiley.
College-level text on principles of stratigraphy and environments of deposition. This chapter deals with sedimentary processes.

Leopold, A. S., and the Editors of *Life.* 1962. *The Desert,* chap. 2, pp. 27–50. Life Nature Library. New York: Time Inc.
A well-illustrated nontechnical presentation written for the layman, which includes the geologic work of both wind and water in the desert environment.

Leopold, L. B., and Langbein, W. B. 1960. *A Primer on Water.* U.S. Geological Survey. Washington, D.C.: Government Printing Office. 50 pp.
A nontechnical publication for the layman that provides general information about hydrology, the development of water supplies, and the use of water.

Leopold, L. B., and Langbein, W. B. 1966. River meanders. *Scientific American,* vol. 214, no. 6, pp. 60–70. (Offprint No. 869. San Francisco: Freeman.)
A report on the geometry and mechanics of river meanders and why they develop.

Matthes, G. H. 1951. Paradoxes of the Mississippi. *Scientific American,* vol. 184, no. 4, pp. 18–23. (Offprint No. 836. San Francisco: Freeman.)
Shows why a number of generally accepted assumptions about the Mississippi River are not true.

Morisawa, M. 1968. *Streams: Their Dynamics and Morphology.* New York: McGraw-Hill. 175 pp. (Paperback.)
An excellent comprehensive treatment of the work of running water and the resulting landforms, written in terms comprehensible to those with little previous knowledge of the subject.

Tuttle, S. D. 1970. *Landforms and Landscapes,* chaps. 3, 4, and 8, pp. 25–47 and 107–121. Dubuque, Iowa: W. C. Brown. (Paperback.)
Chapters 3 and 4 deal with stream erosion and deposition, and the geomorphic features that result from the work of running water. Chapter 8 covers wind action, waves, and shorelines.

# 8 SEDIMENTARY ROCKS AND PROCESSES

8-1  The Nature of Sediments

8-2  Clastic Rocks

8-3  Organic Rocks

8-4  Chemically Formed Rocks

8-5  Features of Sedimentary Rocks

8-6  The Importance of Sedimentary Rocks

8-7  Summary

Viewed as a component of the whole lithosphere, the sedimentary rocks may be thought of as a thin veneer lying on the granitic rocks that make up the bulk of the continental crust (Section 2-3). This veneer is variable in thickness and patchy in distribution, in places flat-lying and elsewhere warped or broken. Nevertheless, sedimentary rocks form the bedrock over some three-fourths of the earth's land surface.

If you arm yourself with a geologic hammer and set out on a field trip, the chances are therefore good that the first bedrock you come to will be some kind of sedimentary rock. How will you be able to tell that the rock you see is sedimentary, and how can you identify it as to specific variety?

We have seen that the chief products of rock weathering, which are fed to streams by downslope movements and eventually washed off the continents into the sea, are sand, clay, and various chemical compounds in solution. The sand is made up of grains of quartz, feldspar, mica, and other minerals, but by far the most plentiful is quartz. This mineral is so abundant that when we use the term *sand* we customarily mean quartz sand. The clay consists of microscopically small particles of silicate minerals, including flakes of micalike minerals, and is so fine-grained that its individual minerals can be identified only by special laboratory techniques. The salts in solution include calcium carbonate, magnesium carbonate, calcium sulfate, sodium chloride, silica, and a number of other substances in smaller amounts.

What happens to these land-derived materials when they reach the sea? If we put a mixture of water, sand, and clay into a glass container, shake them thoroughly, and place the container at rest, we find, predictably, that the solid material settles to the bottom. (The Latin *sedimentum* means "a settling.") But we also find that the coarser, heavier grains of sand settle first, making a layer on the bottom, and that the clay settles more slowly, finally coming to rest as an upper layer, separated more or less distinctly from the sand below. Standing water evidently exerts a sorting action on suspended sediment. We should expect to find a similar result in natural water bodies, including the sea; and a large number of layered rocks do indicate just such a mode of formation as outlined above. They differ from sand and clay only in being more or less consolidated. Geologists speak of layered rocks as being **bedded,** or **stratified,** or as occurring in **beds,** or **strata** (Figure 2-7).

If we dissolve ordinary salt in water and let a pan of its solution evaporate to dryness, the salt will reappear as a precipitate on the bottom of the pan. That this same event has occurred in nature in past ages, on an enormously greater scale, is shown by the presence of beds of salt among the sedimentary rocks of the crust. A similar origin is indicated for deposits of anhydrite (calcium sulfate).

Oysters, clams, corals, and other marine animals build hard shells or skeletons from material that they obtain from the waters in which they live. The material is calcium carbonate, and it is present in seawater by virtue of having been brought to the sea in solution from the lands. Many clear shallow sea floors are today carpeted with oyster beds, coral colonies, or shell fragments broken up by the waves and distributed about. Once again, we find just such beds of shells and shell fragments, in fossil form, in the solid rocks.

Thus each of the chief kinds of material contributed to the sea over the ages has found its way into sedimentary rocks of some kind—by settling out, by chemical precipitation, or by organic processes. Let us now enumerate the principal kinds of sedimentary rocks, and then see what we can deduce from them in the way of geologic history. You should refer to Table 8-1 as you study the remainder of this chapter.

**8-2**
**CLASTIC**
**ROCKS**

It is convenient to consider the sedimentary rocks in three groups, of which the first is the **clastic** rocks. The term means fragmental or broken; the clastic rocks consist of fragments, grains, and particles of mineral and rock material, which accumulated as sediment by settling out of water, and later became **lithified** (consolidated into rock). From coarsest-grained to finest-grained, and also from least abundant to most abundant, the main clastic rocks are conglomerate, sandstone, siltstone, and shale.

### Conglomerate

**Conglomerate** is lithified gravel (Figure 8-1). It consists of rounded, water-worn fragments, from pebbles a fraction of an inch in diameter up to boulders in size, mixed with more or less coarse sand. There are all gradations, from coarse bouldery conglomerate, such as might be found in an alluvial fan, through sand-and-pebble rock of the kind that British geologists aptly call "pudding-stone," to sandstone with occasional layers of small pebbles. A coarse clastic rock in which the large fragments are sharp and angular instead of rounded is termed **breccia** (Figure 8-2).

No generalization can be made as to the composition of conglomerates beyond saying that they tend to be composed of rock and mineral fragments that are resistant to solution and abrasion. Each conglomerate reflects the

Table 8-1  Classification of the Sedimentary Rocks

| Group | Material | Process of Lithification | Rocks | Abbreviation | Pattern on Cross Sections |
|---|---|---|---|---|---|
| CLASTIC | Gravel: boulders, pebbles, quartz sand Larger than 2 mm | Cementation | Conglomerate | cong. | |
| CLASTIC | Sand: chiefly quartz grains 1/16 – 2 mm | Cementation | Sandstone | ss. | |
| CLASTIC | Silt: very fine quartz and mica grains 1/256 – 1/16 mm | Cementation | Siltstone | slst. | |
| CLASTIC | Clay, mud: clay minerals, organic matter Smaller than 1/256 mm | Compaction | Shale | sh. | |
| ORGANIC | Limy mud and shell fragments | Cementation, compaction, crystallization | Limestone | ls. | |
| ORGANIC | Peat: partly decayed plant materials | Compaction | Bituminous Coal | | |
| CHEMICAL | $SiO_2$ in solution | Chemical precipitation | Chert | cht. | |
| CHEMICAL | NaCl in solution | Chemical precipitation (Evaporation) | Salt | | |
| CHEMICAL | $CaSO_4$ in solution | Chemical precipitation (Evaporation) | Anhydrite, Gypsum | anhy., gyp. | |
| CHEMICAL | $CaMg(CO_3)_2$ in solution (in ground water) | Replacement (of limestone) | Dolostone | dol. | |

147

**Figure 8-1**
A coarse conglomerate. The poor sorting, absence of stratification, and rounded nature of fragments indicate deposition in a stream channel. Upper Cretaceous of eastern Utah. (Photo by E. M. Spieker.)

**Figure 8-2**
Specimen of breccia. The light-colored fragments are dolostone; they were once part of a continuous bed above a bed of gypsum. When the gypsum was dissolved away in weathering, the dolostone collapsed into angular fragments. Later the intervening spaces were filled with fine red sand, which became firmly cemented to form the dark matrix of the breccia. Pennsylvanian of eastern Wyoming. The specimen is about 5 inches long.

kinds of minerals and rocks that were present in the area that was eroded to produce it.

Although some gravels reach the sea and become spread out in beds, more commonly they are restricted to high-energy environments of deposition on land, such as river channels, small deltas, and alluvial fans. Thus their chances of becoming covered up, lithified, and preserved in the rock record are relatively poor, and conglomerates are neither as thick nor as widespread as most of the other kinds of sedimentary rock. Gravels become conglomerates by means described under sandstone in the next section.

148

## Sandstone

A clastic rock in which the grains range in size from 1/16 to 2 mm is **sandstone.** Sandstone has a much less mixed-up composition than conglomerate; quartz is the dominant mineral, and it is very nearly the only one in some unusually pure sandstones. Grains of feldspar and flakes of muscovite are perhaps the commonest clastic constituents next to quartz. Many other minerals are present in small amounts.

The process by which gravel is converted to conglomerate, and sand to sandstone, is **cementation.** This simply means the filling of the void spaces among the grains, thus solidifying the mass. The four common cementing materials are calcium carbonate (calcite), silica (quartz), iron oxide (limonite or hematite), and clay. The nature of the cementing material can usually be told without difficulty. Sandstones cemented with calcite effervesce in dilute hydrochloric acid; those with siliceous cement are very hard and tough, since they consist of quartz grains cemented by quartz; those containing iron oxide are yellow, red, or brown in color. A sandstone having none of these characteristics is probably cemented by clay. Calcium carbonate, silica, and iron oxide are mineral compounds that were introduced by percolating waters into the sand at some time after it was deposited. Clay cement was laid down at the same time as the clastic grains, becoming a binding agent when the rock dried out and was compressed by the weight of overlying sediments. Thus sandstone typically consists of two components: the clastic grains, and the cement that holds them together.

**Figure 8-3**
Interbedded sandstone and shale. Sandstone makes the steep cliff at the top of the mesa and the one part way down; shale makes the intervening slopes, which are partly obscured by talus material. Upper Cretaceous rocks in the Book Cliffs of western Colorado. (Photo by E. M. Spieker.)

Although some sandstones bear evidence of having accumulated in river channels, on floodplains, or even in dunes heaped up by the wind, the majority of sandstones were laid down in the sea, and form stratified deposits interbedded with other marine sedimentary rocks (Figure 8-3). Many sandstones formed along the shorelines of ancient seas, in the same manner as the wide sandy beaches of today, and became covered with younger sediments when the sea advanced slowly inland.

## Siltstone

Intermediate in texture and composition between sandstone and shale is **siltstone.** This rock is made up mainly of clastic particles ranging in size from 1/16 mm down to 1/256 mm; there is also a considerable proportion of clay. The particles of silt consist largely of quartz and mica. The rock is commonly tightly bonded by clay, or by a mineral cement such as silica. Siltstone differs from sandstone in its finer grain size and large content of clay. It differs from shale in being recognizably granular or gritty, and in not splitting into thin flakes parallel to bedding.

Siltstone is generally not found in thick beds, but rather in compact slabby beds a few inches thick, separated by thicker layers of shale. The hard thin beds of siltstone commonly stand out in relief on a weathered exposure. Slabs of siltstone are admirably suited for outside walls and flagstone terraces.

## Shale

**Shale** is lithified mud. Its distinguishing characteristic is that it splits, parallel to the bedding, into thin chips or flakes of a thickness from that of cardboard to that of a wooden shingle. The chief constituents of shale are the several clay minerals, but these are so extremely fine-grained that we cannot identify them by ordinary means. Rather, we describe shales by such aspects as color and the presence of nonshaly material. Many shales are neutral shades of gray, but some are black, owing to an appreciable content of organic matter, or red, from the presence of a little hematite. Most shales are more or less *silty;* we may term a shale *micaceous* if it contains recognizable flakes of mica, *sandy* if sand-size quartz grains are present, *carbonaceous* if the shale is dark with organic matter, or *calcareous* if calcium carbonate is a constituent.

When a deposit of mud is buried by younger deposits, their weight tends to drive the water out of the mud and to force the solid particles into closer and closer contact. This process, termed **compaction,** is what converts mud into shale. The individual clastic particles consist mostly of flakelike clay minerals, and during compaction they come into close contact and assume a more or less parallel position, with their flat sides horizontal (at right angles to the compressive force). Thus shale comes to have its characteristic platy bedding.

Mud and clay, being so very fine-grained, tend to remain in stream suspension longer than the coarser clastics, and to be washed off the lands in greater quantities. That this has been true in the geologic past is indicated by the fact that shale makes up about 80 percent of all the sedimentary rocks, and that most shales give evidence of having been deposited in the sea (Figure 8-3).

Under this heading we place two kinds of sedimentary rock, namely, limestone and coal. All that they have in common is that each came into existence through the agency of living things. In other respects they are entirely different.

**8-3 ORGANIC ROCKS**

### Limestone

A **limestone** is any rock that consists of 50 percent or more of calcite. The great majority of limestones have been produced by marine animals, and a few primitive marine plants, that build their shells and skeletons of calcium carbonate extracted from the sea water. When the organisms die, their hard parts accumulate on the sea floor, where they may be covered and preserved.

Thus we might expect most limestones to consist of well-preserved, complete fossil shells, perhaps with enough mud to act as a cement to hold them together. Such limestones are known but are not the most common kind. Shell-secreting organisms are most abundant in shallow water, where there is light and where food may be extracted from passing currents. But shallow waters are periodically thrown into violent commotion by storm waves, which may tear the shells free and break them into fragments or grind them into limy mud. Thus a great many limestones have originated as layers of broken-up fragments, so-called fossil hash, in which we may recognize bits and pieces but few if any complete shells (Figure 8-4). When such fragmented material is moved about by waves, or swept across the sea floor by submarine currents, it may be washed and sorted in the same way as sand; limestones with a history of this kind are correctly interpreted as both organic and clastic in origin.

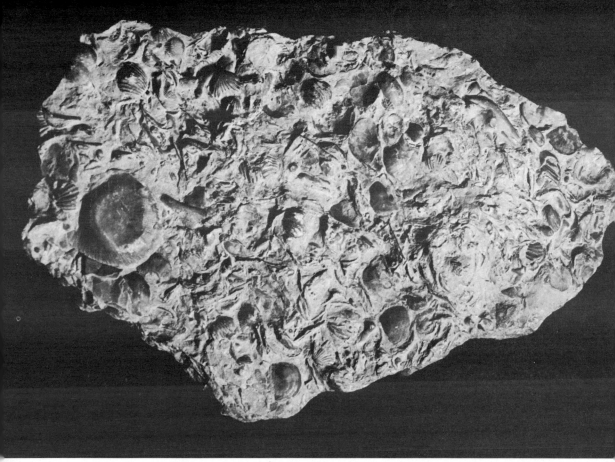

**Figure 8-4**

A "fossil hash" limestone, made up of the whole and fragmented shells of marine invertebrates. An Ordovician limestone of southwestern Ohio.

Once a limestone is formed and is raised above sea level along with the associated strata to become part of the continental bedrock, its calcite may be dissolved and reprecipitated by waters that percolate through it. This **recrystallization** has the effect of altering the original fabric of shell fragments and mud to a fabric of interlocking calcite grains, with plainly visible cleavage surfaces. Fossils are partly or entirely obliterated in such a recrystallized, or "crystalline," limestone.

Added to these factors are several others that make limestones a highly variable group of rocks. Limestones may be gray or black, owing to disseminated organic matter or they may be nearly white, or various shades of tan and buff. Some occur in massive beds several feet thick, whereas others are thin-bedded and slabby (Figure 8-5). Scattered quartz grains may be present. Thus it is clear that in describing a given limestone, it is wise to use several appropriate adjectives—for example, "light gray fossiliferous thin-bedded sandy limestone."

Some limestones are not even stratified, but occur in moundlike masses ranging in size from that of an overturned bushel basket to masses hundreds of feet thick and miles in length. These are fossil **reefs,** similar in origin to today's coral reefs (although they contain the remains of many groups of organisms besides corals) (Figure 8-6).

**Figure 8-5**
A contrast in limestones. The light gray rock is a high-purity limestone (nearly all calcite), which is quarried for the manufacture of lime (calcium oxide). The dark rock is also limestone, but it is thin-bedded and contains shaly layers and nodules of chert. Ordovician strata, northern Virginia.

**Figure 8-6**
A reef in Mississippian strata, Sacramento Mountains, New Mexico. Core of the reef surrounds the tent-shaped limestone mass in center of view. Stratified limestone can be seen dipping away from the reef on both sides.

## Coal

**Coal** occurs as beds with other sedimentary rocks and is organic in origin. In all other respects it is different from the other rocks discussed in this chapter; coal is an exception to practically all the statements we have made.

To begin with, coal is not made up of mineral matter, but of the carbonified remains of plants. Secondly, these plants lived on the continents, not in the sea, so that coal is entirely nonmarine in origin. Third, little or no transportation is involved in the history of coal: the plants lived, died, and were buried in the same place.

Present-day **peat** is a brown porous spongy mass of partly decayed wood, leaves, seeds, bark, and other plant remains, which accumulates in swampy lowlands. At a few places in coal beds, masses of ancient peat have been found that have been preserved from alteration to coal by being impregnated with calcite; and study of these fossil peats has shown that they are identical with modern peat (except for the species of plants, which are different from those living today). Thus it is clear that peat is the parent material of coal. When covered with sediment, peat compacts to a more solid material known as **lignite;** mounting pressure from deeper burial converts lignite into **bituminous coal,** or simply coal. A thickness of about three feet of peat is necessary to form one foot of coal. In the progressive change from peat through lignite to coal, moisture and gases that result from decay of the plant material are gradually driven off, but much of the carbon remains, and carbon increases in proportion. Thus the process of conversion of plant matter to coal is appropriately termed **carbonification.** A typical bituminous coal contains more than 65 percent carbon. (Anthracite, or "hard coal," is a product of metamorphism of bituminous coal (Section 15-4). The carbon content of anthracite is 85 percent or more.)

A tree that is blown over or a branch that falls to the ground, in an ordinary forest or woodlot is attacked by fungi, insects, and air-dwelling bacteria, and in a few years it rots away completely. But woody matter falling into stagnant swamp water is attacked only by bacteria that can live in such waters, and its decay is arrested at an early stage because the action of these bacteria is inhibited by their own waste products. Plant material in the water of peat swamps is thus sealed off from further decay; it is, in effect, pickled.

Bituminous coal is a black rock, finely banded with bright and dull layers, that breaks with a blocky fracture (Figure 8-7). Coal beds range in thickness from a fraction

**Figure 8-7**
Typical bituminous coal, showing
bright and dull banding and blocky
fracture. Pennsylvanian of eastern
Ohio. About three-fourths natural size.
(Photo by J. M. Schopf, U. S.
Geological Survey.)

of an inch to many feet. They generally rest on a layer
of **underclay,** which appears to be the much altered soil
in which the coal-forming plants grew. Tree stumps are
occasionally found in coal beds, with their roots extending
into the underclay.

Coal beds are numerous and important in Pennsylvanian rocks of the eastern United States (Figure 8-8), and
in Cretaceous and Paleogene rocks of the western Great
Plains and Rocky Mountains. The Pennsylvanian coal
beds and stream-deposited sandstones and sandy shales
tend to alternate with beds of shale and thin limestones
that contain marine fossils (Section 22-3). To account for
this cyclical repetition of coal-bearing nonmarine strata
with marine strata is one of the puzzles of sedimentary-rock geology.

**Figure 8-8**
Coal beds exposed in a highway cut. Behind the man is a single bed, the Pittsburgh coal; just above this is a thinner coal bed. The man is standing on underclay about one foot thick. Beneath the underclay is a thin discontinuous bed of limestone, and below this is shaly sandstone to road level. The rock above the coal beds is mostly shale. The openings on either side of the view are old mine entries from which coal was dug many years before the rocks were laid open for the highway. Pennsylvanian strata near Pittsburgh. (Photo by J. M. Schopf, U. S. Geological Survey.)

Rocks formed by chemical means include varieties that are nonclastic and inorganic in origin. The chief types are chert, gypsum, rock salt, and dolostone.

**8-4
CHEMICALLY
FORMED ROCKS**

### Chert

**Chert** is a hard, dense rock made up of extremely fine-grained silica. Most chert is white or gray; black varieties are called **flint.** Chert breaks with conchoidal fracture, giving shallow curved surfaces and sharp edges.

Stratified deposits of chert are relatively rare. Much more commonly the rock occurs as knobby potato-sized masses, or **nodules,** or as discontinuous irregular beds, in limestone and dolostone (Figure 8-9). Some chert apparently originated when jellylike masses of silica settled from seawater to the bottom, where they hardened into nodular form. Other cherts seem to be products of **silicification,** a type of replacement (Section 4-4) in which silica-bearing waters percolating through limestone dissolve a part of the calcite and precipitate silica in its place. Some chert has clearly been formed in this way, as it faithfully reflects the textures and even the fossils of the original limestone.

The siliceous nature of chert means that it is highly

**Figure 8-9**
Nodules and thin beds of chert. These
are more resistant to weathering than
the dolostone in which they occur, and
so they stand out in relief. (Photo
courtesy of New Mexico Bureau of
Mines and Mineral Resources.)

resistant to abrasion (Figure 7-5). It is also much less
soluble in surface waters than the surrounding limestone
or dolostone. On weathered surfaces the chert tends to
stand out in relief, and ledges of cherty limestone or
dolostone tend to have a bumpy, knobby, or ribbed
appearance (Figure 8-9).

## Gypsum and Rock Salt

**Gypsum** consists of the mineral of the same name,
$CaSO_4 \cdot 2H_2O$, and **rock salt** consists of halite, $NaCl$. Thus
each rock is made of a single mineral. Both rocks occur
in thick widespread beds in certain parts of the geologic
column. They are commonly found together, but not
always.

Crusts of precipitated salts are today being deposited
in salt lakes and coastal lagoons in arid regions, and this
fact provides us with the key to the origin of ancient
deposits. Since gypsum and salt were produced through
evaporation of a water body, together they are conveniently referred to as **evaporites.** We conclude that they
were formed by the drying-up of wide, shallow seas on
the continent, cut off from the open ocean by sand bars,
a line of reefs, or some other barrier. Under a climate
so arid that loss by evaporation was greater than addition
by rainfall and stream inflow, such a shallow sea would
in effect become a great evaporating pan, and its contained
salts would be precipitated as the water dried up. The
great thickness of some deposits of gypsum and salt,
several hundreds of feet, indicates that this process must
have been repeated many times, the barrier being

**Figure 8-10**
The prominent light-colored bed in the lower part of the cliff is gypsum. Note the deep solution channels. The overlying strata are shale (covered with sliderock) and sandstone at the top of the cliff. Jurassic strata, northwestern New Mexico. (Photo by R. W. Foster.)

breached periodically to allow more seawater to flow in, and the floor of the basin sinking gradually to accommodate the thickening deposits of evaporites.

Beds of gypsum are exposed at the surface at a number of places (Figure 8-10). Where these same beds are encountered by the drill under a thick cover of younger rocks, they turn out to consist of anhydrite, $CaSO_4$. Apparently the calcium sulfate was originally deposited as anhydrite, and this has been hydrated to gypsum by near-surface waters in the zone of weathering. Rock salt is so soluble that it is removed by these waters, and beds of salt can be found only by drilling wells or sinking shafts. Sandstone, shale, and thin-bedded dolostone are the chief sedimentary associates of gypsum and salt.

### Dolostone

**Dolostone** is a rock made up of the mineral dolomite, $CaMg(CO_3)_2$. It may form as an evaporite, like gypsum and salt, with which it is interbedded at a number of places. This type of dolostone is gray or tan, fine-grained, and uniform in texture.

Much more widespread is the variety of dolostone that is a product of the replacement, or **dolomitization,** of limestone. This process may have occurred soon after deposition of the limestone, perhaps while it was still covered by seawater; or long afterward, through the agency of magnesium-rich underground waters. In any event, the effects of dolomitization are to alter the bulk composition of the rock from calcium carbonate to calcium magnesium carbonate; to rearrange its fabric, through recrystallization, to a more coarsely granular form; to leave a number of irregular voids, especially where fossil shells are dissolved away and not replaced; and thus to mask or obliterate most of the fossils of the original limestone. Though dolostone resembles limestone in a general way, taken as a group dolostones tend to be less variable, as the numerous differences in the original rock are reduced by dolomitization.

## Stratification

Beds of sedimentary rock that are parallel to each other over wide areas are said to exhibit **parallel stratification** (Figures 2-7, 8-3, 8-10). This type of stratification indicates quiet conditions of deposition, with little action by currents or other agents that would interfere with accumulation of sediment in even layers. Some rocks show parallel stratification on a very small scale, individual beds being one-half inch or less in thickness. Such layers are termed **laminae,** and the rock is said to be **laminated** (Figure 10-4).

The observed tendency of most sediments to accumulate as flat-lying strata has led to the **law of original horizontality,** which states that newly deposited strata are horizontal or nearly so. From this law it follows that

8-5
**FEATURES OF SEDIMENTARY ROCKS**

**Figure 8-11**
Cross-stratification in Jurassic sandstone, Zion National Park, Utah. The sandstone is part of a widespread "fossil dune" deposit. (Photo by J. H. Rathbone.)

wherever we find sedimentary rocks that depart appreciably from the horizontal, we know they have been tilted out of their original position by great earth forces (Figure 14-1).

Clastic sediments that are deposited from moving rather than standing water accumulate with stratification inclined in the downcurrent direction, at an angle to the top and bottom of the layer in which it occurs. Changeable winds invariably cross-stratify dune sands (Figure 8-11). The shifting currents in a stream channel cross-stratify the sands and gravels that they deposit. Marine sandstones and clastic fossil-hash limestones commonly show cross-stratification, produced by shifting submarine currents.

Rocks that show cross-stratification obviously did not accumulate in quiet water, but in an environment characterized by high energy. Studies of cross-stratification give us information about the agent that produced it, the direction from which the currents came, and the intensity with which they moved.

### Ripple Marks and Mud Cracks

The surface of a bed of sand that is agitated by waves or currents takes on a rippled form (Figure 8-12), and **ripple marks** are common features in sandstones (Figure 8-13). The size and configuration of ripple marks, and the cross section of the rippled bed where this can be observed, tell us whether the marks were produced by waves or by currents, and, if the latter, in which direction the currents were moving.

You may have seen **mud cracks** (Figure 8-14) in a puddle or roadside ditch from which the water has receded. The cracks are formed as the mud shrinks in drying. This phenomenon has occurred at numerous times and places in the past, as is shown by the presence of mud cracks in shale or as casts in the bottom part of the bed of sandstone or other rock that covered the cracked surface. The presence of mud cracks is clear evidence that the bed in which they occur was exposed to the atmosphere before being buried. Likely sites for mud-crack formation are floodplains and shallow coastal lagoons.

### Concretions and Geodes

A **concretion** is a sharply defined mass of mineral matter in sedimentary rocks, which has been precipitated from solution around a nucleus (Figure 8-15). Concretions range in diameter from less than an inch to several feet, and in shape from spherical to oval, flattened, and irregular. They are commonly composed of material different from

Figure 8-12
Modern ripple marks. Tidal flat, Ganges Delta, India. Scale is in inches. (Photo by E. D. McKee, U. S. Geological Survey.)

Figure 8-13
Ripple marks on a Cretaceous sandstone, near Denver, Colorado. (Photo by J. R. Stacy, U. S. Geological Survey.)

Figure 8-14
Mud cracks on the floor of a desert basin, New Mexico.

the rock in which they occur. Thus we may find concretions of pyrite in shale, of nodular chert in limestone, or of hematite in sandstone. The nucleus at the center of a concretion may be a sand grain, a twig, or a shell fragment. Some concretions evidently formed as the enclosing sediment was being deposited, but others accumulated long after deposition, through precipitation of mineral matter by circulating underground waters.

A **geode** is a hollow rounded body lined with crystals that project into the central cavity (Figure 8-16). Geodes form after the enclosing rock does, probably in cavities dissolved by underground water. Later the cavities become partially filled with mineral matter, commonly calcite or quartz. Since these substances had open space in which to grow, well formed crystals are the rule. For this reason geodes are often seen in mineral collections.

## Fossils

Fossils, which are found in sedimentary rocks almost exclusively, are of central significance in the interpretation of earth history. They are discussed separately in Chapter 9.

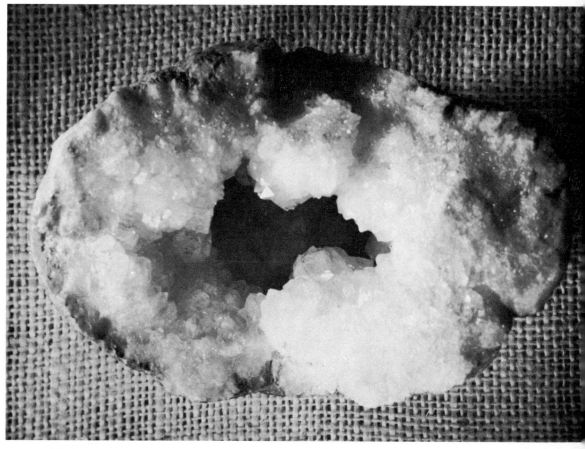

**Figure 8-16**
Part of a geode lined with quartz
crystals. Natural size.

## Scientific Importance

Among the three great classes of rocks, the sedimentary rocks constitute by far the most useful records of earth history. The conditions under which igneous and metamorphic rocks were formed are difficult or impossible to study at first hand, because their sites of formation are largely inaccessible; or to reproduce experimentally, because of the high temperatures and pressures required. Consequently much of our interpretation of the history recorded by these rocks is little more than educated guesswork. The processes of sedimentation, on the other hand, are relatively easy to study in nature or to duplicate in the laboratory. Furthermore, the stratified nature of the sedimentary rocks means that they illustrate the law of sequence and hence are inherently chronologic. Such features as cross-stratification, ripple marks, and mud cracks, to say nothing of fossils, make the sedimentary rocks highly rewarding records of geologic history.

**8-6**
**THE IMPORTANCE
OF SEDIMENTARY
ROCKS**

163

The fact that the vast bulk of sedimentary rocks accumulated in shallow waters shows clearly that their sites of deposition were confined to the continental shelf (Figure 2-2), or to shallow inland seas that flooded parts of the continent in times past. (Hudson Bay of northern Canada, and the Baltic Sea of northwestern Europe, are present-day examples of such shallow seas.) By careful mapping of the nature and distribution of sedimentary rocks of a given age, we can reconstruct the extent of lands and seas as they existed at that time. We can also reconstruct with some confidence the nature of climates in past times, through, for example, a study of evaporites (dry, hot), coals (warm, humid), and limestone reefs (mild, equable). We can deduce that the crust has remained relatively stable where the sedimentary section is thin, but that it has sunk progressively where the section is unusually thick. The fact that in places the much-metamorphosed roots of ancient mountains are now exposed at the surface raises the question: Where did the material eroded from these mountains go? Often we can find the answer in exceptionally thick piles of coarse sedimentary rock that must have been derived from such a source. Owing to their stratified nature, sedimentary rocks readily show the effects of crustal tilting, folding, and other deformation. Finally, in the form of fossils the sedimentary rocks contain the entire record of the development of life on earth.

## Economic Importance

The sedimentary rocks are an indispensable storehouse of useful materials. For one thing, they contain our "fossil fuels": coal, oil, and gas. Coal, once the mainstay fuel for heating homes and factories and for operating railroads, is still used in making coke for steel manufacture, and as a raw material for a great variety of chemical products, including aspirin, dyes, and plastics. The primary use of coal today, however, is as fuel in steam plants that generate electricity. The country's insatiable demand for electric power keeps surface and underground coal mining at a high level of activity.

Oil and gas are fluids of organic origin, which are found in some regions in underground accumulations or "pools." These are not pools in the ordinary sense, but are volumes of porous sedimentary rock that are saturated with oil or gas, generally under pressure. Wells drilled to reach them range in depth from 1,000 feet or less to more than 30,000 feet, or 5.7 miles. Products refined from crude oil provide power and lubrication for our means of transportation, and heat many of our buildings. Gas, which produces fewer atmospheric pollutants than either coal or oil, is in strong demand as a heating fuel.

The immense deposits of iron ore of the Lake Superior region are sedimentary in origin, having come into existence in shallow Precambrian seas. Iron oxide or iron carbonate was precipitated, apparently by chemical or biochemical means, along with great quantities of chert-like silica. The rich hematite ores that supplied the steel industry of the eastern United States for decades were the result of enrichment in iron by intense and long-continued weathering, which removed much of the silica in solution. These rich ores are now largely depleted, but technical advances in metallurgy allow the steel industry of today to make use of unweathered iron-bearing siliceous rock, or **taconite.** There are very large reserves of this material.

If the raw materials just mentioned were the only ones contributed by the sedimentary rocks, these rocks would be valuable enough; but to enumerate these materials is only a part of the story. Limestone and sandstone are quarried, cut, and shaped for architectural use. Limestone and shale are manufactured into cement, and this in turn is mixed with sand, gravel, or crushed stone to make concrete. Clay is used for construction materials like brick and tile, and for ceramic products like porcelain and chinaware. Gypsum goes into plasterboard, and limestone and sand into glass. Salt, limestone, and coal are mainstays of the chemical industry. Thus there are as many practical as scientific reasons for being interested in the sedimentary rocks.

## 8-7 SUMMARY

Sand, clay, and various compounds in solution, washed off the lands, are the main parent materials of the sedimentary rocks. These rocks accumulate in beds or layers, and are said to be stratified. The clastic sedimentary rocks, which are made of fragments and grains, include conglomerate, sandstone, siltstone, and shale. Conglomerate, which is lithified gravel, is generally stream-laid and tends to be local in distribution. Sandstone consists of sand-size grains, mainly of quartz, cemented together by calcite, silica, iron oxide, or clay. Sandstones are widespread, and many are marine in origin. Siltstone, intermediate in grain size between sandstone and shale, generally occurs in thin slabby beds between thicker beds of shale. Shale is formed when mud or clay is lithified by compaction, owing to the weight of overlying sediments. The most abundant of the sedimentary rocks, shale is largely of marine origin.

Limestone is a sedimentary rock that consists of at least

50 percent calcite. Most limestones are of marine origin, and, being made up of fossil shells or their broken-up remains, are classified as both organic and clastic. Limestones have a wide range of color, thickness of beds, and content of noncarbonate mineral matter such as quartz grains and clay. Some limestones have been changed in fabric, or "recrystallized," by percolating underground water. Another rock of organic origin, but revealing an entirely different history, is bituminous coal. It consists of the carbonified remains of plants that grew in swamps on land, were partially decayed, and have been much compacted since accumulating where they grew.

Chert (or flint, if black) is a hard siliceous rock that customarily occurs as nodules and thin beds in limestone and dolostone. It is notably resistant to weathering. Chert may form as masses of silica at the same time as the enclosing rock, or later as a result of replacement. Gypsum and rock salt are evaporites, formed when isolated bodies of seawater, frequently replenished, repeatedly dried up. Dolostone, formed mainly of the mineral dolomite, is produced chiefly through replacement of limestone, either closely following deposition or long thereafter.

Most sedimentary rocks exhibit parallel stratification, which records deposition in relatively quiet water. According to the law of original horizontality, sedimentary strata are horizontal or nearly so when deposited. Clastic sediments that show cross-stratification were deposited from currents of running water or moving air, in an environment of high energy. Ripple marks in sandstone, and mud cracks in shale, give information on the history of these rocks. Concretions and geodes are found in a variety of rocks. Fossils are of great importance as records of the development of life.

Sedimentary rocks are relatively good records of past events and reveal much about earth history. Since most of them accumulated in the sea, their distribution tells us about the past extent of seaways and bordering land areas. Such rocks as coal and salt yield information on past climates; thickness and composition of clastic rocks indicate something of the source of sediments and the site in which they were laid down. Sedimentary rocks furnish all our fossil fuels and much of our iron ore, and they provide raw material for cement, bricks, chinaware, plaster, glass, and chemicals.

## SUGGESTED READINGS

Ernst, W. G. 1969. *Earth Materials,* chap. 6, pp. 110–125. Englewood Cliffs, N. J.: Prentice-Hall. (Paperback.)
A good presentation of the classification and description of the common sedimentary rocks.

Krumbein, W. C., and Sloss, L. L. 1963. *Stratigraphy and Sedimentation,* 2nd ed., chaps. 4 and 5, pp. 93–189. San Francisco; Freeman.
A college-level text that presents the principles of stratigraphy and sedimentation in detail. These chapters deal with sedimentary rocks and processes.

Kuenen, P. H. 1960. Sand. *Scientific American,* vol. 202, no. 4, pp. 94–106. (Offprint No. 803. San Francisco: Freeman.)
Discusses how the shape of wind- and water-transported sand grains provides clues to their history.

Laporte, L. F. 1968. *Ancient Environments,* chap. 2, pp. 7–29. Englewood Cliffs, N. J.: Prentice-Hall. (Paperback.)
Considers the environmental influences of source areas, transportation, deposition, and lithification on the formation of sedimentary rocks.

Spock, L. E. 1962. *Guide to the Study of Rocks,* 2nd ed., chap. 8, pp. 174–227. New York: Harper & Row.
An authoritative and fairly complete coverage of the great variety of sedimentary rocks.

Stetson, H. C. 1955. The continental shelf. *Scientific American,* vol. 192, no. 3, pp. 82–86. (Offprint No. 808. San Francisco: Freeman.)
Discusses the continental shelf as a natural laboratory where geologists may study the processes of formation of sedimentary rocks.

# 9 FOSSILS AND FOSSILIZATION

9-1  Conditions Favoring Fossilization
9-2  The Preservation of Fossils
9-3  The Interpretation of Fossils
9-4  The Significance of Fossils
9-5  Summary

Sedimentary rocks accumulate in the limited zone in which atmosphere, hydrosphere, and lithosphere mingle. This zone is also inhabited by the living things that compose the biosphere. Not surprisingly, then, the bones and shelly vestiges of animals and the more resistant parts of once living plants are important components of many sedimentary rocks. These remains, preserved in many different ways and in rocks of many ages, are **fossils,** which may be defined as *direct indications of life in the geologic past.*

## Possession of Resistant Parts

Although some remarkable examples are known in which entire organisms have been preserved, most fossils are incomplete in that they represent only the more resistant parts of ancient animals and plants. Upon death, the soft, fleshy tissue of living things is quickly attacked by bacteria, dried out in the air, or eaten by scavengers. Fate is kinder to shells, bones (Figure 9-1), and the resistant coatings of plant spores, pollen grains, and seeds, and it is these that make up most of the fossil record. Obviously, plants and animals that lack resistant structures stand a very poor chance of leaving a fossil record. Or stated positively, a prime requirement for fossilization is the possession of resistant parts.

## 9-1 CONDITIONS FAVORING FOSSILIZATION

Figure 9-1
Excavating dinosaur bones from sandstone of the Morrison formation (Jurassic) at Dinosaur National Park, Utah. White gashes on the rock are chisel marks. (Photo by E. L. Shay.)

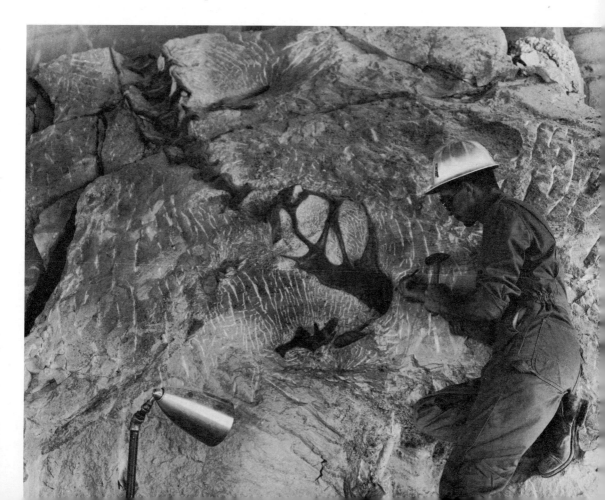

## Rapid Burial

Hard parts alone do not guarantee an organism a place in posterity, for even these may eventually be dissolved in water or weak acids, disintegrated during weathering, or partly eaten by other organisms. Thus, a second requisite for fossilization is that the organism be buried soon after death, so that at least its more resistant parts are protected from destruction.

Volcanic ash is fine-grained and settles quickly to form extensive beds of tuff, many of which contain exquisitely preserved fossils. Nearly 30 million years ago, for example, a volcanic explosion in what is now south central Colorado produced a cloud of ash, part of which settled down into a nearby lake. Not only did this rain of ash smother the inhabitants of the lake and quickly bury their remains, but it brought down with it insects from the air and leaves stripped from neighboring trees. As a consequence, the thinly laminated ash beds that accumulated on the lake floor now contain fish skeletons, well-preserved fossil insects, and plant leaves in which nearly all the original details are clearly visible.

Organic matter may also be protected from the destructive activities of air, bacteria, and scavengers through quick burial in stream-deposited muds, in swampy bogs, in pits of tar, in brine- or oil-soaked regolith, in ice or permanently frozen ground, or in mounds of wind-blown sand or dust. Fossil representatives of both animals and plants are known from rocks that accumulated in all these places or under all these conditions. Even so, it is on the floor of the sea and on the bottom of lakes that conditions for rapid burial persist almost continuously for long intervals of time. Not surprisingly, the sedimentary rocks produced in these places yield the bulk of the fossil record.

From what we have said thus far, it is clear that preservation of organisms as fossils, or **fossilization,** is a process that takes place only under certain conditions and in certain places. Because this is so, it is probable that only a very few of the many millions of animals and plants living at any one time find their way into the fossil record. Organisms that live where little sediment accumulates, or where sediment is deposited very slowly, are poorly represented as fossils. On the other hand, those that live (or die) in places where sediment accumulates rapidly, or where remains are protected from decay, are well represented. Thus, fossils of animals and plants that lived on land, where erosion mostly dominates over deposition and where remains are often exposed to air long after death, are far fewer than those of organisms that inhabited large lakes or the sea, where life is abundant and where sediment accumulates more or less continuously.

## Entire Preservation

Perhaps the most instructive of all fossils, as well as the rarest, are those preserved in such a way that both the soft and hard parts are essentially unchanged. In northern Siberia and Alaska, many complete animals have been discovered in regolith that has been frozen continuously for the past several thousand years. The best known of these represent two extinct types of wooly mammoths (distant cousins of modern elephants) and an extinct species of furry *Rhinoceros*. The nature of the material in which these shaggy fossils are found suggests that the living animals became mired in streams of soft, sticky mud, which then built up around their massive bodies and ultimately covered and suffocated them. The entire mass then congealed quickly in the deepfreeze of an Arctic winter and has remained frozen for thousands of years.

Similarly well-preserved carcasses of wooly rhinos have been discovered in oil- and brine-soaked regolith in Poland, and human remains not old enough to be fossils are preserved entire in peat deposits of northern Europe. Interesting and instructive as these well-preserved forms are, however, neither the unusual circumstances of burial nor the antiseptic conditions of preservation have been widespread enough in the past to be of particular importance in the preservation of fossils.

## Desiccated Remains

Fossils in which soft tissues are preserved in altered form are far more numerous than those preserved in the more exciting, but rarer, ways we have just discussed. In a very dry climate, for example, fleshy parts may dry out so quickly that they are incompletely destroyed by bacteria and are rapidly made unpalatable to marauding scavengers. Dried-out or mummified remains of extinct ground sloths, to which some desiccated skin and tough muscular tissue are still attached, have been found in dry caves in Argentina and in an old volcanic crater in New Mexico.

## Carbonified Remains and Impressions

The soft tissues of plants, and less commonly those of animals, may be altered and fossilized by carbonification, that is, by conversion to coal (Section 8-3). In this process

## 9-2
## THE PRESERVATION OF FOSSILS

tissues are compressed and reduced in volume so that their carbonified remains are usually thin, fragile, coaly films. These films are the only vestiges of the original substance of many organisms (Figure 9-2); but they surround, or are found beneath, the more resistant parts of others (Figure 9-4). If the carbonified film is lost from fossils preserved in fine-grained rocks, a replica of the surface, or an **impression,** may still show much fossil detail.

**Figure 9-2**
Leaves of the fern *Pecopteris* preserved by carbonification in fine-grained Middle Pennsylvanian shale from Clinton, Missouri. Natural size. (Photo by J. M. Schopf, U. S. Geological Survey.)

Carbonification is most widely operative where fine-grained sediment accumulates rapidly enough to bury organisms before there is much decay. The process is largely responsible for the beautiful plant fossils found in shales above coal beds the world over (Figure 9-2), and for remarkable assemblages of fossil worms, jellyfish, and arthropods like those known from black Cambrian shales in British Columbia.

## Unaltered, Permineralized, and Replaced Hard Parts

Most fossils represent just the hard parts of organisms. These resistant structures may remain unchanged for millions of years after their burial (Figure 9-3); more commonly, however, they have been either slightly or completely altered in the course of time.

Pore spaces in the original bone or shell may be filled in after burial by mineral matter precipitated from water percolating through the enclosing sediment. Such fossils have been altered by being "added to" and are said to

**Figure 9-3**
Although some of the brachiopod shells on this slab of Ordovician limestone have been partly recrystallized, many of the shells have remained unaltered. From muscle scars and other delicate features plainly visible on the specimen in the center of the view, skilled paleontologists can reconstruct many parts of the soft-bodied animal, even though the shell represents a species extinct for more than 400 million years. Natural size.

be **permineralized.** As a rule, permineralized bones and shells are heavier and denser than the original structures. Consequently, they are generally better fitted to withstand the ravages of weathering and erosion than unaltered hard parts are (Figure 9-4).

The bones and shells of most animals dissolve slowly in the slightly acid water that circulates through many

**Figure 9-4**
The bony skeleton of this Paleogene fish is permineralized, and an almost perfect outline of the fleshy body has been preserved by carbonification. Natural size.

**Figure 9-5**
This upright trunk in Yellowstone National Park is all that remains of an Early Noogene tree that was buried in siliceous volcanic ash and has been exhumed by erosion. While buried, the wood was replaced by silica introduced by circulating ground water.

rocks, and it is not unusual to find that the original substance of many fossils has been removed in solution. This phenomenon is characteristic also of the change from limestone to dolostone (Section 8-4). Commonly, however, the part removed has been **replaced** by another substance precipitated from circulating underground water. In some of these fossils, microscopic structures are preserved in faithful detail, even though the original wood, shell, or bone has been completely replaced by mineral matter. The nature of such exact replicas (Figure 9-5) implies that the process of replacement was slow and orderly—probably at the ionic level—so that fine details such as cell walls were not disturbed. In other examples of replacement, only the gross form of the initial bone or shell remains, implying that the entire structure was removed by solution before the resulting rock cavity was filled by precipitated mineral matter.

Walls of the rock cavities produced by solution of fossils are marked by impressions that are mirror images of irregularities on the surfaces of bones and shells. Because of the similarity of such cavities to the sand or clay molds constructed in a foundry, they are termed **natural molds** (Figure 9-6). The filling of natural molds by new mineral matter corresponds roughly to the casting of molten metal, plaster, or concrete in foundry molds. By analogy, replicas of fossils formed in this way are termed **natural casts** (Figure 9-7).

**Figure 9-6**
A snail shell that was originally
embedded in Silurian limestone was
removed by ground-water solution
when the limestone was converted to
dolostone. The cavity that remains is a
mold, and the smooth fragment
occupying its lower third is dolostone
that once filled the inside of the shell.
The outer surface of this fragment is a
mold of the internal surface of the
snail shell. Natural size.

**Figure 9-7**
Brachiopods on the surface of this
Mississippian sandstone are preserved
as casts (raised areas) and molds
(depressed areas). Although none of the
original shell material remains, nearly
all specimens can be identified and
their age determined. Slightly less than
natural size.

## Burrows, Tracks, Coprolites, Gastroliths

In addition to their own remains, animals and plants of
the past have left behind a variety of somewhat less
distinct records. In some rocks, for example, we find the

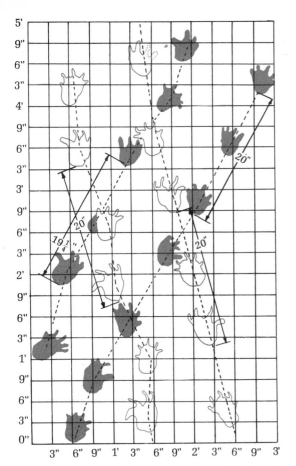

**Figure 9-8**
Casts of amphibian footprints from Pennsylvanian rocks in Ohio. The size and shape of the footprints, which form two intersecting trackways, show that the animal had a stride of about 20 inches, was 6 to 8 feet long, and probably weighed about 100 pounds. About one-fourteenth natural size. (From J. E. Carman, 1927.)

filled-in burrows of ancient animals; in others, recognizable footprints, tracks, and trails of various sorts (Figures 9-8, 9-9). **Coprolites,** permineralized or replaced animal excrement, are common fossils in some strata; they are of considerable interest because of the light they shed on the feeding habits and digestive processes of long vanished groups. Polished pebbles in many Mesozoic rocks are **gastroliths,** the gizzard stones of extinct land and aquatic reptiles. Although it stretches our definition of fossils somewhat to include them, they are certainly valuable in the information they contribute about the size, shape, and patterns of migration of the animals that used them.

**9-3**
**THE**
**INTERPRETATION**
**OF FOSSILS**

Although an interest in fossils is nearly as old as man himself, general awareness of their nature and significance is relatively recent. A few Greek scholars (Aristotle and Theophrastus, for example) recognized that fossils were the remains of once living animals and plants. However, their roundabout explanations of how such things came to be entombed in solid rock serve to indicate that they failed to appreciate the significance of fossils.

**Figure 9-9**
Dinosaur footprint in the Dakota
sandstone (Cretaceous), near Denver,
Colorado. (Photo by J. R. Stacy, U. S.
Geological Survey.)

Other Greeks not only understood that fossils were
vestiges of past life, but also perceived their significance
in reconstructing the history of the rocks that enclose
them. For example, Herodotus (484?–425 B.C.) pointed out
that the fossil shells of marine animals found far inland
in Egypt indicated former submergence of that part of
Africa beneath the sea. Sea shells found in the interiors
of Malta, Armenia, and Sardis caused Xenophanes and
Xanthus of Sardis to reach similar conclusions about these
areas some 500 years before the birth of Christ.

During the Dark and Middle Ages, the thoughtful observations of the Greeks were generally ignored or forgot-
ten. Fossils were regarded by some as "sports of nature,"
whereas others held they were the work of the devil,
placed on earth to terrify or confuse mankind. Still others
argued that fossils were the abortive attempts of a "life
force" to produce living things from rocks. Some believed
fossils to be merely animal- or plantlike shapes produced
inorganically by the growth of minerals. These divergent

views led to long and vigorous arguments about the nature of fossils. But the arguments were fruitless and unproductive until the fifteenth century.

A change in opinion about fossils did not come about overnight. The great Leonardo da Vinci (1452–1519) collected and studied fossils, and correctly regarded them as the remains of once living organisms, buried in rock when it was soft sediment. But Leonardo's genius was largely lost on his contemporaries. Even George Bauer (1494–1555; better known as Agricola), the one who coined the word *fossil,* was prepared to admit that only a few of the ones he had seen were once alive.

During the fifteenth and sixteenth centuries, ideas as to the nature of fossils gradually changed. The very weight of observations, and a growing perception of how rocks are formed, made it evident that fossils are remains of once living organisms entombed in rock that was previously soft sediment. This interpretation of fossils has not been seriously challenged since the beginning of the eighteenth century.

9-4
THE
SIGNIFICANCE
OF FOSSILS

A few early Greek scholars, as well as Leonardo, recognized the significance of fossils and interpreted them as mute evidence of the former submergence of lands in which they occur. But fossils can tell us much more than this about the world we live in. Robert Hooke (1635–1703), a brilliant English experimental physicist, made the novel suggestion that rock layers might be recognized as the same in different places because they contain fossils of the same types of animals. He also pointed out that a tropical climate must once have existed in southern England, because of tho kinds uf fussils found there in Mesozoic rocks.

Hooke's first suggestion was considerably elaborated a century later by another Englishman, William Smith (Section 0-5). Smith noted that fossils in a sequence of rock layers changed upward and downward from layer to layer, but that they could be counted on to maintain their general characters as a layer was traced laterally. Using fossils in this way, Smith was able to trace rock strata for great distances laterally, and by 1815 he had assembled a geologic map of England, the first of its kind.

Since Smith's time, a growing number of geologists and biologists have turned their talents to a more intensive and more systematic study of fossils and the fossil record. These studies have served conclusively to demonstrate the significance of fossils and their unique importance in the business of piecing together earth history.

## Fossils and Paleogeography

Herodotus was perhaps the first to suggest that fossils could be used in charting the boundaries of ancient lands and seas; that is, that they would be helpful in what has come to be called **paleogeography.** Although it has not always been clear, it seems obvious today that rocks containing fossils of marine organisms were deposited in the sea; the fact that many such rocks are now far from the sea certainly implies that past geography differed greatly from that of the present.

## Fossils and Ancient Climates

Fossils are also very useful in reconstructing the climatic conditions under which the rocks containing them were deposited. Modern reef-forming corals, for example, grow most profusely in the shallow warm waters of tropical seas, and it has long seemed logical to assume that their long-extinct cousins had similar preferences and tolerances. If this is so, however, fossil corals testify that areas in now temperate or arctic regions were once submerged beneath shallow tropical seas. Likewise fronds of fossil palms, plants now at home only in tropical and subtropical latitudes, have been found as far north as Spitzbergen, and fossil bones record the fact that alligators like those now most comfortable in Florida not long ago disported themselves on the plains of Alberta and Saskatchewan. Presumably, then, climatic patterns have changed with the passage of time; and fossils are of considerable significance not only in suggesting this conclusion, but in working out the patterns themselves.

## Fossils and Geologic Time

Recognition of the fact that successive layers of rock contain different associations of fossils, but that the sequence of fossil assemblages is much the same in all parts of the earth, has been important in dividing the rock record into divisions, and geologic time into eras, periods, and epochs. In short, fossils are of paramount significance in geologic chronology.

## Fossils and Evolution

The succession of fossils established by careful study of fossiliferous rock layers had a very great impact on scientific, social, and philosophical thought during the past century. The succession itself is irrefutable evidence that animal and plant communities have multiplied in number, increased both in size and in diversity, and changed slowly in character with the passage of time. This fact was underscored in 1859 by Darwin, who recognized that the fossil record of the biosphere embodies tangible evidence to support his conclusion that life has evolved from simple beginnings to its present complexity and diversity.

The study of fossils, or **paleontology** (literally a "discourse on ancient beings"), has long been an integral part of geology. For many years, geologists studied fossils primarily to date rocks in the geologic record, and they paid scant attention to the biologic aspects of successive fossil communities. The use of fossils in dating and correlating rocks is still important, but geologists of more recent vintage have broadened the scope of their studies to include the entire spectrum of animal and plant relations. Thus paleontology today is actually an overlapping field (Figure 1-1), in which zoologic and botanic principles are applied to the interpretation of ancient animals and plants and in which the important dimension of time is added to our understanding of the biosphere.

## 9-5 SUMMARY

Fossils, or direct indications of life in the geologic past, generally consist of shells, bones, seed coats, and other resistant parts of animals and plants. To persist as fossils, these hard parts must be buried soon after the organism dies. Rapid burial may take place under a variety of conditions on land, but is most effective on the floors of shallow seas. Only a minority of the animals and plants that lived at any time in the past became preserved in the fossil record.

Rarely, complete animals may be preserved; for example, mammoths have been discovered frozen in the Arctic tundra. Other means of fossilization include desiccation, or drying out; alteration to coal, or carbonification; preservation of unaltered hard parts; filling of pore spaces by mineral matter, or permineralization; and replacement, either ion by ion or through solution and reprecipitation. A shell dissolved from rock leaves a natural mold; this may be refilled with mineral matter, to form a natural cast. Fossils include such indirect evidences of past life as burrows, tracks, coprolites, and gastroliths.

Fossils have been noted and argued about since the time of the ancient Greeks. The present interpretation of their origin has been generally accepted since about 1700. Fossils are useful

1. to indicate the distribution of ancient lands and seas, in the study of paleogeography;

2. to tell something of past climates and the conditions under which some sedimentary rocks were formed; and

3. to date rocks by giving their relative ages and hence their place in the geologic time scale. Modern paleontology also includes the biological aspects of the fossil record, and thus overlaps the fields of zoology and botany.

## SUGGESTED READINGS

Clark, D. L. 1968. *Fossils, Paleontology and Evolution,* chaps. 1, 2, and 3, pp. 1–23. Dubuque, Iowa: W. C. Brown. (Paperback.)
An introduction to life of the past. These chapters cover paleontology, fossils and fossilization, and evolution.

Collinson, C. W. 1959. *Guide for Beginning Fossil Hunters.* Illinois State Geological Survey, Educational Series 4, 37 pp.
A useful pamphlet for amateurs that gives simple, well-written instructions on how to find and collect fossils. Good figures and nontechnical descriptions of the principal fossil groups are included.

Ericson, D. B., and Wollin, G. 1962. Micropaleontology. *Scientific American,* vol. 207, no. 1, pp. 96–106. (Offprint No. 856. San Francisco: Freeman.)
Fossils so small that they can be identified only with magnification provide excellent clues to ancient changes in climate, and are used to establish zones for correlation.

McAlester, A. L. 1968. *The History of Life,* chaps. 3, 4, and 5, pp. 39–99. Englewood Cliffs, N.J.: Prentice-Hall. (Paperback.)
Includes a discussion of the modes of preservation of fossil organisms, both marine and nonmarine.

Matthews, W. H., III. 1962. *Fossils: An Introduction to Prehistoric Life.* New York: Barnes & Noble. 337 pp. (Paperback.)
An introduction to the study of fossils, written chiefly for the amateur collector.

# 10 IDENTIFICATION AND CORRELATION

10-1 Twofold Nature of the Sedimentary Record

10-2 Identification

10-3 Correlation

10-4 Lithologic versus Chronologic Equivalence

10-5 Cambrian Rocks in Northern Arizona: An Example

10-6 Summary

Because geology is concerned with the evolution of the earth and its inhabitants through time, every geologist is necessarily a historian. Like other historians, he must derive the sequence of past events from the records they left behind. Earth history is recorded in a bewildering variety of rock types, of which some are exposed at the surface and others are encountered by drilling into the crust. The completeness of the history that we reconstruct depends on our ability to unravel this complex rock record, deduce the events that produced it, and arrange these events in chronological order. Since a large and relatively readable share of earth history is recorded in fossiliferous sedimentary rocks, we need to be familiar with some of the principles and processes involved in interpreting this sedimentary record.

A sequence of fossiliferous sedimentary rocks exposed in the walls of a canyon or along a road is a document of the history of that place. Depending on the way in which we wish to read it, this documentary record can be divided into rock units, or formations, or into biologic units, or zones. In making a complete study, we generally use both divisions.

A **formation** is a subdivision of the geologic record based solely on type of rock. In Figure 10-1, three distinct rock units are shown, one above the other. Each is a formation. The limestone at the top, the shale below it, and the interbedded sandstone and shale at the bottom severally constitute easily recognizable, distinct parts of the sequence. A formation is given a geographic name, derived from the name of a place in or near which it is typically exposed or best developed. The Columbus Limestone and the Olentangy Shale are Ohio formations that are best exposed near the city of Columbus and along the Olentangy River.

In contrast to formations, **zones** are divisions of the geologic record based on the fossils that occur in a

10-1
**TWOFOLD NATURE OF THE SEDIMENTARY RECORD**

Figure 10-1
Formations and zones contrasted.
Contacts are the boundaries of
formations; time lines bound zones.

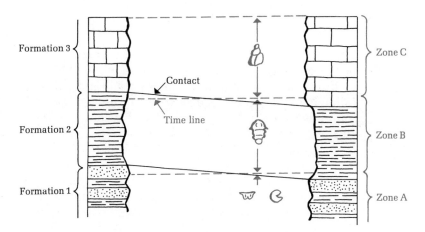

sequence of sedimentary strata. Obviously a zone, like a formation, consists of a body of rock, but in defining its boundaries no attention is paid to the nature of the rock. The zone is a rock unit containing certain distinctive fossils or assemblages of fossils. Three assemblages of distinctive fossils characterize the rock sequence shown in Figure 10-1, and each of these defines a separate zone. Note that the boundaries of zones and formations into which this section has been divided do not coincide. This is not always the case in dividing the rock record, but it is a very common one.

Zones derive their names from that of one of their characteristic fossils. The zone of *Olenellus,* for example, includes all rocks in the world, regardless of their type, that contain this distinctive Cambrian trilobite and its customary associates.

Only a small fraction of the earth's history is recorded by the rocks exposed in a single canyon wall or along one highway. A comprehensive, worldwide history can be written only if the many local histories can be compiled into a single volume. Such a compilation, of course, requires that there be ways to "match up," or collate, the rock record from canyon wall to canyon wall, from region to region, and from continent to continent.

There are two distinct ways of fitting together the rock records of separate areas. Matching up rock units, or formations, is **identification,** which involves a determination of lithologic equivalence between the rocks of two or more regions. Matching up zones, or other rock units known to have been deposited during the same time intervals, is **correlation,** which, for the most part, requires establishment of faunal (or floral) equivalence between the rocks in two or more regions. Because methods of determining equivalence differ somewhat in each of these two types of regional collation, they are best discussed separately. In practice, geologists usually match up rocks in both ways at the same time.

## 10-2 IDENTIFICATION

The lithologic equivalence of rocks continuously exposed between two areas can be demonstrated by "walking out," or by visually tracing, their outcrop: that is, by directly following their exposed edges from one area to the other. Individual beds exposed in areas like that shown in Figure 10-2 can be traced visually or actually followed out, and there is no question that beds in one part of the area are the lithologic equivalents of those in other parts.

Walking out a formation fails if the rock record between two areas is covered by water, soil, or vegetation, or if it has been removed by erosion. In situations like this,

which are very common indeed, lithologic equivalence can often be reasonably inferred if a series of strata occurs in similar or identical sequence in separate areas. For example, the invariable occurrence in isolated exposures of superposed beds of black limestone, green shale, and red sandstone would certainly suggest identity of these distinctive beds from exposure to exposure.

Rock strata that are not exposed at the surface can be identified by several techniques that have been developed in the exploration for oil. When a well is drilled, the cores and the rock chips or "cuttings" brought to the surface are faithful indicators of the formations penetrated by the bit. When these are described by a geologist, and the results are plotted to scale on a strip of paper representing the well, an accurate record, or **lithologic log,** of the formations penetrated is produced. Comparison is then made with the lithologic logs of nearby wells, and formations are identified from well to well by rock type and position in sequence.

Subsurface formations can also be identified by their electrical properties. For example, the electrical resistance of the various strata penetrated by a well may be measured by electrodes moving slowly up the well, and recorded mechanically on paper strips, or **electric logs.** Because each rock unit tends to make characteristic curves on such logs, these curves can be used for identification. When lithologic and electric logs for a number of wells are

**Figure 10-2**
Individual beds of sedimentary rock can be either "walked out" or visually identified over great distances in country like this. Paleogene lake beds near Rifle, Colorado. (Photo by J. H. Rathbone.)

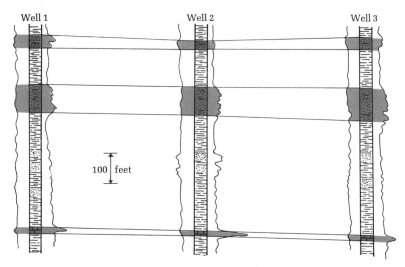

**Figure 10-3**
Strata encountered underground in wells drilled for oil are readily identified from well to well. The central column for each well is the lithologic log. It shows by standard patterns the rocks penetrated, as determined by geological examination of cuttings or cores brought up during drilling. The wiggly lines beside each column are records of the rocks' electrical behavior, as told by instruments lowered into the well. Even though these wells may be miles apart, the combination of lithologic log and electric log makes identification precise and accurate.

placed side by side, the various formations can be readily identified (Figure 10-3).

**10-3**
**CORRELATION**

The business of correlating the rock records between two areas is largely that of establishing the existence of the same zones in both rock sections. Because zones are rock units characterized by a distinctive fossil or assemblage of fossils, the procedures of correlation are directed primarily toward recognition of the same sequence of fossils in rocks of two or more regions. If it can be established that rocks in two sections can be divided into zones on the basis of the same fossils, the rocks in one zone can then be correlated with those in the same zone in the other section.

To a limited extent, correlation of rock records may also be done without fossils, for some types of rock are known to have been deposited over extensive areas at the same time. Tuff, for example, accumulates as the result of ash falling over great areas within relatively short periods of time. Thus the identification of a tuff bed as the same in two sections also implies correlation, for it was clearly deposited at about the same time everywhere it occurs.

Some rhythmically or cyclically bedded sequences of limestone, dolostone, gypsum, and rock salt can also be correlated from place to place *within the same basin of deposition,* for sequences like this are the results of cyclical changes in depositional conditions that probably affected the entire basin at the same times (Figure 10-4). Certain old lake clays can also be correlated if they can be identified as the same beds from place to place, for they consist of light and dark bands that result from

A.                         B.                         C.

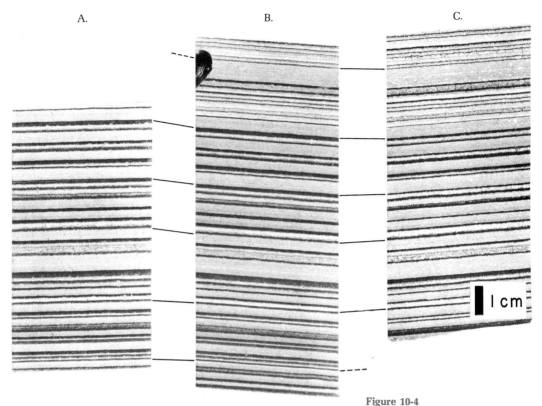

1 cm

**Figure 10-4**
Laminations of anhydrite (light) and calcite (dark), Upper Permian rocks in the Delaware Basin of west Texas. Lithologic equivalence is obvious; chronologic equivalence is highly probable. It is 20 miles between wells A and B, and 9 miles between B and C. (Courtesy of D. W. Kirkland.)

cyclical changes in climate (daily, seasonal) that affected the whole lake at the same time. In general, however, these few special types of rock are of limited areal extent. They are of little use in regional correlation and of no use in correlation from continent to continent.

It follows from the principle of uniformity of process that, regardless of time, similar or identical combinations of place and process result in similar rock records. Consequently, matching two or more rock records by lithologic equivalence (identification) implies only that the rocks in the different regions were deposited as a result of similar processes operating in similar places. The identified equivalents may be, but are not necessarily, the same age (Figure 10-1).

On the other hand, repeated observation of fossiliferous strata in many widely separated parts of the earth has established two striking facts. First, there is in all continents a gradual change in the character and appearance of fossils from the oldest fossiliferous rocks to the youngest. Second, the succession of fossils in all parts of the earth represents animals and plants that went through the same sequence of developmental stages. That is, in

## 10-4
## LITHOLOGIC
## VERSUS
## CHRONOLOGIC
## EQUIVALENCE

every sequence known, fossil fish precede fossil amphibians, which in turn precede fossil reptiles and birds. Nowhere is this sequence reversed or scrambled.

From these two facts it has been inferred that the biosphere has changed progressively and irreversibly with time; and that during any given interval of the past, lands and seas over all the earth were populated either by identical animals and plants, or more reasonably, by communities of organisms similar in their stage of development. Such an inference leads naturally to the conclusion that rocks containing identical fossils, or fossils of organisms at the same general level of development, were deposited *at the same time.* Thus, correlation of rocks, based primarily on the fact of biologic equivalence, implies that rocks correlated are the same age.

## 10-5
## CAMBRIAN ROCKS IN NORTHERN ARIZONA: AN EXAMPLE

The sequence of sedimentary rocks cut through by the Grand Canyon of the Colorado River affords an almost unparalleled opportunity to make a practical test of the matters just discussed. Several thousand feet of brightly colored, nearly horizontal strata are exposed to view in the walls of this mile-deep chasm (Figure 2-11), and from a point in the canyon or on its rim, individual beds can be walked out or visually traced for many miles to the east or west. Collecting and identifying fossils through this thick sequence, and thus dividing the strata into zones, as well as tracing individual beds from one end of the canyon to the other, is arduous work and can hardly be accomplished from a point on the rim. However, careful work of this sort has been done, and the results are pertinent to our discussion of identification and correlation.

**Figure 10-5**

Diagrammatic cross section showing three Cambrian formations in the walls of the Grand Canyon, northern Arizona. The Little Colorado River enters the canyon at its east end; the Grand Wash Cliffs are at the west end. The two ends of the section are about 120 miles apart. The rows of dots represent levels at which particularly distinctive fossils are found. Vertical exaggeration about 64 times. (After McKee, 1945.)

In the lower part of the canyon walls, three distinctive formations can readily be distinguished. These contain fossils that show them to be of Cambrian age, and each of them has been walked out from west to east for nearly 230 miles along the sinuous course of the Colorado River. Figure 10-5 is a graphic cross section showing these strata

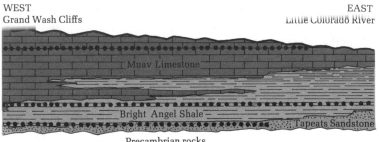

WEST
Grand Wash Cliffs

EAST
Little Colorado River

Precambrian rocks

between the Grand Wash Cliffs on the west and the Little Colorado River on the east.

The lowest formation, the Tapeats Sandstone, forms cliffs of medium- to coarse-grained pebbly brown sandstone with prominent cross-stratification and occasional beds of conglomerate and shale. It increases in thickness from 180 feet at the west end of the Grand Canyon to nearly 300 feet at the east end. Above the Tapeats Sandstone is the Bright Angel Shale, 350–450 feet of dull greenish shale, crumbly micaceous siltstone, and silty dolostone that forms a bench above the Tapeats cliff. The Bright Angel Shale is overlain by 150–800 feet of gray and buff Muav Limestone, which in most parts of the canyon forms very steep slopes or cliffs.

Fossils are not abundant in every bed of these three Cambrian formations, but especially distinctive fossil assemblages have been discovered in thin intervals at many places. The positions of three of these assemblages are shown by rows of dots in Figure 10-5. Because of the very small scale of this figure, each row of dots appears as a thin line; actually, each row is a zone and represents a rock interval several feet thick. Presumably, other zones intermediate between the three shown could be recognized if the rocks were uniformly fossiliferous.

Just a glance at Figure 10-5 is sufficient to demonstrate that in the Grand Canyon area the boundaries of Cambrian formations are not parallel to the boundaries of Cambrian zones (the rows of dots). Subdivision of these strata into formations and zones has resulted in two very different kinds of rock units.

What is the significance of all the hard work that has gone into establishing the facts just mentioned? If we recall that zones are bodies of rock deposited during the same time intervals, it is clear that each of the Cambrian formations is younger at the east end of the Grand Canyon than at the west end. Equally apparent from Figure 10-5 is the fact that during each successive interval of time, Tapeats Sandstone was accumulating in one part of the Grand Canyon area, Bright Angel Shale in another, and Muav Limestone in yet another. Thus parts of all three formations are the same age. To be more specific, it is clear that the Tapeats Sandstone at the east end of the Canyon is the same age as the Bright Angel Shale at the west end; that the Bright Angel Shale at the east end of the Canyon is almost entirely equivalent to the Muav Limestone in the Grand Wash Cliffs at the west end.

If we keep in mind the fact that each type of sedimentary rock is the result of a special combination of processes operating in specific places, we can carry our discussion a step further. Each of the three Cambrian formations in the Grand Canyon area is the result of sediment deposition in a specific environment. Marine fossils entombed in all three formations tell us that all three environments were on the sea bottom. Because rocks formed in all three environments occur side by side within the same zone and in the same geographic order, we may safely deduce that while these formations were accumulating there was an eastern area of sand deposition, a western area of limestone accumulation, and an intermediate area in which shale and siltstone beds were being deposited.

The fact that the boundaries between formations "rise" eastward with respect to zonal boundaries implies that the three marine depositional environments gradually shifted eastward with the passage of Cambrian time. Because the coarsest clastic sediments accumulate nearest the shore of modern seas, and finer materials settle farther seaward, it is logical to assume that the same environmental pattern characterized Cambrian seas. If this is so, the eastward migration of the Tapeats Sandstone, Bright Angel Shale, and Muav Limestone clearly implies an eastward migration of the sea in which they were deposited. These three formations are the record of an eastward transgression of the Cambrian sea in northern Arizona.

Thus, the principal result of all the climbing, walking, and fossil collecting among the Cambrian rocks of the Grand Canyon area is the knowledge that during the Cambrian Period, what is now northern Arizona was invaded from the west by a sea that gradually flooded eastward and ultimately submerged all the region through which the Colorado River has since cut its canyon. It is from careful studies like this in many parts of the earth that earth history is written.

**10-6 SUMMARY** A sequence of sedimentary rocks constitutes a record of geologic history. A formation is a subdivision of the record made on the basis of rock type, or lithology. A zone is a subdivision of the geologic record that contains a distinctive assemblage of fossils. A formation is named for a geographic locality or feature, a zone for a characteristic fossil. Both are used in fitting together the rock record of many separate areas into a coherent story.

Identification is the lateral matching-up of formations. It can be done by walking out or tracing an exposed outcrop, or by matching similar sequences of strata that are either exposed at the surface or encountered in wells. Correlation, on the other hand, involves the lateral tracing of zones, with the object of showing that two or more rock sections are of the same geologic age. Correlation means time equivalence because of the worldwide, irreversible evolution of organisms that has taken place during geologic time. Local correlation, within a basin of deposition, can sometimes be done by comparing cyclically bedded sequences of strata.

Three Cambrian formations can be traced along the walls of the Grand Canyon for some 230 miles. The lowest, a sandstone, is believed to represent a nearshore environment, probably a sandy beach; the middle formation, a shale, a belt of finer sediment some distance from shore; and the highest formation, a limestone, a sea floor still farther from land. Characteristic fossil assemblages are found in three zones. When these relations are plotted on a cross section, the formations are seen to "rise" eastward with respect to the zones, which represent time lines. We therefore deduce that the sea in which these sediments accumulated must have migrated eastward with the passage of Cambrian time.

## SUGGESTED READINGS

Dunbar, C. O., and Rodgers, J. 1957. *Principles of Stratigraphy,* chap. 16, pp. 271–88. New York: Wiley.
A college-level stratigraphy text that is a good reference on principles. This chapter deals with correlation.

Eicher, D. L. 1968. *Geologic Time,* chaps. 4 and 5, pp. 67–116. Englewood Cliffs, N.J.: Prentice-Hall. (Paperback.)
A fairly comprehensive coverage of correlation including sections on paleomagnetic and modern quantitative techniques.

Harbaugh, J. W. 1968. *Stratigraphy and Geologic Time,* chap. 2, pp. 23–33. Dubuque, Iowa: W. C. Brown. (Paperback.)
Chapter 2 deals with stratigraphic correlation.

Krumbein, W. C., and Sloss, L. L. 1963. *Stratigraphy and Sedimentation,* 2nd ed., chap. 10, pp. 332–89. San Francisco: Freeman.
A college-level text that presents the subjects of stratigraphy and sedimentation in detail. Chapter 10 is concerned with the principles of correlation.

Weller, J. M. 1960. *Stratigraphic Principles and Practice,* chap. 15, pp. 540–69. New York: Harper & Row.
A comprehensive college text that deals with the fundamentals of stratigraphy, with emphasis on interpretation of sedimentary rocks as a means of constructing geologic history. Chapter 15 deals with correlation.

# 11 LAND SCULPTURE BY STREAMS

11-1 A Contrast in Valleys
11-2 Valley Development
11-3 Regional Reduction
11-4 Interruptions in the Cycle
11-5 Drainage Patterns
11-6 Summary

If we stand on the rim of the Yellowstone River canyon in Yellowstone National Park, we see before us a deep gorge whose walls are composed of cliffs of bare rock (Figure 7-4). At the bottom the river foams along in a series of white-water rapids. We should not need the services of a park ranger to inform us that this has been, and still is, the site of powerful stream erosion and rapid valley cutting. Now, by way of contrast, consider the view of the Ohio River valley shown in Figure 7-14. Here we see a wide floodplain, which is featureless except for low willow-covered natural levees along the channel. The walls of the valley form subdued tree-clad slopes. Altogether lacking are the spectacular features to be seen at Yellowstone Canyon.

Yet these two utterly different landscapes are both stream valleys, which owe their characteristics to the work of running water and downslope movements. No doubt you can call to mind a picture of a valley that you know from your own experience, which quite likely does not resemble either of the two we have chosen as examples. Taking the earth's land surface as a whole, it seems obvious that there are great differences from region to region in the way in which streams have etched the landscape.

In Chapter 7 we discussed the work that streams do in eroding, transporting, and depositing rock material. In this chapter we shall inquire into the ways streams carve and sculpture the face of the solid earth.

## 11-1 A CONTRAST IN VALLEYS

The millions of stream valleys that furrow the land surface of the globe may seem at first to be infinitely variable. In size they range from gullies a few feet wide and a few yards long to regional features miles in width and hundreds of miles in length. Some valleys contain water only in wet seasons, whereas others hold perennial streams. Some are cut into bedrock, others are floored with alluvium. How may stream valleys be classified for orderly study?

Fortunately it has been found that valleys, like most other natural features, undergo progressive development, or evolution with time; indeed, such a conclusion naturally follows from our consideration of erosion and deposition. By observing certain of the characteristics of a given valley, we can readily determine its position in the **cycle of valley development.**

## 11-2 VALLEY DEVELOPMENT

### Young Valleys

A large proportion of valleys are found to have these aspects in common: (1) a steep gradient, (2) a V-shaped cross profile, and (3) a floor occupied entirely by a stream. Such valleys are said to be **young,** or **youthful.**

A steep gradient imparts high velocity and abundant energy to the stream. Hence young valleys are typically undergoing active erosion. Though they may start as small gullies in loose regolith (Figure 11-1), most of them are rapidly cut down into bedrock (Figure 11-2). Rapids, falls (Figure 11-3), plunge pools, and potholes (Figure 7-6) are characteristic of many young valleys.

**Figure 11-1**
Gullies, or valleys in extreme youth, forming in loess, western Iowa. (Photo by R. L. Handy.)

**Figure 11-2**
Young valley of the Rio Grande, in northern New Mexico.

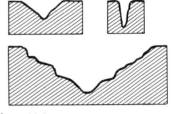

Figure 11-3
Small waterfall in the channel of Taughannock Creek, central New York. The creek flows in a young valley cut into flat-lying sedimentary rocks of Devonian age.

The walls of young valleys typically show evidence of slump, talus formation, rock-slides, and other downslope movements. Valley bottoms may become temporarily choked with boulders and other debris contributed by these gravity-actuated processes (Figure 6-1).

Because the stream in a young valley is occupied with erosion, not with deposition, no floodplains form and there are few bars or other deposits of more than seasonal permanence. In most young valleys, the valley bottom and the stream channel are one and the same thing (Figure 11-4).

It should be clear from this description that the canyon of the Yellowstone River (Figure 7-4) is a young valley. Its gradient is some 56 feet per mile, its profile is a typical V, and there is no room along the valley bottom for anything but the rushing waters of the river. Many, perhaps most, young valleys approach this form very closely (Figures 11-5, 11-6). Others are exceptional on one

Figure 11-4
Characteristic profiles of young valleys.

Figure 11-5
Rapid stream erosion and slow valley widening have produced this deep canyon in east central Utah. The rock is massive sandstone of Jurassic age. (Photo by E. M. Spieker.)

count or another. We find, for example, valleys in which abrasion by the stream has outstripped the contribution of rock waste from the walls. Examples are the deep slotlike gorges of upstate New York, of which Watkins Glen (Figure 7-6) is perhaps the best known; and many vertical-walled canyons of the desert regions (Figure 11-5). At the Grand Canyon of the Colorado River (Figure 2-11), downslope movements have widened the valley greatly, and the stratified nature of the rocks has given the walls a steplike profile; yet even in such a valley the cross profile is that of an opened-out V (Figure 11-4).

We add two items of caution as to what the term "young valley" does not mean. It does not refer to age in years (even if that could be determined), and it does not refer to any particular size of valley. It applies to a stage of valley development. To sum up—if a valley, new or ancient, large or small, has the characteristics outlined above, it is a young valley.

### Base Level

Every stream ends somewhere, generally by flowing into another stream, a lake, or the sea. Obviously a stream cannot erode the floor of its valley appreciably below the elevation of the water body into which it flows. This elevation, projected in imagination beneath the stream's valley, is its **base level.** It marks the limit of downward erosion by the stream.

The **ultimate base level** is sea level. Theoretically, if a continental landmass remained unaffected by any up-building forces for a long enough time, stream erosion would eventually reduce it to a nearly horizontal surface just high enough above sea level to allow streams to flow off the land. We may envision ultimate base level as an imaginary surface, at approximately the elevation of the sea, extending beneath all land surfaces.

But since most streams do not flow directly into the ocean, their downward erosion is limited by **local base levels** of one sort or another. Small streams near Columbus, Ohio, can deepen their valleys only to the level of their junctions with the Scioto River; Niagara River can cut its gorge no deeper than the elevation of its entrance into Lake Ontario. Streams in arid regions (Section 7-6) can erode downward only to the level of the desert basins where they end. Another kind of local base level is produced when a downcutting stream encounters an exceptionally resistant rock layer in its channel, which slows erosion and acts as a base level for the upstream segment of the valley (Section 7-2; Figure 11-7).

Viewed from the perspective of geologic time, all local base levels are temporary. Major streams like the Scioto River lower their channels very slowly toward ultimate base level; lakes such as Ontario eventually become filled with sediment washed in by streams, or are drained by downward erosion of the outlet. Desert basins are filled,

**Figure 11-7**
Floodplain and meanders of Fontenelle Creek, western Wyoming. Resistant strata farther downstream produce a local base level.

and are finally excavated by through-flowing streams; and resistant rock masses in the path of streams are ultimately worn through by abrasion. But from the short-range point of view, most streams in young valleys are working toward a local base level.

### Mature Valleys

A stream that has cut its valley down nearly to base level has but a short vertical distance to go before reaching its lower end; in other words, its gradient is low. Hence the stream loses that part of its energy that depends on steep gradient. Yet its energy may still be high, because as the valley has been deepened and widened by erosion and downslope movements, the stream has come to drain a much larger area than formerly, and its discharge is greatly increased. Furthermore, it may be expected to carry a considerable load of debris derived from the enlarged drainage area. Once base level is approached, most of the available energy that the stream possesses is applied not to further downcutting but to debris transport and to lateral abrasion against the valley walls, in short, to valley widening. This transition from downward to sideward erosion produces a pronounced change in the characteristics of stream valleys, and serves as a fairly distinct mark of difference between the two stages of valley development that we recognize.

A valley whose floor is near base level is said to be **mature.** A mature valley is characterized by a pan-shaped cross profile, a floor wider than the stream channel, and a laterally swinging stream with a low gradient. At flood stages the stream leaves its channel and spreads across the valley floor, depositing sediment and producing a floodplain. The term *mature* applies to a broad range of valley development. In valleys in early maturity, the flat floor is narrow, the walls rise from it abruptly, and the stream is deflected from wall to wall in wide gentle arcs (Figures 11-8, 11-9, 11-10). In late-mature valleys, on the other hand, the walls are low and gently sloping, the floodplain is miles across, the stream flows in a series of great meander bends with point bars and natural levees and the gradient is very low, perhaps only a fraction of a foot per mile. The part of the Ohio River Valley that is portrayed in Section 11-1 is clearly in late maturity (Figure 7-14).

Look back at the long profile of a typical stream and the accompanying discussion, in Section 7-2. We can now put this into terms of valley development. Most young valleys are confined to the upper reaches of river systems, where gradients are steep and base level is a long way

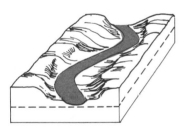

**Figure 11 0**
A valley in early maturity. Note the pan-shaped cross profile and the stream swinging laterally on a narrow floodplain.

**Figure 11-9**
Early-mature valley of the Potomac River between West Virginia (on left) and Maryland. The light-colored bands of trees clearly show the narrow floodplain. The river passes through the distant mountain in a water gap. (Photo by J. H. Rathbone.)

**Figure 11-10**
Mature valley in southern Peru. Irrigated farmland (dark) outlines the floodplain. The stream is braided because it is heavily loaded with sediment from melting glaciers in the Andes Mountains. (Photo by M. P. Weiss.)

down; most mature valleys are found in rivers' lower reaches, where gradient is gentle because base level is being approached. Thus it is normal for a major stream valley to change from youth to early maturity and then to late maturity progressively downstream. Interruptions in this normal pattern, such as the one described in Section 7-2 for the Yellowstone River, can now be explained on the basis of local base level. Above the falls the valley is in maturity, downcutting by the river being inhibited by the resistant rock that makes the falls; below, the valley is the typically young gorge mentioned at the beginning of this chapter.

**11-3
REGIONAL
REDUCTION**

What we have said so far in this chapter relates to the stages of development of individual valleys. Any sizable part of the earth's land surface, however, is likely to include not just one valley but whole systems of them, together with the high areas, or **divides,** that separate one valley from the next. In time, these stream-eroded regions change, and in so doing go through typical stages that are recognizable. Because the effect of stream erosion is to reduce the land to lower levels, we speak of the **cycle of regional reduction.**

Suppose we start with several thousand square miles of brand new land surface, which has never before been exposed to weathering or erosion. Such a surface might be produced by the uplift of a previously submerged part of the continental shelf or interior; or by the outpouring of plateau basalts (Chapter 4); or by the melting away of a widespread ice sheet (Chapter 12). Let us say that this new surface is nearly flat and stands some hundreds of feet above sea level. Under these circumstances, any changes that the region undergoes through stream erosion will take place between two vertical limits: the elevation of the new land surface, and base level (sea level projected inland beneath the region). The events that take place and the landscapes produced can be readily understood if these two governing limits are kept in mind.

**Regional Youth**

When rains first fall on the new land, the water that runs off will naturally collect in the low places, and spilling to ever lower levels, will at length find a few paths by which it will flow out of the region toward the sea. At first all channels will be shallow and ill-defined, and swamps or ponds may dot the landscape. Those few streams that flow down somewhat steeper slopes than the average, or across more easily eroded rock, will cut their channels down relatively fast. Thus a few main or trunk streams will develop, with the others as tributaries. Stream erosion will be accompanied by weathering and downslope movements.

At an early stage of **regional youth**[1] (Figure 11-11 A), small streams have started to cut into, or *dissect,* the original surface; but most of this surface remains as undrained, undissected flat areas between the small young valleys. In Figure 11-11 B, dissection has gone somewhat further, and one stream is even approaching base level, but the largest part of the land is still at its original elevation, interstream areas are wide and flat, and there are only a few valleys.

There are numerous examples of regions in youth. Much of Ohio, Indiana, and Illinois consists of a low-relief surface that emerged from beneath melting ice late in Pleistocene time; in the 10,000 years or so since disappearance of the ice, streams have made only a small start at dissecting the surface. Extensive plains and plateau areas in western North America are also in regional youth (Figures 11-12, 11-13, 18-3).

## Regional Maturity

As the streams deepen their valleys, they also lengthen them by eroding at the upstream end (Figure 11-1); small gullies eventually develop into young tributary valleys. Thus evolves a pattern of many branching valleys, which grow headward at the expense of the original land surface. Downslope movements, widening adjacent valleys, tend to narrow and sharpen the divides.

The stage of **early maturity** is reached when:

1. the original surface has been so dissected that divides are narrow and nearly all the land slopes sharply toward streams;
2. there is an intricate, well-developed drainage network; and
3. the flat land, if any, is in the form of narrow floodplains along the major streams (Figures 11-11 C, 11-14, 11-15).

In this stage the region is at *maximum relief;* a few remnants of the original surface may still exist on the highest divides as the major streams approach base level. All the rest of the surface slopes from the upper of these limiting elevations toward the lower.

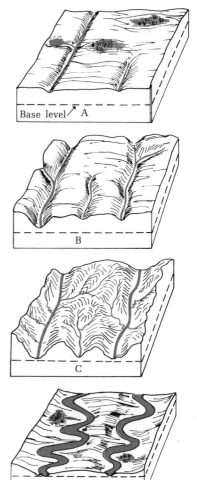

**Figure 11-11**
Stages of regional reduction, from early youth in A to a regional erosion surface in D.

---

[1]The terms *regional youth* and *regional maturity* must not be confused with the terms *young* and *mature* as applied to valleys. You will save yourself much trouble if you will learn thoroughly the distinction between these two sets of terms.

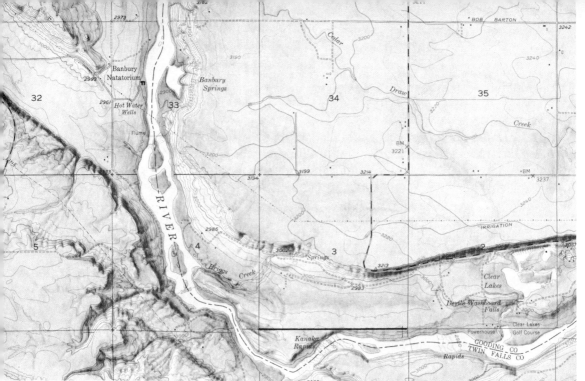

**Figure 11-12**
An area in topographic youth, southern Idaho. The main stream is the Snake River. Bedrock consists of horizontal lava beds; this is a part of the Columbia River basalt plateau. Note the slight amount of stream dissection away from the main valley. Black line represents one mile. (Part of the Thousand Springs quadrangle, Idaho, shaded-relief edition. U. S. Geological Survey.)

**Figure 11-13**
Surface of the Alberta Plateau, western Canada, being dissected by stream erosion. Streams are in young valleys; the drainage pattern is dendritic. A resistant bed of sandstone caps the plateau. (Photo courtesy of Royal Canadian Air Force.)

**Figure 11-14**
A maturely dissected land surface. Near Beaumont, southern California. (Photo by J. R. Balsley, U. S. Geological Survey.)

**Figure 11-15**
A topographic surface in maturity. Divides are sharp and valleys narrow, and practically the whole area is in slope. Black line represents one mile. (Part of the Juanita Arch quadrangle, Colorado, shaded-relief edition. U. S. Geological Survey.)

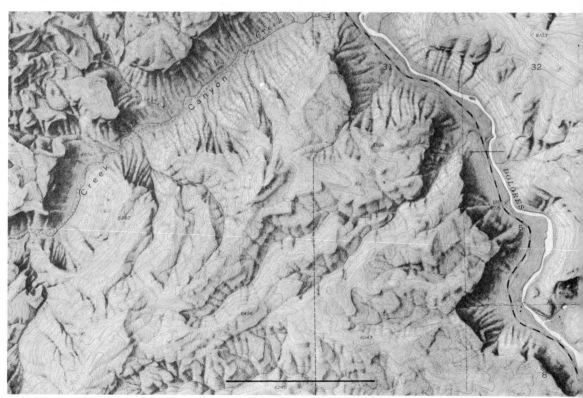

Later stages of maturity are characterized by progressively lower relief, as more of the high lands are destroyed, more streams attain base level, and wider floodplains develop. If continued long enough, stream erosion eventually produces a late-mature **regional erosion surface,** close to ultimate base level (Figure 11-11 D). Features of such a landscape are valleys with wide floodplains, large meandering streams with very low gradients, and subdued divides with thick residual soil, no rock exposures, and very gentle slopes.

Thus stream erosion has reduced a low-relief region several hundred feet above sea level to a low-relief region almost at sea level. The cycle of regional reduction is complete.

### Residual Mountains; Mesas and Buttes

The Catskill Mountains of New York, the Cumberland Mountains of eastern Kentucky and adjacent areas, and the Absaroka Mountains east of Yellowstone Park in Wyoming are all underlain by thick piles of essentially flat-lying stratified rock. The rocks of the first two are sedimentary; those of the Absarokas are lava beds and sheets of pyroclastics. Each of these mountainous areas has been sculptured by streams to an early-mature stage in the cycle of regional reduction. Each is the remains of a plateau that was once much more extensive, and hence these mountains and others like them may appropriately be termed **residual mountains.** In all such areas the terrain is mountainous solely because of deep erosion of thick masses of elevated stratified rocks, not because of any inherent crustal deformation.

Topographic features analogous to residual mountains, but on a smaller scale, are mesas and buttes. A **mesa** (Spanish, "table") is a broad flat-topped feature that stands above the surrounding countryside because an exceptionally resistant rock layer is underlain by weak, easily eroded strata. Common rocks that cap, or "hold up," mesas are sandstone and lava; generally the underlying formation is shale. The edge of a typical mesa is shown in Figure 8-3. A **butte** (French, "mound") is a small isolated remnant of formerly extensive flat-lying rocks (Figure 18-9). As indicated by the angular profiles of mesas and buttes, and the thinness or absence of soil, these topographic features are particularly characteristic of arid and semiarid regions.

**11-4
INTERRUPTIONS
IN THE CYCLE**

Although we have no trouble in citing examples of regions in youth and in early and middle maturity, we find that late-mature regional erosion surfaces are very rare. Since these surfaces are the end products of the cycle of regional reduction, it seems strange that examples should be hard to find. Why is this?

For one thing, the length of time required to reduce a region through the stages of maturity to a widespread erosion surface is tremendously longer than that necessary for the development of a youthful or early mature topography. As the region progressively loses elevation, as stream gradients decrease, and as a blanket of vegetation and soil forms on thick residual regolith, the rate of erosion is markedly slower. To produce D from C in Figure 11-11 takes many times longer than to produce C from earlier stages. Hence it may require an appreciable period of time, even in geological terms, for a regional erosion surface to develop.

Now one of the assumptions that was implicit in our account of the reduction of a land surface was that the region was not affected by downward or upward crustal movement while it was being eroded; in other words, that it held still from start to finish of the erosion cycle. But we have seen that vertical movements of the lithosphere are by no means unusual, and are even to be expected over appreciable periods of geologic time. Thus it is altogether likely that, at some time in its long-drawn-out later stages, the cycle of erosion is interrupted because regional base level is radically altered by lowering of the land (or rise of sea level), or by uplift of the land (or retreat of the sea).

## Burial by Sediment

A land surface like that in Figure 11-11 D has little relief and stands but a few feet above sea level. Therefore a lowering of the land (or rise of the sea) of only a few tens of feet would be enough to flood the area completely, converting it to shallow sea bottom. We should expect that, in the ordinary course of events, layers of sediment would accumulate in this shallow sea, burying the former erosion surface and preserving it in the record of the rocks. Just such a series of events is indicated at many places in the rock record. Repeatedly we find sequences of sedimentary strata that were deposited on surfaces of low relief that cut across older rocks and give every evidence of having originated through stream erosion. Attention is given to such surfaces (unconformities) in Chapter 14.

## Dissection by Erosion

Earth forces may be positive as well as negative. Suppose that a regional erosion surface, instead of being lowered, is bodily uplifted several hundred feet, or that sea level declines by a like amount. Referring again to Figure 11-11 D, it is clear that such an event would have the effect of lowering base level. So the streams now have several hundred feet to fall before reaching the sea; in other words, their gradients are greatly increased. Provided with new energy, the rejuvenated streams start downcutting their valleys, and the whole cycle of regional reduction starts over again. The end product of a former cycle becomes the starting surface of a new one. Since there are already late-mature streams on the land surface, inherited from the old cycle, these will cut young valleys on the uplifted surface. The occurrence of valleys with strongly meandering courses and youthful V profiles is clear indication of uplift of an erosion surface and rejuvenation of its drainage. Examples are shown in Figures 11-16 and 11-17.

Thus late-stage surfaces of regional reduction are seldom preserved for very long. They are either lowered and buried by sediments or raised and cut away by streams.

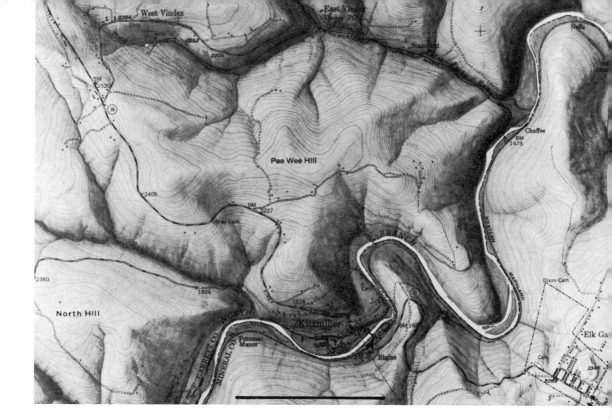

**Figure 11-17**
North Branch of the Potomac River once flowed in meander bends on a wide floodplain. Regional uplift has caused the stream to cut the V-shaped young valley that it now occupies. The earlier floodplain has been almost entirely dissected. Black line represents one mile; contour interval 20 feet. (Part of the Kitzmiller quadrangle, Maryland–West Virginia, shaded-relief edition. U. S. Geological Survey.)

**Figure 11-16**
Goosenecks of the San Juan River, a spectacular example of entrenched meanders. The river and its tributaries have been rejuvenated by regional uplift of the land. Bedrock is Permian sandstone and shale. Southeastern Utah. (Photo by John S. Shelton.)

**11-5
DRAINAGE
PATTERNS**

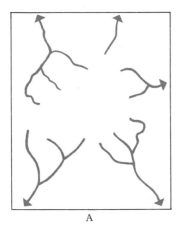

A

B

**Figure 11-18**
Radial (A) and dendritic (B) drainage
patterns. Compare B with the pattern
shown by the streams of Figure 11-13.

Often we may learn something of the geology of an area by looking at the patterns taken by the main streams and their tributaries. Here we are referring to the arrangement of stream courses on the land surface, as viewed from above and plotted on an aerial photograph or a topographic map (generally the latter).

A **radial** drainage pattern (Figure 11-18 A) shows plainly that there is a high point in the mapped area—a hill, volcanic cone, or other isolated peak—from which streams flow outward in all directions. A treelike, or **dendritic,** pattern (Figures 11-14 and 11-18 B) develops where the rocks have similar resistance to weathering, downslope movements, and stream erosion over a considerable area. A dendritic pattern tells us that the rock beneath the surface is uniform or massive.

A **trellised** drainage pattern develops in regions where stratified rocks have been tilted by earth forces and then cut across, or *truncated,* by erosion (Frontispiece; Figures 11-19 A and 18-5). Some of the strata are resistant to erosion and stand up as long ridges, whereas others are relatively weak and form belts of low land between the ridges. The main streams flow across the edges of the tilted strata toward the sea, probably having inherited their courses from an earlier cycle, in the manner discussed above; these main streams cut through the ridges in water gaps. Subsidiary streams excavate their valleys in the belts of weak rock between the ridges, and a characteristic right-angle pattern develops (Figure 11-19 B). Resemblance to a trellis is furthered in some areas by the formation of short streams that flow down the sides of the ridges. A trellised pattern tells us immediately that we are dealing with a region of folded or tilted strata of unequal resistance to erosion.

**11-6
SUMMARY**

Stream valleys may be classified according to stage of development. Valleys with a sharp gradient, with a V-shaped cross profile in which the stream occupies the whole valley bottom, are said to be young. Young valleys undergo active erosion by the stream and active widening by downslope movements.

Downward cutting is inhibited when a stream reaches its base level, the level of the water body into which it flows, below which it cannot erode its valley. There are many local base levels, all of which are temporary when considered from the viewpoint of geologic time. Ultimate base level is sea level. Valleys with a flat or pan-shaped cross profile are said to be mature. This signifies that the stream is approaching or has reached base level, and is using its energy to transport sediments and to erode laterally rather than downward. Floodplains, meanders, point bars, and low gradients characterize mature valleys. The term is applied to a wide range of stages.

The cycle of regional reduction includes the changes that an area undergoes through stream erosion between (1) the elevation of the surface on which the erosion started, and (2) base level. In regional youth, the original surface has been only slightly dissected, valleys are small, and divides are wide and flat. In early maturity, divides are narrow, most of the land is in slope, the stream network is well developed, and floodplains may have started to form along the principal rivers. This is the stage of maximum relief. Relief is progressively lower in later stages of maturity; eventually a regional erosion surface, near sea level, is produced. Residual mountains are examples of areas in the early mature stage of regional reduction.

A regional erosion surface takes a long time to form. The cycle may be interrupted at a late stage by encroachment of the sea and burial of the erosion surface under a layer of sediment; or by uplift of the land and dissection of the surface in a new cycle of regional reduction.

Stream systems generally take patterns reflecting the arrangement of the underlying rocks. A radial pattern forms on a volcano or other isolated high spot; a dendritic pattern on large areas of uniform rock; and a trellised pattern on tilted rocks of uneven resistance to erosion. Water gaps are characteristic of areas of trellised drainage.

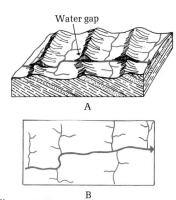

**Figure 11-19**
A, a block diagram showing a land surface underlain by tilted rock layers of differing resistance to erosion. The main stream cuts through the sandstone ridges in water gaps; its tributaries flow in valleys underlain by shale. B, map of the same area, showing the trellised drainage pattern.

## SUGGESTED READINGS

Beiser, A., and the Editors of *Life*. 1962. *The Earth,* chap. 5, pp. 105–29. Life Nature Library. New York: Time Inc.
Sculpture of the landscape is presented for the layman using a nonscientific approach. Good color photographs.

Bloom, A. L. 1969. *The Surface of the Earth,* chap. 5, pp. 81–102. Englewood Cliffs, N.J.: Prentice-Hall. (Paperback.)
Presentation of the sequential evolution of valley and regional landscapes through time.

Shimer, J. A. 1959. *This Sculptured Earth: the Landscape of America.* New York: Columbia University Press. 256 pp.
This story of features of the landscape is written in popular language, and should be of special interest to the traveler who has visited the scenic localities described.

Tuttle, S. D. 1970. *Landforms and Landscapes,* chaps. 5 and 6, pp. 48–86. Dubuque, Iowa: W. C. Brown. (Paperback.)
Covers the development and classification of stream-eroded landscapes and the theory and philosophy of geomorphic systems.

# 12 THE WORK OF GLACIERS

12-1 Ice on the Lands
12-2 Glaciers and Glacier Motion
12-3 Valley Glaciers
12-4 Ice Sheets
12-5 Some Effects of Glaciation
12-6 Pre-Pleistocene Glaciations
12-7 Summary

About 10 percent of the earth's land surface is covered with glacier ice. The area includes the continent of Antarctica, the large island of Greenland, and the high parts of major mountain ranges all over the world. The ice of these regions is about one percent of the total hydrosphere. Since this moisture is locked up in solid form on the land, sea level is lower (by some 300 feet) than it would otherwise be. Glaciers appreciably modify the earth's climates. Most significant from our point of view, glaciers can be observed in the process of doing geological work: eroding, transporting, and depositing rock materials.

There is clear evidence that glaciers were much more extensive in the recent past than they are today. During the Pleistocene Epoch nearly one-third of the lands were ice-covered. Important regions, including the north central and northeastern United States and central and northern Europe, directly or indirectly owe their water supply, their soil, their scenery, and the very configuration of their lands and waters to now vanished sheets of ice. Hence the geologist is much interested in the work of ice that has long since disappeared. Applying the principle of uniformity of process, he uses what he can learn about the geological work of modern glaciers in interpreting the work of past ones.

## 12-1 ICE ON THE LANDS

A **glacier** is a thick mass of ice that moves slowly on a land surface. We recognize two types. **Valley glaciers** are confined to valleys in mountainous terrain like the Alps, the Alaskan ranges, and Glacier National Park, Montana. **Ice sheets** are broad glaciers that blanket whole regions, covering plains, plateaus, and mountains with ice that may reach thicknesses of 5,000 feet or more. Modern examples are the ice sheets of Antarctica and Greenland.

Glaciers depend for their sustenance on an annual increment of snow, and so they form only where there is a carry-over of snow from one season to the next. Abundant precipitation is more important than extreme cold. The southeastern coast of Alaska, which is the warmest part of the state, has the greatest concentration of glaciers because it receives the heaviest snowfall. Large land areas around the Arctic Ocean are very cold, but are barren of glaciers because there is not enough snow.

The fragile six-sided crystals that are snowflakes are readily crushed and compacted on a glacier surface by the weight of new snowfalls, and are shortly altered to a loose aggregate of rounded granules of ice. With deeper burial these granules are deformed, locally melted and refrozen, and recrystallized, to produce a solid mass

## 12-2 GLACIERS AND GLACIER MOTION

consisting of individual ice grains that interlock tightly like a three-dimensional jigsaw puzzle.

Ice, you will recall, is listed among the minerals; and ice in large enough quantities to make up a glacier is a rock. It is a one-mineral rock, like pure limestone (calcite) or rock salt (halite). Despite its interlocking crystalline fabric, ice is a weak rock, for it exists close to its melting temperature. As snowfalls continue, ice accumulates to such a thickness that its bottom layers start to yield under the weight of the ice and snow above; thus beginning to move, the ice mass becomes a glacier.

Glacier motion, though not apparent like that of a stream, is nonetheless real. If a straight row of stakes is placed directly across the ice of a valley glacier, and the site is revisited a few days or weeks later, two facts are immediately apparent. First, all the stakes are several feet downvalley from their original position, obviously showing that the glacier is in motion. Second, the row of stakes is now curved, the stakes in the middle part of the glacier being farther downvalley than those near the edges. Thus we must conclude that the central part of the glacier moves faster than the marginal parts, undoubtedly because of decreased friction away from the valley walls. By measuring the distance covered and the time elapsed, we can readily compute the rate of ice movement at any point along the row of stakes, or an average rate for the whole mass. We shall probably find that the maximum surface velocity near the center is somewhere between one and two feet per day.

Although the mechanism of glacier motion is complex, two factors seem to be dominant. One we may call **basal slip:** the motion of the whole mass, as it grinds and slides across a rocky floor. Grooved and striated bedrock surfaces beneath glaciers are evidence of this kind of movement. Added to basal slip is **internal flow,** which allows parts of the ice to move faster than other parts. Internal flow seems to take place chiefly as a result of microscopic slippage along planes within individual grains of ice. This slippage within grains does not destroy the solidity or coherence of the ice, or the internal arrangement of atoms in the crystalline grains. Hence the internal flow of a glacier is the sum total of billions of tiny movements within its constituent ice grains. This nonturbulent motion is in marked contrast to that of streams.

As suggested by its name, internal flow is restricted to the deeper parts of a glacier, which are under confining pressure from the weight of overlying snow and ice. Here the ice develops shear planes and foliation (Section 15-3), and becomes in effect a metamorphic rock (Figure 12-1). The upper ice, not being under strong pressure, is brittle

Figure 12-1
Edge of the Greenland ice sheet. The streaked, dark material in the lower third of the cliff is ice, heavily loaded with rock debris; the streaks mark shear planes and foliation. Till (ground moraine) in the foreground. (Photo by R. P. Goldthwait.)

Figure 12-2
Crevasse in the Matanuska Glacier, Alaska. Deep within the glacier the ice moves by flowage, but it cracks like this at the surface where it is brittle. (Photo by J. R. Williams, U. S. Geological Survey.)

and develops cracks, or **crevasses** (Figure 12-2), as it is carried along more or less unevenly on the lower ice.

When movements of the crust raise a region to mountainous elevations, or when cooling of the climate affects mountains that already stand high, more snow falls on

12-3
VALLEY
GLACIERS

213

the summit areas each winter than is lost through melting in the warm seasons. Patches of permanent ice, termed **icefields,** begin to accumulate in the low places, where the snow drifts deeply and there is some protection from sunlight. In the thickest icefields, especially those at the heads of former stream valleys, enough ice accumulates to flow; and the masses begin to move downward under gravity, following preexisting stream valleys. Thus are valley glaciers born.

### Erosion and Its Effects

As anyone knows who has tried to dislodge a stick or stone frozen in ice, its adhesion to foreign objects is strong indeed. The ice in the depths of icefields freezes tightly to the rock on and against which it rests, and when the ice starts to move it pulls along whatever fractured rock can be pried loose. This process, called **plucking,** allows the ice at the upper end of a valley glacier to "dig in," converting its original place of accumulation to a scoop-shaped hollow, or **cirque,** in the mountainside (Figures 12-3, 12-4).

The supply of ice in a cirque is replenished by fresh snowfalls, by drifting, and by avalanching from the surrounding heights. Daytime melting may allow water to percolate down between the ice and the cirque walls, where it freezes and promotes frost wedging, a highly significant weathering process in glaciated regions. Plucking and frost wedging are especially effective at the upper end of the cirque, where they produce a steep to vertical **headwall.**

Imagine a mountain range with a row of valley glaciers in cirques on either side, each glacier eroding its cirque headward. Eventually the two rows of cirque headwalls will meet back-to-back, converting the mountain crest to a knife-edge ridge, or **arête,** of bare rock. The Continental Divide in Glacier National Park follows a series of arêtes for many miles. Three or more glaciers pushing their cirques headward into an isolated mountain may at length produce a **horn,** or high jagged spire of bare rock. Horns and arêtes may be recognized from afar, and they are invariable indicators of present or past glaciation (Figure 12-5).

A second process of glacier erosion is rock grinding, or **abrasion.** Rock fragments, contributed by plucking and frost wedging and frozen tightly in the sides and bottom of the ice, make a valley glacier a powerful agent of abrasion. One of the effects is to rasp and scrape the walls and the floor of the valley, often polishing and grooving the bedrock (Figure 12-6). Another is to straighten the

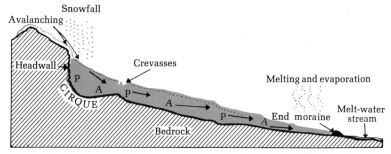

**Figure 12-3**
Longitudinal section of a valley glacier. A, abrasion; P, plucking. Arrows within the glacier indicate direction of ice movement. Whether the lower end advances, stands still, or retreats depends on the balance between supply (snowfall) and wastage (melting and evaporation).

valley, by grinding away irregularities on opposite sides. What was formerly a somewhat winding, narrow stream valley with a V profile is "reamed out" into a much straighter, **U-shaped valley** that typically has a troughlike form (Figure 12-7).

Intensity of abrasion depends on thickness of the ice. Therefore main, or trunk, glaciers cut their valleys deeper than their tributaries are able to do. The latter flow in U-shaped troughs far above the floor of the main valley. Exposed to view when the ice melts away, these subsidiary troughs are appropriately termed **hanging valleys** (Figure 12-8).

**Figure 12-4**
Valley glaciers in action. Cirque occupied by ice in left center; note avalanche scars on the headwall, the narrow crack between headwall and glacier, and numerous crevasses. Sharp ridge in left foreground is an arête; it leads up to a horn. Baffin Island, Canadian Arctic. (Photo by R. P. Goldthwait.)

**Figure 12-5**
An ice-carved
landscape: the Sierra
Nevada, California.
Winter snows fill the
cirques and glaciated
valleys; high winds
blow the knife-edged
arêtes free of snow.
Many horns are visible,
including the high peak
on the skyline (Mt.
Lyell, 13,090 feet).
(Photo by Marshall
Moxom.)

**Figure 12-6**
Glacier-abraded surface
with erratics. Yosemite
Valley, California.
(Photo by E. L. Shay.)

**Figure 12-7**
Glacier-eroded valley, with characteristic U-shaped cross profile and unobstructed extent. Glaciated tributary valleys "hang" far above the main valley floor. Western Norway.

**Figure 12-8**
The magnificent glaciated valley of Yosemite National Park, California. The waterfall at right cascades from a hanging valley. (Photo by Mary R. Hill.)

**Figure 12-9**
Valley glaciation in action. Note the streaks of rock debris in the main glacier, and the small tributary glaciers on the left side of the valley. Were the ice to disappear, the valley would look much like the one shown in Figure 12-7. Tiedemann Glacier, British Columbia. (Photo courtesy of British Columbia Department of Mines and Petroleum Resources.)

A mountain region actively undergoing glaciation is not very inviting to most of us, and perhaps is best viewed in photographs or from the insulated comfort of an airliner. The features of erosion that we have just enumerated are largely buried in ice and snow; barren masses of rock not yet reduced by the glaciers project through the snowfields and between the ice-filled valleys (Figure 12-9). But a mountain region once glaciated and now freed of ice presents an altogether different landscape, which invites first-hand exploration on the ground. Miles of unobstructed view, waterfalls pouring from hanging valleys, sheer cliffs leading up to arêtes and horns, cirque-cupped ponds, or **tarns**—all these make glacially eroded mountains the most scenically striking of any on earth. Fine examples in North America include the high Sierra Nevada of California (Figure 12-5), the Teton and Wind River ranges of Wyoming, and the northern Rocky Mountains of the United States and Canada as displayed in Glacier, Waterton, Banff, and Jasper national parks.

## Transportation

The process of plucking yields angular rock fragments, and abrasion yields ground-up **rock flour** of silt size. These

218

materials are carried within the ice, especially at the bottom and sides. Coarse rock debris may also be borne on the glacier's upper surface, where it accumulates as a result of avalanching or landsliding from adjacent cliffs.

Glacier transportation differs from stream transportation in several ways. In the first place, glacier ice can move fragments far bigger than those moved in streams. Masses of rock the size of a house trailer are not uncommon (Figure 12-10). Second, ice exerts no sorting action on the particles that it transports, with the result that fragments ranging in size from silt grains to great boulders are all mixed together and carried alike. Third, some of the load travels on top of the ice. Fourth, little if any rounding or grain-against-grain abrasion affects the silt- and sand-size particles, which remain angular and fresh throughout their journey. Larger pieces that are frozen in the marginal ice may be flattened and scratched on one or more sides as they are dragged against the bedrock.

**Figure 12-10**
Large glacier-transported boulder (erratic) in Yellowstone National Park.

## Products of Deposition

The lower end of a glacier, or the **ice front,** stands at that point where the supply of ice from up the valley is just equal to loss through melting and evaporation. If supply is greater than loss, the ice front moves down the valley; if loss exceeds supply, the ice front retreats. Note that it is the front or end of the glacier that may advance, stand still, or retreat; the ice itself is always moving forward, downvalley, under gravity.

Rock debris carried in or on the glacier may simply be dumped at the ice front when the ice melts. The resulting material is termed **till.** It consists of an unstratified, unsorted mixture of rock flour, pebbles, cobblestones, and boulders, all jumbled together. A ridge of till extending across the valley at the ice front is an **end moraine** (Figure 12-11).

Often some of the larger fragments in an end moraine are found to be quite different from the bedrock in the vicinity, and to be derived from ledges of bedrock some distance up the valley. Such "foreign" boulders deposited from glaciers are termed **erratics.** They may occur by themselves as well as in moraines (Figures 12-6, 12-10).

A prominent feature at the lower end of a valley is a stream of meltwater, which gushes from beneath the ice, cuts through the end moraine, and takes a course down the valley (Figure 12-12). As the lower end of a typical glacier is full of rock debris, and the moraine consists of nothing else, it is not surprising to find that meltwater streams are loaded to capacity with rock flour, sand, and gravel. The channels of such streams that are

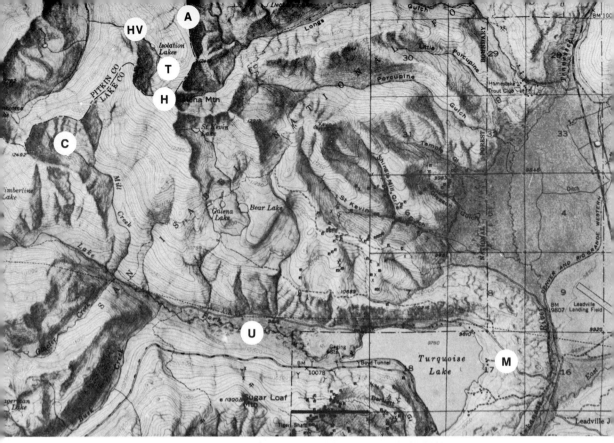

**Figure 12-11**

An end moraine, M, marks the terminus of a former valley glacier that moved down the valley, U. Turquoise Lake is a moraine-dammed lake. Other features of valley glaciation include an arête, A; cirque, C; horn, H; hanging valley, HV; and tarns, T. There are numerous other examples of each in addition to the ones lettered. Elevations range from 9,500 to nearly 13,000 feet.

The black line represents one mile; contour interval 50 feet. (Part of the Holy Cross quadrangle, Colorado, shaded-relief edition. U. S. Geological Survey.)

**Figure 12-12**

The Nisqually Glacier, Mount Rainier, Washington, in 1964. High ridges of morainal material along valley sides, from one of which the picture was taken, show that the glacier was formerly much larger. Lower end of the glacier is heavily loaded with debris, which is dark where moist. Stream of meltwater is braided. (Photo by E. L. Shay.)

downvalley from the glacier are likely to be a maze of sand and gravel bars, in other words, to exhibit a characteristically braided pattern.

Unlike till, the sediments carried by meltwater streams are washed and sorted, and come to rest in more or less well-defined layers. All glacier-derived sediments that are sorted and layered are termed **stratified drift.** The stratified drift—mostly cross-bedded sand and gravel—that a stream spreads down the valley from an ice front constitutes **outwash,** and the deposit that it makes is a **valley train** (Figure 12-13). The finer sand and the rock flour tend to stay in stream suspension, and to come to rest as evenly stratified beds in the lake or sea into which the stream flows.

At this point the preceding paragraphs should be reviewed, so as to get clearly in mind the terms that apply to glacial *erosion* as distinct from those that apply to *deposition.* Cirque, horn, and hanging valley are examples of features produced by the erosional activities of valley glaciers; several other terms have been mentioned and should be added to the list. As for glacial deposits, till and stratified drift are the *classes of material* of which these deposits are composed, whereas end moraines and valley trains are among the *features* produced.

In summary, a valley glacier acquires the bulk of its load by plucking and abrading the bedrock of its cirque and valley; carries this load on and in the ice, greatly modifying its valley as it moves; and deposits it either

**Figure 12-13**
Braided channels of the Nelchina River, which drains the glacier in the distance. Broad flat area is a valley train. Copper River region, Alaska. (Photo by J. R. Williams, U. S. Geological Survey.)

directly or through the agency of meltwater. With their knowledge of present-day valley glaciers, geologists can reconstruct with fair accuracy the history of those that have disappeared.

**12-4**
**ICE SHEETS**

The island of Greenland, with an area about one-fourth that of the mainland of the United States, is almost entirely covered by an ice sheet. The interior of the island is a monotonous expanse of snow-covered ice, with a maximum elevation of about 10,000 feet. Geophysical surveys made by expeditions crossing the ice sheet reveal that locally it is at least 10,000 feet thick, and hence must extend down to or even somewhat below sea level. At places around the coast, high mountains partially hem in the ice, and great valley glaciers squeeze between mountain masses as they move from the interior to the sea.

The Antarctic ice sheet is of comparable thickness and elevation, but its area is more than seven times as great, about that of the United States and Mexico combined. In places along the Antarctic coast are mountains through which the ice spills in immense valley glaciers, but around much of the continent the ice overrides the coastline and extends into the sea, merging with the floating "shelf ice" formed by freezing of ocean water.

The ice of an ice sheet moves outward from one or more central areas of accumulation. This has been verified by observations in Greenland and Antarctica, but it was known before the present program of intensive studies in these regions began. When the directions of the grooves and striations left behind on the bedrock by the Pleistocene ice sheet of North America are plotted on a map, they clearly show that the ice radiated from two centers, one in central Canada and the other far to the west in British Columbia (Figure 24-4). Similar evidence in Europe shows that there was a center of ice radiation in Scandinavia and another in northern Russia. Thus an accumulation of ice starts to move outward when it becomes so thick that the pull of gravity on the mass exceeds the strength of the ice. A downslope gradient is not required; indeed, in their gradual flooding of an uneven land surface, ice sheets may locally move uphill.

**Erosion and Its Effects**

At places in once glaciated regions we find knobs or humps of resistant bedrock with profiles like that shown in Figure 12-14. One side is gently sloping, smooth, and polished or grooved; clearly it has undergone abrasion.

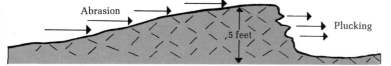

Abrasion

Plucking

.5 feet

**Figure 12-14**
Effects of ice erosion on this small exposure of bedrock show clearly that the ice moved from left to right.

The opposite side is steep and angular; just as clearly, it was produced by plucking. Such glacially eroded masses of rock are as good as signboards in pointing the direction of former ice movement.

Although plucking may be locally notable, in the cross-country movement of ice sheets abrasion is the far more significant process. The ability of a thick ice sheet, well shod with rock debris, to abrade its rocky floor is hard to overstate. For example, as the ice moved southward across central New York State, it gouged deep bedrock trenches parallel to the direction of ice flow (Figure 12-15). These trenches, a mile or two in width, as much as 40 miles long, and at least 2,000 feet deep, represent chiefly the work of abrasion in the relatively soft sedimentary rocks of the region. Partly filled by water after the ice disappeared, these basins today hold the Finger Lakes: Cayuga, Seneca, Keuka, and others.

**Figure 12-15**
Canadice Lake (left) and Hemlock Lake, near Rochester, New York, looking south. The lakes occupy deep bedrock trenches produced by ice-sheet abrasion. Note the smooth, molded topography. Little valleys on the lake sides are postglacial; several have deltas. (Photo by Wahl's Photographic Service, Inc.)

Figure 12-16
Deep grooves cut into limestone by glacier abrasion, Kelleys Island, Ohio. The bottom ice probably carried a heavy load of rock debris from the scouring-out of the Lake Erie basin just to the north.

Whole mountain ranges were surmounted by the ice sheets. The Laurentian Highlands of Quebec, the Adirondack Mountains of New York, and the Green Mountains and White Mountains of New England are examples of such ranges. The effects of abrasion can be seen in a number of features, including the smooth, rounded mountain profiles; the steep-sided, trenchlike character of many of the major valleys; and the numerous abraded surfaces of bare rock (Figure 12-6).

Glacier grooving of bedrock surfaces is exceptionally pronounced at a few places. A famous example is on Kelleys Island near the south shore of Lake Erie (Figure 12-16). Here the ice sheet, moving southward heavily laden with rock debris, abraded the limestone bedrock in deep parallel furrows that look as though they had been milled by a machine.

**Figure 12-17**
Glacier front with end moraine. Note the mixed coarse and fine material in the till making up the moraine. The light ground at left center is outwash from an earlier stage; the ice is now pushing forward over this material. Northwest Greenland. (Photo by R. P. Goldthwait.)

## Deposits of Till

Ice sheets make extensive deposits of material dumped directly from the ice (Figure 12-17). End moraines mark positions at which the glacier edge stood for some time.

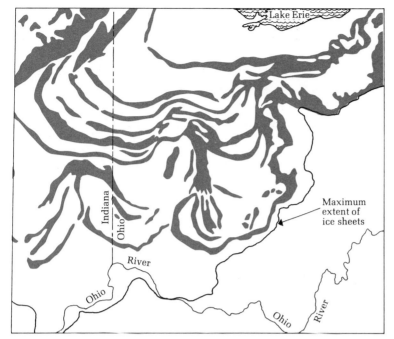

Figure 12-18
End moraines left by the last ice sheet in Ohio and eastern Indiana as the ice front retreated by stages. The lobate pattern is characteristic of such moraines.

They are long ridges, or belts of low hills, that extend across country for many miles. Viewed from above, end moraines are characteristically curved in a series of lobes, like a looped-back curtain (Figure 12-18). These lobes are concave toward the direction from which the ice came, and they demonstrate that an ice sheet does not push forward uniformly along a straight front, but rather in a series of lobes, or tonguelike projections.

Large areas on the glaciated side of end moraines commonly received a sheet of till plastered on bedrock. This **ground moraine** was left behind when a heavily loaded ice sheet melted away, leaving its rock debris spread widely. Ground moraine underlies tens of thousands of square miles of nearly flat **till plain,** notably in Ohio, Indiana, Illinois, Michigan, and Wisconsin. Weathering since deposition has converted the till into fertile cropland.

Certain regions of ground moraine contain many **drumlins** (Figures 7-24, 12–19). These are streamlined hills, some 50 to 150 feet in average height, elongated parallel with the direction of ice flow. Drumlins are generally made up of clayey till. Apparently these till masses could neither be carried away by the ice nor spread out as ground moraine, and instead were simply overridden. Drumlins occur by the scores in some areas of ground moraine but are absent from others.

Erratics are common in all morainal deposits. They range in size from pebbles to massive boulders weighing many tons. Sometimes erratics of a distinctive rock type

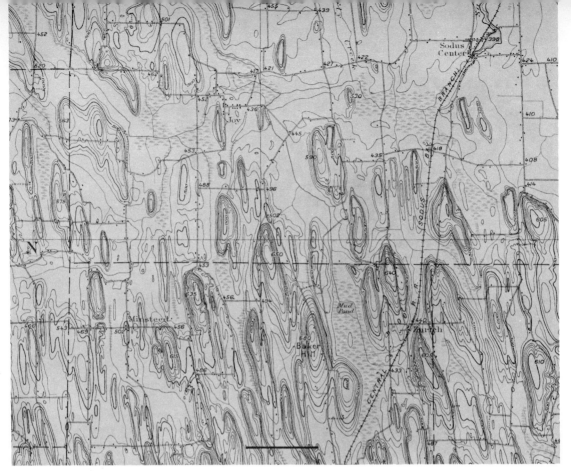

**Figure 12-19**

Drumlins, as shown on a topographic map. The black line represents one mile; contour interval 20 feet. Compare Figure 7-23. (Part of the Palmyra quadrangle, New York. U. S. Geological Survey.)

can be traced back to the parent ledge from which they were plucked by the ice, thus giving us a clue as to the distance and direction of ice movement.

## Deposits of Stratified Drift

Streams of meltwater leaving an ice sheet may flow down valleys that lead away from the glacier front, depositing outwash in the form of valley trains. Or where the land surface slopes evenly away from the edge of the ice sheet, meltwater streams may spread over the countryside in branching braided patterns, forming widespread **outwash plains.** Composed of relatively well-sorted, evenly bedded sand and gravel, these outwash plains may be many square miles in area and scores of feet thick.

In areas formerly occupied by ice sheets, we find narrow winding ridges of sand and gravel, a few tens of feet in height and width and sometimes several miles long (Figure 12-20). These are called **eskers.** They are interpreted as deposits made by streams that flowed in tunnels beneath slowly moving or stagnant ice. When the ice finally disappeared, the stream-laid deposits remained in the esker form we see today.

**Kames** are small knobby hills of sand and gravel (Figure 12-21), most of which originated as these sediments were

washed into holes and crevasses in glacier ice or between masses of melting ice. They were left as steep-sided hills when the ice melted away. Groups of kames commonly formed between lobes of ice that advanced from different directions.

Stratified drift derived from ice sheets is of great economic importance. Urban and industrialized regions demand very large quantities of concrete, for bridges, airports, foundations, and highways; and concrete cannot be made without sand, gravel, or some other aggregate. In the midwestern and northeastern United States, which were glaciated during the Pleistocene Epoch, hundreds of plants process millions of tons of sand and gravel every year. Every type of stratified-drift deposit is used; valley

**Figure 12-20**
The curving ridge is an esker. It is as much as 70 feet high and can be traced for 4½ miles. Northeastern North Dakota. (Photo by J. R. Reid.)

**Figure 12-21**
A kame. Kettle Moraine State Forest, southeastern Wisconsin, (Photo by Dean Tvedt, Wisconsin Conservation Department.)

trains and outwash plains are probably most intensively exploited. Thus we find that much metropolitan existence is literally based on sand grains and pebbles that were washed from melting ice sheets. Where outwash gravels are buried, they provide another resource of great value, namely water. The occurrence of underground water is discussed in the next chapter. Some of the features produced by ice sheets are shown in Figure 12-22.

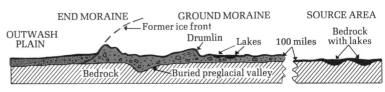

**Figure 12-22**
Diagrammatic cross section through some of the features produced by ice sheets.

## Repeated Glaciation

In many road cuts where ground moraine is exposed, it is found that the soil merges downward into fresh, unweathered till, in which the rock fragments are angular and hard; and that this till, instead of resting on bedrock, lies on an older till layer. In this older layer, decomposition is far advanced. The upper few feet are weathered to clay, there is abundant iron oxide staining, and all boulders except siliceous ones are so deeply decomposed as to crumble in the fingers. Clearly the lower till is the result of deposition by an earlier ice sheet, its deeply decomposed character recording a long period of weathering when the region was free of ice. The fresh upper till was laid down by a later ice sheet, which spread across the old till (ground moraine) without removing it. The existing soil layer reflects weathering since disappearance of the last ice sheet.

Putting the records together from many exposures, geologists have found that there were four important glaciations during the Pleistocene Epoch (Section 24-3) separated by three long interglacial periods. (We are in a postglacial period now; it may turn out to be an interglacial one.) The earliest ice sheet advanced southward about two million years ago, and the last one disappeared less than 10,000 years ago, as shown by radiocarbon dating of materials that it left behind.

**12-5**
**SOME EFFECTS**
**OF GLACIATION**

## Redistribution of Regolith

The source area of an ice sheet, where the ice is thickest and from which all movement is outward, undergoes intense erosion. Soil and residual regolith are scraped off, and the ice bites more or less deeply into the underlying bedrock. The marginal regions of an ice sheet, on the other hand, are sites of deposition, and the ground receives till and stratified drift in the forms we have enumerated.

To cite an example, central Canada from Labrador to the Yukon is a vast region of bare rock, lake basins scoured by the ice, and patches of thin soil that have developed since the last ice disappeared. By contrast, the southern fringes of Canada and the northern United States have thick fertile soils, the bulk of which have developed by weathering on till that was transferred to its present position from northern sources by the same Pleistocene ice sheets that scraped central Canada clean. Similarly in Europe: the barren mountains of Norway and Sweden, where ice sheets centered, contrast sharply with the outwash plains, ground moraine, and other crop-producing flatland deposits of Denmark and northern Germany. On each continent, what the north has lost the south has gained.

Other materials to be mentioned here are those deposits of loess (Section 7-9) that resulted when winds picked up and redistributed large amounts of glacier-produced rock flour.

## Rearrangement of Drainage

A thick sheet of ice moving across a land surface completely disrupts the drainage system. Valleys disappear under the ice and are filled with till or gravel. Their stream waters are ponded against the ice front, and the overflow finds new channels. Meltwater streams appear and establish themselves in new courses, only to cease flowing with disappearance of the ice sheet. The till or outwash surface left by the ice may be so flat that new drainage can be established only with difficulty. Low places develop into swamps or ponds, which are connected by sluggish streams wandering in a seemingly aimless way.

After a period of time, streams establish a more normal drainage pattern, commonly dendritic (Section 11-5). But this pattern is quite likely to be entirely different from the one in existence before glaciation. Southern Ohio, for example, now drains southward into the Ohio River. In pre-Pleistocene time, however, there was no Ohio River valley, and drainage in the region was to the northwest, by a river flowing into northern Indiana and thence southwest across Indiana and Illinois to the Mississippi River (Figure 12-23). Evidence for this statement comes from water-well records; the buried gravels that fill the old river valley make good reservoirs for underground water, and so have been the object of much drilling. Present-day drainage systems in glaciated regions, then, no matter how well developed and firmly established they

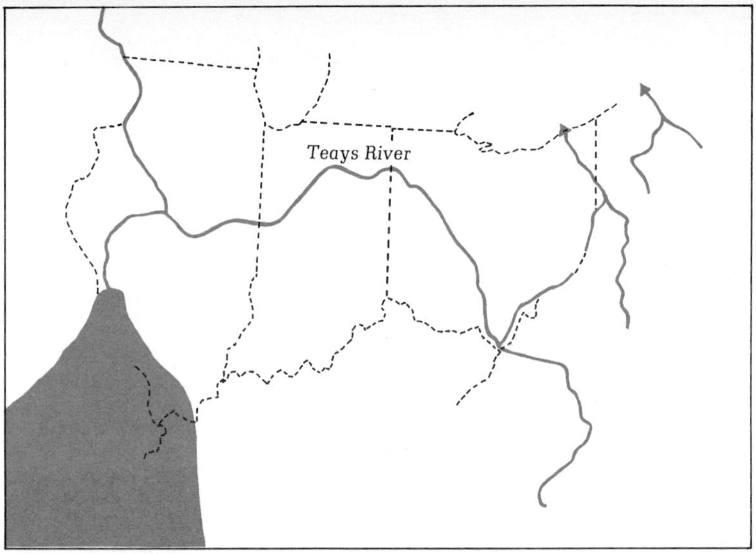

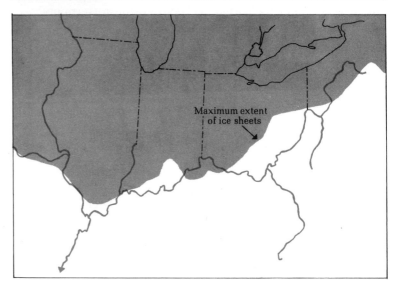

**Figure 12-23**
Drainage of the Illinois-Indiana-Ohio region before the Pleistocene ice sheets. The sea extended far up what is now the lower valley of the Mississippi River. The Teays River Valley is now buried beneath glacier-deposited regolith.

Teays River

Maximum extent of ice sheets

**Figure 12-24**
Major streams of the present day. The Ohio River, the lower Missouri, and part of the Mississippi owe their courses in large part to the ice sheets.

appear, may bear little relation to those that existed before glaciation (Figure 12-24)

### Formation of Lake Basins

Perhaps the most distinctive aspect of formerly glaciated landscapes is the presence of lakes. The basins that hold these lakes are formed by valley glaciers and by ice sheets, by the processes of erosion and by those of deposition. Lake basins are produced in other ways, but glaciation is the parent of far more of them than any other group of processes.

Small lakes in bedrock basins are typical of formerly glaciated high mountains. Gemlike little tarns in cirque basins are beloved of hikers and trout fishermen. Some glaciated valley floors hold a series of lakes, where glacier erosion produced a sort of stair-step long profile, abraded basins alternating with plucked ledges of bedrock. Hundreds of tarns and valley lakes are shown on topographic maps of quadrangles in the high Sierra Nevada, Glacier National Park, and similar terrains.

A second group of lakes that owe their origin to valley glaciers includes those that are dammed behind end moraines. These lakes, which are found farther down the valleys than the ones just mentioned, tend to be considerably bigger. Among many examples are Jenny Lake in Grand Teton National Park, St. Mary Lake in Glacier National Park, and Turquoise Lake in the Sawatch Range of central Colorado (Figure 12-11).

Lake basins formed by ice sheets are numbered in the hundreds of thousands. The vast lowland of central Canada, from which the North American ice sheets came, is pock-marked with innumerable lakes, which lie in bedrock basins abraded by the ice. They range greatly in size and depth, but all have in common their origin through erosion. Far to the southward, the ice sheets found a stream-eroded lowland and from its bedrock they scoured out the basins that now hold the Great Lakes. The origin of the Finger Lakes of New York is also erosional.

Most of Minnesota's famous "ten thousand lakes," as well as those of nearby areas, are in shallow basins, or **kettles,** in ground moraine and outwash. In the waning stages of glaciation, when the ice sheet was stagnant, the zone of melting was many miles wide. Masses of ice became detached from the main front and, on finally melting, left depressions that now hold lakes. Large numbers of these lakes, together with the poorly integrated drainage system of which they are a part, can be studied from topographic maps of many Minnesota and Wisconsin quadrangles.

## 12-6 PRE-PLEISTOCENE GLACIATIONS

Because we have referred thus far only to features associated with present-day glaciers or their Pleistocene predecessors, it might be concluded that glaciation is associated only with fairly recent times. This is far from the case. Lithified tills (termed **tillites**) are known to occur in Precambrian rocks more than two billion years old in Africa and North America (Section 19-7); similar deposits of much later Precambrian age have been identified in all the continents except Antarctica.

In 1961, petroleum geologists making field studies in the Sahara Desert found wide areas of well-preserved parallel striations on Upper Ordovician sandstones. The striations were clearly produced before the sandstones were buried by overlying Silurian strata. The announcement that there must have been an ice sheet in northern Africa in Ordovician time was greeted with skepticism by many geologists; but an international field conference in 1970 seems to have confirmed its existence. The striations are accompanied by tillites, outwash deposits, and erratics. Movement of the ice sheet was toward the north and northwest.

Tillites of late Paleozoic age are found in India, Australia, Africa, South America, and Antarctica. Thus there were at least four intervals of widespread glaciation before the most recent one, the Pleistocene.

The identification of certain sedimentary rocks as tillites (and thus as products of glacier deposition) is not always easy. Striations on bedrock and boulders, as well as jumbled texture and heterogeneous composition like that of till, can all be produced in mudflows, landslides, and submarine slumps. None of these features, however, commonly contains erratic boulders of exotic rock type, nor are these deposits of such widespread areal extent as most of the pre-Pleistocene strata identified as tillites.

**12-7**
**SUMMARY**

Glacier ice covers 10 percent of the earth's land surface today, chiefly in Greenland and Antarctica, and occupied some 30 percent in the recent past. Ice greatly modifies the lands that it covers.

Glaciers form where there is heavy snowfall, a portion of which carries over from one season to the next. Under the weight of new snowfalls, old snow compacts and recrystallizes to form ice. Though a crystalline rock, ice is weak and may yield under its own weight. Glaciers flow by a combination of basal slip, or motion of the whole mass, and internal flow, or the sum of a multitude of tiny movements within individual ice grains. Brittle ice at the surface of a glacier develops fractures, or crevasses, owing to movement of the ice underneath.

A valley glacier originates in an icefield in high mountains, and moves down a preexisting stream valley. By the process of plucking, it produces a cirque, with a vertical headwall, at its upper end. Glaciers eroding their cirques headward on opposite sides of a ridge or isolated peak produce sharp bedrock features called arêtes and horns. By abrasion, glaciers ream out their valleys to a troughlike form with a characteristic U profile. Recently glaciated high mountains make spectacular scenery.

Valley glaciers transport debris on and in the ice. As ice exerts no sorting action, material dumped directly by a glacier is unsorted and unstratified. Such material, termed till, makes up end moraines at the ice terminus. Erratics, or foreign boulders, are other products of deposition. Streams of meltwater, commonly braided, deposit stratified drift, or outwash, consisting mostly of cross-bedded sand and gravel, as a valley train extending downvalley from the ice front.

Ice sheets thousands of feet thick are exemplified by the ones in Greenland and Antarctica. Pleistocene ice sheets radiated outward from centers in central and western Canada, and in Scandinavia and northern Russia. Ice sheets move when the pull of gravity exceeds the strength of the ice. They gouge deeply into bedrock, to form such features as the basins of the Finger Lakes of New York, and they abrade overridden mountains like the Adirondacks and the mountains of New England. Deposits of till include end moraines, ground moraine, and drumlins. Among deposits of stratified drift are outwash plains, eskers, and kames. Sand and gravel from these features is valuable as aggregate for concrete.

Successive sheets of till, with deeply weathered upper surfaces, show that there were four major glaciations during the Pleistocene Epoch, separated by long interglacial periods. The last ice disappeared less than 10,000 years ago. There are records of earlier glaciations in the late Paleozoic, in the Ordovician, and at two intervals in the Precambrian.

Ice sheets remove soil and regolith from one region and deposit them in another; rearrange the drainage on the land surface; and produce thousands of lake basins. These basins result from erosion of bedrock in the source area of the ice sheet, and from the melting of detached ice blocks, to produce kettles, or shallow basins in ground moraine or outwash, in the area of deposition.

## SUGGESTED READINGS

Bloom, A. L. 1969. *The Surface of the Earth,* chap. 7, pp. 128–145. Englewood Cliffs, N.J.: Prentice-Hall. (Paperback.)
A presentation of the story of ice on the land, which includes a section on Pleistocene climatic change.
Dyson, J. L. 1962. *The World of Ice.* New York: Knopf. 292 pp.
A highly readable nontechnical yet authoritative account of glaciers and glaciation.
Field, W. O. 1955. Glaciers. *Scientific American,* vol. 193, no. 3, pp. 84–86. (Offprint No. 809. San Francisco: Freeman.)
Considers the critical role ice plays in the water economy of the earth, and the profound influence it exerts on the weather in all parts of the world.
Flint, R. F. 1971. *Glacial and Quaternary Geology.* New York: Wiley. 892 pp.
A college text and reference book containing detailed information on all topics relating to glaciers and glaciation.
Janssen, R. E. 1952. The history of a river. *Scientific American,* vol. 186, no. 6, pp. 74–80. (Offprint No. 826. San Francisco: Freeman.)
Presents the story of the Teays River, the main pre-Pleistocene drainage system of the Illinois-Indiana-Ohio region.
Tuttle, S. D. 1970. *Landforms and Landscapes,* chap. 7, pp. 87-106. Dubuque, Iowa: W. C. Brown. (Paperback.)
A concise treatment of glaciers and glaciation written at a level understandable to the introductory student.

# 13 GROUND WATER

13-1   Infiltration versus Runoff

13-2   Porosity and Permeability.
       Aquifers

13-3   Distribution

13-4   Wells and Springs

13-5   Geologic Work of Ground Water

13-6   Man and Ground Water

13-7   Summary

That part of the rainfall that does not immediately flow off across the land surface but sinks into the ground constitutes underground water, or simply **ground water.** It is a significant part of the hydrologic cycle (Chapter 2). Ground water, like the water in streams, is on its way back to the ocean from which it originated. In most regions, only a part of this return trip is made underground; ordinarily ground water emerges as springs, or seeps into streams, long before reaching the sea.

Ground water does a great deal of geologic work; for example, it is an agent in precipitating mineral cement around grains of quartz so as to convert sand to sandstone, and in dissolving limestone to form enormous voids like Mammoth Cave. Furthermore, ground water is a mineral resource of prime importance. Many cities and industries, and also farms that use irrigation, depend on ground water for their very existence. Thus it behooves us to learn something about the occurrence and habits of this earth material.

## 13-1 INFILTRATION VERSUS RUNOFF

What proportion of the precipitation that falls on the land will sink in rather than run off? When rain falls on loose, open material like sand or gravel, the water can readily pass into the ground, whereas soil that is tight and clayey will allow little water to sink in and will force most of it to run off. Clearly, the size of the openings in the regolith or rock is one factor that controls the amount of **infiltration** (sinking-in). Another factor, and an even more evident one, is the type and amount of vegetation. A thick blotter-like layer of sod in a field of grass or alfalfa will absorb several times as much water as the mostly bare soil of a cornfield. A third factor is the character of the precipitation. Long, gentle rains put much more water below ground than heavy showers, in which the water comes too fast to soak in deeply. Finally there is topography, or attitude of the land surface. Acre for acre, a flat surface in regional youth can be expected to absorb more water than a region in maturity (Figure 11-11).

## 13-2 POROSITY AND PERMEABILITY. AQUIFERS

Rock and regolith possess two properties that bear directly on the movement and storage of ground water. **Porosity** is the portion of the total volume that is not occupied by solid matter; it determines how much water can be held. **Permeability** is the ability to transmit ground water, and thus it is a measure of the extent to which the water can move. Clay and shale are made up of rather loosely packed particles; hence their porosity is relatively high. Their permeability, on the other hand, is extremely low, because the pore spaces between the ultrafine particles are so small that water cannot move through them. Thus clay and shale are porous but not permeable. Gravel, by way of contrast, is highly porous and very permeable, as there are many pore spaces and they are so large that water can pass through them with ease.

Any body of regolith or rock that contains and transmits water, and thus is a source of ground water for our use, is an **aquifer.** An aquifer must have both porosity and permeability. The most common aquifers are gravel, sand, loosely cemented sandstone, limestone full of solution cavities, and rock of any kind that is crisscrossed by fractures. Most aquifers, like the gravels in a buried glacial valley, are local and limited in extent; a few, like widespread sheet sandstones, may underlie several states and yield water at thousands of wells. Ground water moves through aquifers at rates ranging from 5 feet or more per day to as little as 5 feet per year.

**13-3 DISTRIBUTION**

On the basis of their water content, earth materials below the surface may be readily divided into two zones. The upper one, in which regolith and rock may be moist but are not saturated, is the **zone of aeration**; the lower, in which all pore spaces are full of water, is the **zone of saturation**. The boundary between these zones, or the upper surface of the zone of saturation, is the **water table.** On entering the ground, a drop of water filters down through the zone of aeration to the water table, at which level it joins the ground water that fills all the pore spaces in the rock.

The water table tends to reflect in a subdued way the topographic relief, rising under hills and being depressed under valleys (Figure 13-1). In valleys and other low areas, it may intersect the land surface; streams, swamps, and lakes are surface expressions of the water table. Depth to the water table, then, ranges from nothing at all in a swamp to, let us say, 10 feet part way up a nearby hill and 30 feet at the hilltop. As we should expect, the water table rises in wet seasons and falls in dry ones.

Ground water in the zone of saturation is not static, but moves slowly under gravity. Laboratory experiments using scale models have shown that the paths of ground-

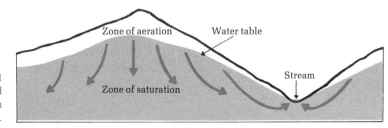

**Figure 13-1**
Distribution and movement of ground water in an area underlain by uniform rock.

Zone of aeration    Water table

Zone of saturation

Stream

water motion in uniform earth material are as indicated by the arrows in Figure 13-1. These demonstrate that much ground water seeps into streams, and here we have the explanation of how certain streams continue to flow even in dry seasons. Position of the arrows on the figure also shows that if the ground water were not regularly replenished, the water table would eventually flatten out; water "piles up" under hills because of its slow rate of lateral movement. If the drought persisted, the water table of Figure 13-1 might even fall below the level of the valley bottom, in which case the stream would go dry. In desert regions the water table is often hundreds of feet below the surface, far too deep to intersect the surface anywhere. Here the movement is not from ground water into streams, but from streams into the ground. Recall what happens to the waters of a "flash flood" in a desert basin (Section 7-6).

The bottom of the zone of saturation lies at depths of several thousand feet to several miles, where earth pressures are so intense as to close all pore spaces in the rocks. Waters from deep in the zone of saturation, as revealed in wells drilled for oil, are hot and are heavily laden with mineral matter.

Figure 13-2 is a cross section through some dipping beds of sandstone and shale, with soil and regolith omitted. The surface is hilly and the water table reflects its relief. The sandstones are porous and permeable, and hence are aquifers. The shales are impermeable, and even where they are saturated, no movement of ground water can take place.

## 13-4
## WELLS AND SPRINGS

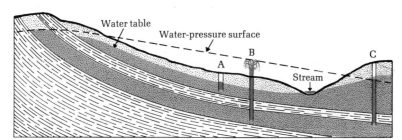

**Figure 13-2**
Area underlain by tilted beds of permeable sandstone and impermeable shale. Well A, which enters only the upper sandstone, is a water-table well. Wells B and C penetrate the lower sandstone, in which the water is under pressure. In such wells, the water will rise to the water-pressure surface. Well B is a flowing artesian well and well C is a nonflowing artesian well.

### Water-Table Wells

Consider first only that part of Figure 13-2 that lies above the upper shale. Here there is only one rock, a sandstone, which is uniform throughout the area. Suppose we decide to drill a well for water at point A. From the surface to the water table, the walls of the well would be wet but no free water would accumulate in the well. Below the water table, however, water would stand in the well, since all open spaces are saturated. To be sure of a supply of water even in dry seasons, we would drill the well many feet below the water table. In a **water-table well** such as this, the water stands at the level at which it was encountered, that is, at the water table. This type of well must always be pumped.

A water-table well could be obtained by drilling into the upper sandstone at any place shown in Figure 13-2. Any well drilled from the surface shown in Figure 13-1 would also be a water-table well.

### Artesian Wells

Now look at the part of Figure 13-2 that includes the two shale beds and the sandstone between them. The shale beds, being impermeable, seal off the lower sandstone and make it a separate and distinct aquifer, open only where it reaches the ground surface (left end of the section). The water held in this sandstone cannot escape and hence is under pressure from its own weight. If a well is drilled into the sandstone, the water will tend to seek its own level by rising in the well to the water-pressure surface. This surface is the highest level at which the water stands in the sandstone, minus the loss of elevation resulting from friction in the system.

Let us assume we are to drill a well at B to the lower sandstone. (Water encountered in the upper sandstone will be kept out of the well by a steel pipe, or "casing.") As soon as the drill passes through the thin shale bed into the lower sandstone, water will pour into the well under pressure, and when the bit is removed, will flow out the top. It will be under enough pressure to reach the water-pressure surface.

Any well in which the water rises above the point at which it was encountered is termed an **artesian well,** named from the French province of Artois (Roman name, *Artesium*), where such conditions exist. The well we have just described is a **flowing** artesian well (Figure 13-3). A well drilled at point C would also encounter water under pressure in the lower sandstone. The water would rise in the well, but not to the surface, because at this place the water-pressure surface is below ground level. The well at C is a **nonflowing** artesian well. Like a water-table well, it must be pumped.

Some artesian aquifers are of wide extent and of great importance to farms, cities, and industries. An example is a Cambrian sandstone that underlies much of southern Wisconsin, northern Illinois, and Iowa. The sandstone is

**Figure 13-3**
A flowing artesian well on the floodplain of the Pecos River near Roswell, New Mexico. The aquifer is a Permian limestone that is exposed in mountains many miles to the west. The pipe joint at the wellhead deflects the flow sideways instead of allowing it to rise into the air as in well B, Figure 13-2.

exposed at the surface in a broad belt in central Wisconsin, and passes beneath younger strata toward the south and west, in which directions it becomes progressively more deeply buried. Many cities obtain their water from wells drilled to this aquifer. Since all the water in the sandstone is derived from rain that falls on its area of outcrop, the water drunk by people living in Rockford, Illinois, all fell as rain in central Wisconsin some 225 miles away. Probably several centuries have elapsed since the time when the water consumed in Illinois today fell as rain in Wisconsin.

## Springs

The arrangement of rocks in the earth's outer crust is extremely variable. Hence there are circumstances in which the normal movement of ground water may be interrupted, and a part of the water diverted to emerge as a natural surface outflow, or **spring**. In Figure 13-4, normal downward movement of the ground water is prevented by an impermeable stratum, and some of the overlying rock becomes saturated. Where such a **perched water table** intersects the land surface, a spring or a row of springs is found. In Figure 11-12, for example, the many

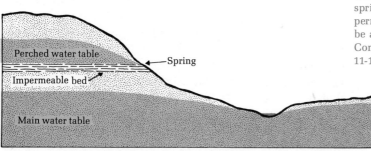

**Figure 13-4**
One of the numerous ways in which springs may be produced. If the permeable bed is persistent, there will be a row of springs along the slope. Compare Banbury Springs in Figure 11-12.

springs along the sides of the valley result from the presence of impermeable zones in the horizontal plateau basalts of the area.

Springs are common in limestone regions, where water may enter a surface depression, travel below ground for a distance, and then emerge from a solution channel. There are numerous other ways in which springs may form. Water subject to magmatic heating may issue as warm or hot springs, exceptionally as geysers (Figure 4-22).

**13-5
GEOLOGIC WORK
OF GROUND
WATER**

The water that soaks into rocks is important in frost wedging and in the various processes of chemical weathering, so important that neither residual regolith nor soil could form without it. Water is also crucially important in rock slides, slump, and the other forms of downslope movement. In its capacity as ground water, it has another significance, as an agent of transportation, removing rock material from some subsurface sites and depositing it at others. Most of the material carried is in the form of dissolved compounds.

**Solution**

Of all the common rocks the most soluble is rock salt. It is exposed at the surface only in regions of extreme dryness; elsewhere any salt that once was near the surface has been removed in solution. Its former position is commonly marked by a zone of broken rock where the overlying strata have collapsed. Thick salt beds exist underground, however, even in the zone of saturation, because ground water cannot enter them. The tightly interlocking fabric of the halite grains that make up the rock gives it an extremely low porosity; furthermore, halite is a weak mineral, and rock salt yields by internal flow under the weight of overlying rocks, like the ice deep in a glacier, so that cracks and crevices cannot form. Salt mines hundreds of feet below the water table are quite dry.

Anhydrite is another easily soluble rock. Where soaked by near-surface waters it is generally hydrated to gypsum, a conversion that involves a volume increase of some 30 percent. Gypsum may remain exposed, even in regions of moderate humidity, but ordinarily it disappears in ground-water solution, leaving collapse zones like those that mark former salt beds. At depth, anhydrite resists entry by ground water, for the same reasons that salt does.

Limestone, though less soluble than the evaporites just mentioned, is still readily attacked by ground water, especially when the water contains abundant carbon dioxide (Chapter 5). Limestone is brittle enough to be cut by fractures, and strong enough to allow these fractures to stand open. Together with permeable zones along bedding surfaces, these fractures are natural avenues through which ground water may move. They are commonly enlarged by solution, and very large caverns may be formed under favorable conditions of temperature and water composition. Many caverns bear evidence of having been produced by currents of moving water that filled them, and it is believed that most big caverns were formed in the zone of saturation. It is possible for us to enter them today because the water table has receded since the caverns were formed.

The most spectacular evidence of solution by ground water, then, is found where the bedrock is limestone (or dolostone), and where the climate is humid, or has recently been so. In such regions the limestone is honeycombed with solution passages, some of which may be very large (Figure 13-5). As we walk through one of the great caverns, such as Mammoth Cave in Kentucky or Carlsbad Caverns in New Mexico, we must remember that what we can see is only a small part—cleared of obstructions, and fitted with lights and walks—of a tremendous underground labyrinth. At least 150 miles of subterranean passageways are known in the vicinity of Mammoth Cave; at Carlsbad there are three levels, at 754 feet, 900 feet, and 1320 feet below the surface, connected with many vertical shafts and steep tunnels.

**Figure 13-5**
Southern end of Natural Tunnel, southwestern Virginia. This is an underground passage some 900 feet long, whose roof averages 75 feet above the stream. It furnishes a valuable grade for the railroad. The rock is Cambrian dolostone. (Photo courtesy of Commonwealth of Virginia, Division of Parks.)

**Figure 13-6**

Limestone region, much of which is pockmarked with sinks. Rain falling on such an area percolates downward instead of flowing off the surface in streams. The higher ground in the north part of the area is also underlain by limestone: Mammoth Cave occurs in these rocks just to the north. Black line represents one mile. (Part of the Mammoth Cave quadrangle, Kentucky, shaded-relief edition. U. S. Geological Survey.)

The surface of the ground in a limestone region may reflect this underground condition, being pitted and pock-marked by closed depressions, called **sinks**. These are solution cavities, a few tens to a few hundreds of feet across (Figures 13-6, 13-7), that lead downward into the system of underground passages. Many sinks result from roof collapse of near-surface caverns; others, from concentration of downward-percolating water along the intersection of vertical features. If a sink becomes choked with residual soil and vegetation, it may hold a pond. Rain falling on a land surface marked by many sinks is funneled downward, so that limestone regions in humid climates are likely to be without normal streams and valleys. That ground-water solution had the same effect on ancient limestones as it has on modern ones is shown by the pre-Pennsylvanian limestone surface pictured in Figure 14-27.

In speaking of water conditions below the surface, laymen often refer to "underground rivers." Streams of water flowing freely in channelways below ground are not uncommon in limestone regions. Elsewhere, however, we have seen that ground water is held in the pore spaces of aquifers, through which it moves slowly as a body, not as a stream. This is true even where the aquifer is the gravel of a buried stream channel. We should not confuse the long-vanished surface stream that left the gravel with the ground water that seeps through this gravel today. Underground rivers in rocks other than limestone are likely to be more imaginary than real.

## Deposition

Every year ground water removes great quantities of calcite from limestone regions. Where does this material go? The bulk of it stays in the soluble form, calcium bicarbonate, and is taken by ground water to streams and by streams to the sea. There some remains in solution and some is used by organisms in building their shells and bones.

But an appreciable fraction of the calcium bicarbonate dissolved in ground water is taken out of solution, or precipitated, before it is very far along on its journey through the rocks. Consider, for example, rain water that trickles from a sink down a crevice leading into a cavern. On its way down, the water acquires a solution load from the walls of the passage. Presently the water emerges on a cavern ceiling, from which it drips to the floor. Each drop of water, as it hesitates a moment before falling, loses a little carbon dioxide. This loss converts some of the soluble calcium bicarbonate into the insoluble

**Figure 13-7**
This area, in southern Indiana, is underlain by limestone, solution of which has produced the sinks that dot the landscape. (Photo by John S. Shelton.)

**Figure 13-8**
Dripstone in Carlsbad Caverns, New Mexico. (Photo by E. L. Shay.)

carbonate, and a film of the latter is deposited. If such a process is long continued, it leads to the formation of a **stalactite**, or icicle-shaped deposit extending from the ceiling; directly below, where the water drops to the floor, a low mound or dome **(stalagmite)** may grow slowly upward. Often a stalactite merges with the stalagmite below it, to form a column (Figure 13-8). Where water drips from many points along a crevice, a row of stalactites may form and eventually join to resemble the folds of a curtain; or a row of columns may coalesce to look like organ pipes. All such features are collectively termed **dripstone**. Dripstone is an impressive feature of most large caverns.

Thus a cavern, which was a site of solution when it was being formed below the water table, becomes a site of deposition when it is in the zone of aeration.

Dripstone is but one result of deposition by ground water. Geodes (Section 8-5) owe their crystal linings to the precipitation of mineral matter, generally quartz or calcite, from percolating ground water in the zone of saturation. The substance of many concretions is similarly derived. The mineral matter that fills the pore spaces of permineralized fossils (Section 9-2), and binds many sandstones and conglomerates together, was deposited from ground water as it moved through the rocks.

### Replacement

Replacement refers to the removal of one substance in solution and the simultaneous deposition of another, without appreciable change in total volume. We have already met the process several times. In Section 4-4 we noted that limestone in the vicinity of large intrusions is susceptible to replacement by ore minerals, especially sulfides like galena and sphalerite; there, however, the agent is believed to be water derived from magma rather than ordinary ground water. The change of limestone to dolostone, as described in Section 8-4, is probably dependent in large part on ground water. So is the replacement of fossil shells (Section 9-2), whether by solution of the whole shells and later filling of the voids or through replacement a molecule at a time, so as to produce an exact replica.

Such different objects as a stalactite, a concretion, a specimen of conglomerate, a specimen of dolostone, and a piece of petrified wood all have one thing in common: each was formed by ground water, or in some degree affected or modified by it.

### 13-6 MAN AND GROUND WATER

Man's ever expanding activities on the land surface have had numerous effects on the underlying reservoir of ground water. One of the most obvious is his habit of placing large areas under concrete or blacktop pavement, which sheds water rather than allowing it to sink in. The "umbrella effect" is well illustrated by a large airport, with its acres of runways, ramps, parking lots, and buildings. After a rain, the water has no opportunity to infiltrate and add itself to the supply in underground storage.

If the rate at which water is taken out of the ground is greater than nature's rate of input, or recharge, and the reservoir is under simple gravity control, the water table will go down; if the system is artesian, the pressure will decline. It then becomes necessary to deepen old wells, drill new ones deeper than formerly, and pump from greater depths. The only way to arrest such a decline is to decrease the rate of withdrawal until it at least balances that of recharge. Modern conservation measures have been in effect in some areas for a long time; for example, when in the 1930s the artesian aquifer below the Pecos Valley of southeastern New Mexico (Figure 13-3) had been overdrilled for irrigation, tight restrictions were put into effect on the drilling of additional wells. Such restrictions are commonplace today in areas of heavy ground-water use.

Ground water, like surface waters and the atmosphere, is subject to pollution. Suppose that a community, following an accepted modern practice, disposes of its trash in a "sanitary landfill," a pit where each day's contribution is flattened and then covered with a layer of earth. There is no smoke or odor and there are no animal pests. Yet the cover of soil or regolith is not impervious to rain water. In wet seasons, water percolates through the entire deposit, dissolving an assortment of materials from the mixture, some of them perhaps harmful, and adds these to the ground-water body. Or as another example, it is found profitable to produce oil from a group of wells only if a way can be found to dispose of the brine that must be pumped up with the oil. Several thousand gallons of brine per day are stored in shallow pits, in the expecta- tion that the water will evaporate and the salt be precipi- tated. Instead, much brine sinks in and seriously contami- nates the ground water. Or an industry, restricted by law from disposing of waste liquids on the surface, drills a deep well and pumps the effluent into the rocks several thousand feet below. Although the results may not be of immediate concern, we have no way of knowing for sure that these wastes will not eventually appear in a ground-water supply. Contamination of ground water, however it happens, is especially serious because the water moves so slowly. Contamination may have been taking place for a long time before this fact is discovered; the effects may last for years or decades even after the source has been cleared up.

Geologic knowledge is necessary not only to identify problems but to help in their solution. Bad effects of sanitary landfills can be minimized by placing them in impermeable materials such as clay or shale; oil-field brines can be pumped back into the same formation from which they came. Most other ground-water problems have comparable geologic solutions. But we emphasize that difficulties arising from overuse or misuse of ground water—a substance widely available to individual, industry, and community—can often be solved only by joint geologic, engineering, social, and political action.

**13-7**
**SUMMARY**

Waters that sink into the ground for a part of their journey seaward constitute ground water. The proportion that infiltrates rather than runs off is determined by size of openings in regolith and rock, type and amount of vegetation, character of the precipitation, and slope of the surface. Porosity is the portion of the total volume of rock or regolith that is void space; permeability is the capacity of the material to allow water to move through it. Any body of rock or regolith that contains and transmits water that may be used is an aquifer. Ground water moves through most aquifers very slowly.

The water table is the upper boundary of the zone of saturation, which is that part of the rock and regolith in which all pore spaces are full of water. From the land surface to the water table is the zone of aeration. The water table reflects the irregularities of the land surface. It eventually flattens out unless the supply of ground water is replenished.

Wells in which the water stands at the level encountered by the drill are water-table wells. In such wells the water is under simple gravity control and must be pumped. In artesian wells, the water rises above the level at which it was encountered, because it is in a confined aquifer and hence under pressure from its own weight. The highest level to which the water can rise is the water-pressure surface. If this surface is above ground at the well site, the well will flow; otherwise it will be a nonflowing artesian well. Some artesian aquifers, especially sandstones, are of regional extent.

A spring is a natural surface outflow of ground water. Springs may form where a perched water table, held up by an impermeable stratum, intersects the land surface; where waters flow from an underground solution passage in limestone; or in various other places. Hot springs and geysers are common in volcanic areas.

Ground water dissolves much limestone in humid climates. Solution is chiefly done below the water table, where a complex system of underground passages may be developed. Surface openings, or sinks, lead downward into the limestone. If the water table is lowered, caverns and other openings, having come into the zone of aeration, become the sites of deposition rather than solution. Stalactites, stalagmites, and other forms of dripstone commonly form. Ground water also forms the crystals that line geodes, and deposits the substance of most concretions and the cementing materials in many conglomerates and sandstones. Ground water is the agent of local replacement of limestone by ore minerals, of the conversion of limestone to dolostone, and of the replacement of fossil shell material.

Supplies of ground water for man's use may be decreased by widespread paving of land surfaces, or by overpumping in excess of natural recharge. Contamination may result from unwise disposal of solid wastes, oil-field brines, and industrial effluents. Wise use of ground water has geological, engineering, and political aspects.

## SUGGESTED READINGS

Baldwin, H. L., and McGuinness, C. L. 1963. *A Primer on Ground Water.* U. S. Geological Survey. Washington, D. C.: Government Printing Office. 26 pp.

A nontechnical publication that provides general information on ground water, including distribution and management of resources.

Mohr, C. E., and Poulson, T. L. 1966. *The Life of the Cave.* New York: McGraw-Hill. 232 pp.

An introduction to caves and the organisms that inhabit caves, for the amateur spelunker. Includes a list of caves open to the public, a glossary of speleological terms, and advice for the person who wants to become a spelunker.

Mohr, C. E., and Sloane, H. N. 1955. *Celebrated American Caves.* New Brunswick, N.J.: Rutgers University Press. 339 pp.

An anthology of fascinating cave stories containing 24 contributions by 15 authors. Popular accounts of exploration and history are given for caves of wide geographic distribution.

Moore, G. W., and Nicholas, G. 1964. *Speleology: The Study of Caves.* Boston: D. C. Heath. 128 pp.

Discusses origin of caves, the formation of cave deposits, underground atmosphere, cave organisms, and man's use of caves.

Olson, R. E. 1970. *A Geography of Water,* chap. 7, pp. 102-15. Dubuque, Iowa: W. C. Brown. (Paperback.)

This chapter provides a concise picture of the origin and nature of ground water, and of its economic significance, distribution, and use by man.

Sayre, A. N. 1950. Ground water. *Scientific American,* vol. 183, no. 5, pp. 14–19. (Offprint No. 818. San Francisco: Freeman.)

This report discusses the question of whether or not the large store of water in the subterranean rocks is being permanently depleted.

# 14 ROCK DEFORMATION

14-1  Evidence in the Crust

14-2  Dip and Strike

14-3  Folds

14-4  Domes and Dome Mountains

14-5  Faults

14-6  Fault-Block Mountains

14-7  Earthquakes

14-8  Unconformities

14-9  Summary

The presence of folded and broken rocks in mountain regions gives clear evidence of kinetic energy generated within the earth. From a study of the chief ways in which rocks have been deformed, we can draw some conclusions about the history that such deformation records.

Crustal stresses affect rocks of all three classes, but their effects are most readily recognized in sedimentary rocks. This is true for two reasons. First, the principle of original horizontality tells us that sedimentary rocks lie flat to begin with; any departure from the horizontal must mean later disturbance, by outside forces. Second, the stratified nature of the sedimentary rocks plainly reveals any bending, breakage, offset, or other deviation from the normal (Figure 14-1). Therefore in this chapter we shall be concerned mostly, though not exclusively, with sedimentary rocks.

Figure 14-1
The attitude of these sedimentary rocks tells us immediately that they have been tilted by earth forces since they were deposited. Jurassic strata, western Wyoming.

We find that some sedimentary rocks have been tilted gently and others steeply; some are inclined in one direction and some in another. It is necessary to have a standard means by which the position, or attitude, of inclined rocks can be accurately described.

Suppose that in the field we observe layers of sandstone and shale that are inclined as shown in Figure 14-2 A. Using a compass, we can read the direction toward which the beds are inclined, which is southeast. This we call the **direction of dip.** The direction at right angles to the direction of dip is the **strike;** here it is northeast-southwest. The **amount of dip** is determined by measuring the angle

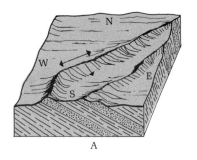

A

between the bedding surface and the horizontal (Figure 14-2 B). It is found to be 25 degrees. The symbol and the numerals in the southern part of the map, Figure 14-2 C, summarize what we have found out. They say in effect, "At this place in the map area there are strata that strike NE-SW and dip SE at 25 degrees." Another symbol on this map, some distance to the northeast, tells nearly the same story.

**Figure 14-2**
Dip and strike. In A, sandstone and shale beds dip to the southeast (single arrow) and strike northeast-southwest (double arrow). B, their angle of dip is 25 degrees. C is a map of the area, showing dip and strike as measured at two places.

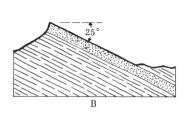

B

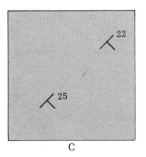

C

In Figure 14-3 a geologist is determining the amount of dip of steeply inclined limestone beds. Determination of dip and strike is not an object in itself. It is merely a means to an end, which is the correct delineation of dipping strata. More accurately, we should say the correct delineation of dipping *surfaces*, because dip and strike may be used to describe the position of any inclined surface: a bedding plane, the surface of a dike, or a fracture in the rocks.

**Figure 14-3**
A geologist measures the angle of dip with an instrument containing a bubble level. (Photo courtesy of Marathon Oil Company.)

It is hard to believe that rock strata can be folded, like rumpled sheets of cloth or paper. Common experience suggests that layers of hard, brittle materials like sandstone or limestone, if subjected to strong compression against their edges, would not bend or fold but simply shatter. And so they would, under ordinary conditions of atmospheric pressure. But when lateral compression is applied to strata that are deep within the crust, confined by the weight of thousands of feet of overlying rocks, the strata are no longer brittle and they yield by folding. Thus folds in rocks, like intrusive igneous masses, were formed at depth in the crust and are exposed to view only through uplift and deep erosion.

For convenience, we distinguish up-folds or **anticlines,** and down-folds, or **synclines** (Figure 14-4). Where the two occur side by side, as they normally do, there is no sharp dividing line; the right side, or **limb,** of the anticline in the figure merges into the left limb of the syncline. Anticlines and synclines range in size from small wrinkles (Figure 14-5) through larger folds (Figures 14-6, 14-7, 14-8) to great arches and troughs whose dimensions are

**14-3**
**FOLDS**

Figure 14-4
Anticline, A, and syncline, S.

Figure 14-5
Small folds in thin-bedded limestone, southwestern Virginia.

Figure 14-6
Anticline and syncline in thick-bedded limestone near Kingston, New York.

**Figure 14-7**
An asymmetrical anticline in Silurian
shale near High Falls, New York.

**Figure 14-8**
Folds in Lower Devonian rocks near
Berkeley Springs, West Virginia. (Photo
by J. H. Rathbone.)

measured in miles. The Rock Springs anticline of southwestern Wyoming, for example, is some 75 miles long
and 40 miles wide; the Massanutten syncline, in the
Appalachian Mountains of northern Virginia, is 80 miles
long by 5 to 10 miles wide. Both these folds, and others
like them, affect several thousand feet of sedimentary
strata. Anticlines and synclines may be broad and open,
as in Figure 14-4, or tightly compressed like accordion
pleats; they may stand upright and symmetrical, or be
inclined so that one limb is considerably steeper than

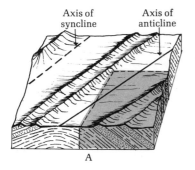
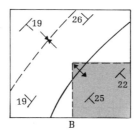

A                    B

**Figure 14-9**
In A, the dipping strata of Figure 14-2 (within the colored area) are found to lie on the southeast limb of a major anticline. The shale and sandstone beds were once continuous over the anticlinal arch, before erosion cut into the fold. In B, the anticline is shown on a map by dip-and-strike symbols and by a line with opposing arrows along the anticlinal crest. The symbols in the northwest corner show that a syncline flanks the anticline; the synclinal axis is shown by a dashed line and converging arrows.

the other (Figure 14-6). When traced in their long dimensions, folds ultimately disappear or die out, like wrinkles in a rug or blanket.

The erosion that removes the rocks above folded strata, thus exposing the latter to our view, also cuts more or less deeply into the folded rocks themselves. Thus we find that the situation depicted in Figure 14-9 A is very common. In the area shown in the figure, the fact that the rocks have been folded into an anticline is not immediately apparent from surface evidence, since the upper part of the fold has been eroded away.

Many folds are so large that they cannot be shown in their entirety on a map at ordinary scales. The strata shown by dip-and-strike symbols on the small map, Figure 14-2 C, all dip to the southeast; it is not until this map is assembled with ones of adjacent territory that the strata of the small map are seen to be on the southeast limb of a large anticline (Figure 14-9 B), which is flanked on the northwest by a syncline. The kind of information needed on each map is dip and strike. Accurate readings, obtained from strata exposed in fields, cliffs, road cuts, and the like, are the basis for our descriptions of the larger anticlines and synclines.

Note that folds are features of the framework or *structure* of the rocks, not of the land surface. Anticlines may stand high and synclines low; but more commonly the reverse is true, anticlines forming valleys and synclines mountains (Figures 14-10, 20-1). (A part of the Massanutten syncline forms a mountain some 45 miles long.) Or folded rocks may be indiscriminately cut across, or *beveled,* as in Figure 14-9 A; in such regions, the edges of the more resistant strata may stand out and help to indicate the attitude of the rocks below the surface. In brief, there is no necessary direct relation between folded strata and the topographic surface that develops upon them.

The geologist is interested in folds for at least two reasons. One, folds record a significant phase in the geological history of a region, in which rocks were compressed by strong earth forces. Since the geologist's task

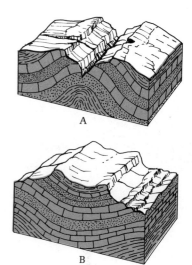

A

B

**Figure 14-10**
Structure and topography "out of phase." In A, an anticline has been breached by stream erosion so that the crest of the structure is marked by a valley. In B, rocks of a syncline stand higher than adjacent rocks, and the syncline forms a double ridge with an intervening shallow valley. (After U. S. Geological Survey.)

is to unravel the geologic history, this interval of crustal stress must be fitted into the chronological record. Secondly, folds are of interest for economic reasons. No rational exploration for earth materials of value can be carried on in a region of folds unless the character of those folds is well understood.

## 14-4
## DOMES AND
## DOME MOUNTAINS

**Figure 14-11**

Maverick Spring Dome, a deeply eroded anticline in northern Wyoming. Crest of the anticline extends from upper left to lower right. Black line represents one mile. (Part of the Maverick Spring quadrangle, Wyoming, shaded-relief edition. U. S. Geological Survey.)

Figure 14-11 shows an uplift that is more nearly equidimensional in plan view than many anticlines, and is thus termed a **dome.** Many other domes are known, some of which are essentially circular. Among the causal mechanisms for the formation of domes are crustal compression; igneous intrusion on a small scale, reaching relatively shallow depths; and vertical uplift owing to intrusion of rock salt. Salt is a weak rock that deforms by flowage, much like glacier ice, at depths of a few thousand feet. Under favorable circumstances it is capable of pushing its way upward and doming the overlying rocks.

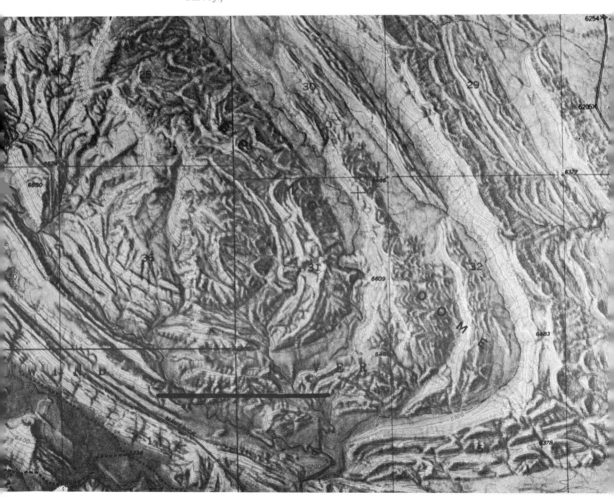

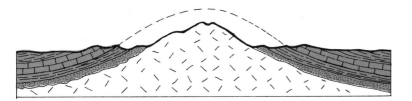

Features of similar form, but much greater size, are **dome mountains.** Well-known examples are the Black Hills of South Dakota, the Ozark Mountains of Missouri, and the Adirondacks of northern New York. Each is a roughly circular uplift of the outer crust, which was at one time covered by a blanket of sedimentary rocks (Figure 14-12). These rocks have been largely removed by erosion from the higher parts of dome mountains, and the mountainous central area is now surrounded by concentric valleys and ridges marking the upturned edges of weak and resistant strata. These marginal strata dip outward beneath the surrounding country.

The nature of the processes that elevate dome mountains is not always clear. They may be arched upward by the same sort of lateral compression that produces folds. Or the arching of the basement rocks with their sedimentary cover may result from the intrusion of a batholith deep within the crust, producing an upward bulge.

**Figure 14-12**
Dome mountains. Sedimentary strata, arched by localized uplift, have been eroded from core of older igneous and metamorphic rock. Strata of different resistance to erosion form ridges and valleys flanking the mountains.

## Displacement along Fractures

Suppose that stresses develop within the lithosphere that tend to move one block of the crust northward and the neighboring block toward the south; or to depress one block and elevate the adjacent one. A fracture may then form between the two blocks, along which movement will take place so as to relieve the stress. Such fractures, termed **faults,** are surfaces along which the rocks on one side are displaced with respect to those on the other. Tiny faults in hand specimens may show displacements of a fraction of an inch; at the opposite extreme are faults along which the rocks on opposite sides are offset for many miles.

Fault surfaces may occupy any position from vertical to horizontal. (Their attitude, like that of dipping rock strata, is described by dip and strike.) Faults may cut any kind of rock and displace any type of geologic structure. Movement may be vertical, horizontal, or a combination of the two. Thus a full classification of fault displacements is a highly involved exercise in solid geometry. For descriptive purposes, however, it is enough to distinguish three chief kinds of faults.

**14-5
FAULTS**

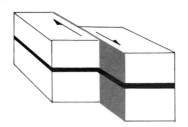

**Figure 14-13**
A strike-slip fault. The movement has been horizontal, parallel to the strike of the fault.

**Figure 14-14**
Aerial view (looking north) of a branch of the San Andreas fault, southern California. The right or east side has moved toward the observer. The hills on opposite sides of the fault are unrelated. Note the folded strata in the hills on the right. (Photo by Spence Air Photo.)

## Strike-Slip Faults

In Figure 14-13 is shown a vertical fault, on which the rocks on opposite sides are displaced horizontally. Because the displacement, or "slip," is parallel with the strike of the fault, the term **strike-slip fault** is appropriate. Not all strike-slip faults are vertical, nor is the slip always 100 percent horizontal; but these are dominant attributes of strike-slip faults and separate them from other types.

Some of the largest faults known are of the strike-slip variety. The San Andreas fault of California (Figure 14-14), nearly 700 miles long, is an important structural feature of western North America. The rocks on the east side of this fault have moved south with respect to those on the west side. Movement has been going on periodically since Paleogene time, and the cumulative offset totals 175 miles or more. Other big strike-slip faults parallel the margin of the Pacific Ocean basin in New Zealand, the Philippine Islands, and Chile. Smaller and shorter ones are known in a number of mountain ranges.

## Normal Faults

The situation shown in Figure 14-15 A is quite different from the one just described. For one thing, the relative movement has been up and down rather than horizontal; for another, the fault surface is not vertical, but inclined. We call the rocks that lie above an inclined fault the **hanging-wall block,** and those that lie below it, the **footwall block.** We use these terms because they make it easy to describe the movement on such faults. In the figure, it is clear that the hanging-wall block has moved down relative to the footwall block. This is the diagnostic feature of a **normal fault.**

Figure 14-15 B is a cross section at right angles to the fault, the dashed line showing the right half of the complete block as it was before faulting. Note that the distance across the block is greater after faulting than before. Normal faulting, it is apparent, indicates crustal stretching. Vertical uplift of the footwall block would produce such a stretched or elongated crustal block.

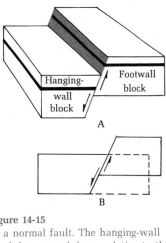

**Figure 14-15**
A, a normal fault. The hanging-wall block has moved down relative to the footwall block. B, cross section showing that the horizontal distance through a block of the crust is greater after normal faulting (solid line) than before (dashed line).

**Figure 14-16**
Fault in Pleistocene sandstone and siltstone beds, Idaho. Visual comparison of strata on opposite sides of the fault shows plainly that the hanging wall (left side) has moved downward relative to the other side, and thus that it is a normal fault. (Photo by H. E. Malde, U. S. Geological Survey.)

**Figure 14-17**
Complex pattern of normal faults in Jurassic strata, southern Utah.

Rocks vertically uplifted and stretched may be offset on individual normal faults, like the one shown in Figure 14-16, or broken into a mosaic of blocks as in Figure 14-17. The rocks shown in the latter figure are soft silty shales, which yield semiplastically to stress, and hence the fault surfaces are tightly compressed. In strong, brittle rocks, such as granite or massive limestone, faults may stand as irregular open fractures. These in turn may be converted into veins, by precipitation of mineral matter from hot magmatic waters. Many "bonanza" ore deposits have formed in networks of veins produced in this manner. In an entirely different kind of geologic setting, where there are thick sections of sedimentary rock, drilling for oil has disclosed many normal faults that are confined to subsurface strata.

### Reverse Faults

A **reverse fault** is one on which the hanging-wall block has moved up relative to the footwall block. This situation is shown in Figure 14-18 A. The dashed line of the cross section, B, gives the outline of the whole crustal block before faulting. Because the ends of the block are closer together after faulting than before, we conclude that reverse faults are produced by compression of crustal rocks. The type of stress that such faults record is similar

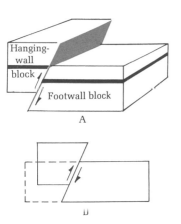

**Figure 14-18**
A, a reverse fault. B, cross section showing that the horizontal distance through a block of the crust is shorter after reverse faulting (solid line) than before (dashed line).

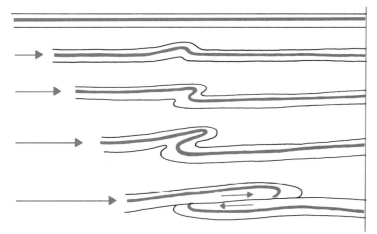

**Figure 14-19**
Development of a thrust fault, or low-angle reverse fault, from a fold. At top, strata are unaffected by stress. Increasing lateral compression, as indicated by arrows, produces an overturned fold, which finally ruptures to form a thrust fault. Note the pronounced crustal shortening.

to the compressive stresses that cause folding. Reverse faults with a low angle of dip, termed **thrust faults,** commonly develop from folds (Figure 14-19).

Reverse faults are most common in regions of crustal compression, especially in the major mountain systems such as the Alps, the Andes, and the Appalachians. The commonest large-scale faults in such regions are low-angle thrusts, on which the rocks of the upper plate (hanging-wall block) may have been pushed for miles across those of the footwall block (Figure 14-20).

**Figure 14-20**
Thrust fault, southwestern Virginia. The light-colored rock is Cambrian dolostone; it has been thrust to the right (northwest) for several thousand feet vertically and several miles laterally, onto dark shales of Devonian-Mississippian age. The dolostone is fractured and the shales contorted, but the fault surface itself is sharp.

**14-6
FAULT-BLOCK
MOUNTAINS**

In Nevada and adjacent parts of Utah and California, there are scores of rugged mountain ranges, trending generally north-south and separated from each other by broad alluvium-floored desert basins (Figure 18-8). Field studies have shown that these mountains have been uplifted along normal faults; hence they are termed **fault-block mountains.** Their steep rocky sides constitute **fault scarps** that have been greatly modified by erosion.

Some of these mountains are simply blocks of the crust upfaulted along one side only, so as to make a range that is asymmetrical in cross section (Figure 14-21 A). Such a range is the Sierra Nevada of California, which is a great mass of granite uplifted by normal faulting on its east side and tilted westward. A smaller but equally famous example is the Teton Range of Wyoming, also a westward-tilted block. The Tetons owe their spectacular scenery to the work of valley glaciers, which have carved deeply into the 6,000-foot fault scarp on their east side. The internal structure of some fault-block mountains is highly complex, and they may be bounded by faults on both sides (Figure 14-21 B). The Slate Range of southeastern California, a part of which is shown on Figure 3-2, is a fault-block range. The fault shown in the figure is a normal fault, on which movement takes place only a few feet at a time, as indicated by the low scarplet that crosses the alluvial fans. Another example of a complex fault-block range is the Panamint Mountains bordering Death Valley (Figure 7-18).

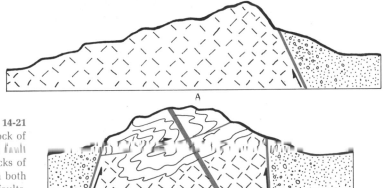

**Figure 14-21**
Fault-block mountains. A, a block of the crust uplifted along a normal fault and tilted to the left. B, rocks of complex structure, uplifted on both sides along normal faults.

**14-7
EARTHQUAKES**

An **earthquake** is a series of vibrations or shock waves passing through rock and regolith. Study shows that earthquake-causing shocks are produced by sudden slippage along faults. Most faulting originates deep within the crust, but some active faults extend to the surface, and after earthquakes, may offset features of the landscape (Figure 14-22). Earthquakes have occurred repeatedly

along major fault zones, and for this and other reasons we infer that movement on faults does not take place all at once, but in many small increments over long periods of time.

The point within the crust where initial rock slippage occurs and the shock waves originate is termed the **focus** of an earthquake; the point on the earth's surface directly above the focus is the **epicenter** (Figure 14-25 A). For a given earthquake, both focus and epicenter are located by scientists using records from a worldwide network of seismographs. Seismograph data also yield significant information on the velocity with which earthquake waves pass through the different zones of the lithosphere, and thus on the nature of the earth's interior.

An abrupt strike-slip movement on the northern part of the San Andreas fault produced the shock waves of the famous San Francisco earthquake in 1906. Roads and fences on the east side of the fault were locally offset as much as 20 feet southward from their continuations on the west side. The Madison Canyon earthquake of 1959 (Section 6-2) resulted from movement along two roughly parallel normal faults, with vertical displacement ranging from 9 to 20 feet. Geologic evidence shows that these faults are long-standing zones of weakness in the crust, and that the 1959 movement was merely the latest in a long series of spasmodic displacements.

At 6:01 A.M. on February 9, 1971, an earthquake struck the heavily populated San Fernando area of southern California. At least 65 lives were lost; had the earthquake come at a later hour the loss of life might have been much greater. Hospitals were badly damaged, reservoirs were

**Figure 14-22**
Small cliff or scarp produced by movement along a fault during the Madison Canyon earthquake of 1959, western Montana. The footwall side, on which the Jeep stands, moved up relative to the other side; hence the fault is a normal fault. (Photo by J. R. Stacy, U. S. Geological Survey.)

**Figure 14-23**
Railroad tracks on the west side of San Fernando were twisted during the earthquake of February 9, 1971. (Photo by Los Angeles Times.)

**Figure 14-24**
This overpass, on the Golden State Freeway near Sylmar, collapsed during the San Fernando earthquake. (Photo by Los Angeles Times.)

in danger of failure, rail lines were twisted (Figure 14-23), and freeway interchanges collapsed (Figure 14-24). More than a thousand landslides were triggered by the earthquake and its aftershocks. The causative fault is a rela-

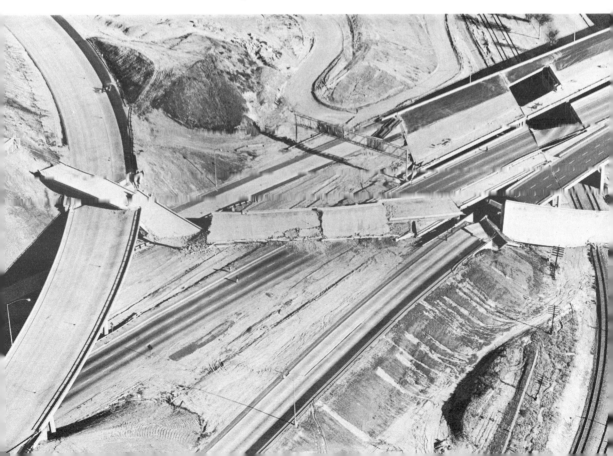

tively minor break, on which movement was apparently upward for some 3 feet and laterally for about 2 feet. Thus, as indicated on Figure 14-25 B, the fault has both reverse and strike-slip components of movement. Displacement occurred along a highly irregular fault surface dipping 40 to 60 degrees northward. The focus of the earthquake (Figure 14-25 A) was at a depth of about 7.5 miles, and the epicenter was 6 miles north of San Fernando.

Earthquake damage results from one or more of four factors. The first and perhaps the most obvious is shaking of the ground, which may produce widespread destruction, including the type shown in Figure 14-24. As a rule, houses or other buildings resting on solid rock undergo much less damage than those on alluvium, or on "made ground" such as might be produced by filling in a swamp. Such unconsolidated material, especially if water-saturated, quivers like jelly when earthquake waves pass through it. The second factor in earthquake damage is landsliding, as illustrated by the rockslide at Madison Canyon (Figure 6-1) and the slump at Anchorage, Alaska (Figures 6-2, 6-3). The third is actual displacement along a surface fault, as shown by the ground in Figure 14-22. The fourth is change in elevation, as when a dock is left high and dry owing to local uplift of the shoreline. Effects of this kind have been noted along the Alaska coast.

News reports of an earthquake commonly state its **magnitude** on the *Richter scale,* by assigning it a number, generally between 3 and 8. The Richter scale is a measure of the energy released. The maximum amplitude, or height, of earthquake waves recorded at a given distance from the epicenter is converted to a numerical figure by a logarithmic scale. An increase of a whole number on the Richter scale represents a tenfold increase in the amount of earth movement recorded by the seismograph. Further, each whole number represents a release of energy about 200 times that of the next smaller number. Hence, although there are nearly 4,000 earthquakes each year, most of the crustal energy released is concentrated in a few large ones.

An earthquake of Richter magnitude 2.5 is felt close by, one of 4.5 may cause a little local damage, one of 6.0 or more is moderately destructive, and one of 7.0 or more is a major earthquake. The San Francisco earthquake had a magnitude of 8.3; Madison Canyon, 7.1; Alaska (1964), 8.5; and San Fernando, 6.6. We might note that damage and loss of life are not necessarily proportional to magnitude. An earthquake of magnitude 6 that affects a densely populated area will be far more destructive than one of magnitude 8 in sparsely inhabited country.

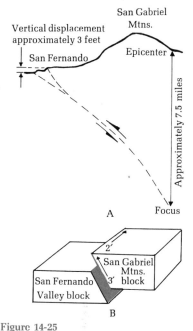

**Figure 14-25**
Fault movement during the San Fernando earthquake of February 9, 1971. A, cross section showing relation of focus, epicenter, and surface faulting. B, relative movement. The San Gabriel Mountains block (the hanging wall) moved some 3 feet upward and 2 feet laterally. Thus the fault shows both reverse and strike-slip aspects. (After California Division of Mines and Geology.)

Another method of measuring an earthquake is to determine its **intensity.** This is a qualitative record, based on personal experiences at the time and on observed damage to man-made structures. One of several scales for recording intensity is known as the *modified Mercalli scale.* On this scale, intensity increases by stages from I (hardly noticed) through VI (felt by everyone, awakens sleepers, frightens many persons) to a maximum of XII (total destruction, general panic). Obviously this is a largely subjective record, and can be arrived at only some time after the event. A single earthquake will yield a range of Mercalli intensities, depending on such factors as distance from the epicenter and types of earth material in the areas affected.

Seismic records over the years suggest that southern California may expect more than 30 shocks of at least San Fernando magnitude (6.6) per century, or roughly one every three years. This is scarcely surprising when we learn that the region is cut by a network of faults, and is geologically young and crustally unstable. Indeed, it is part of the great circum-Pacific belt that we met in Chapter 4 (Figure 4-12). This belt, we now find, is characterized not only by volcanism but also by mountain-building and earthquakes. Most of the earth's great earthquakes occur in this belt, or in its westward extension to the Mediterranean region.

**14-8**
**UNCONFORMITIES**

The end product of the wearing down of a land area by streams is a regional erosion surface. If base level remains at the same elevation for a long enough time, such á surface may be expected to develop irrespective of the kind of rock present in the region. Thus a regional erosion surface may cut indiscriminately across rocks of all types and attitudes—flat-lying, folded, faulted, foliated, massive, and so on.

We are here concerned with those regional surfaces that have been lowered beneath sea level and covered with layers of new sedimentary rock. Such a buried surface is given the general name **unconformity;** three varieties are customarily distinguished: disconformities, angular unconformities, and nonconformities. As is usual with geologic features, unconformities interest us because of the information they afford about past events.

**Disconformity**

An unconformity above and below which the strata are parallel is termed a **disconformity.** The relationship, shown in Figure 14-26 A, records a relatively simple story.

266

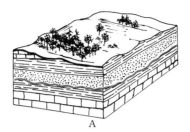
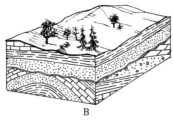
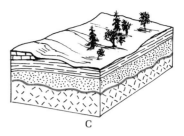

A lower series of strata was deposited, consolidated, exposed, and eroded; at some later time, the sea covered the eroded surface and deposited additional layers of sediment on it. Withdrawal of the sea, lithification of the younger series, and erosion to the present land surface completes the history of the area. Disconformities may be smooth, or show many feet of relief (Figure 14-27). The rocks immediately below some disconformities show evidence of old soils; those immediately above may consist of conglomerate or breccia in which the fragments are composed of the same kind of rock as that below the disconformity. Clearly a disconformity represents an interval during which erosion, not deposition, took place: that is, a pronounced gap in the record of the stratified rocks.

The length of time represented by a disconformity, its "time significance," is not indicated by the characteristics of the surface itself, but rather by the ages of the rocks above and below, as indicated by the fossils they contain. If, for example, the strata below the disconformity of Figure 14-26 A are Upper Cambrian and those above it are Middle Devonian, then the disconformity represents the whole of Ordovician, Silurian, and Early Devonian time, an appreciable fraction of geologic history. We can only guess as to what went on in this region during this span of at least 100 million years. No doubt rocks were deposited here that represented some, possibly much, of

**Figure 14-26A**
Disconformity.

**Figure 14-26B**
Angular unconformity.

**Figure 14-26C**
Nonconformity.

**Figure 14-27**
A disconformity (traced in black) that reveals some geologic history. The light-colored rock, a Mississippian limestone, was exposed at the surface for a long time after deposition, and underwent intense solution by ground water. This produced the rough surface exposed in the cliff to left of center, and also a sink full of broken limestone blocks (right of center), probably produced by roof collapse. Later, red Pennsylvanian sediments infiltrated the rubble on the limestone surface, to form the dark rough-weathering rock. That the surface is a disconformity is shown by the parallelism of Mississippian and Pennsylvanian rocks, best seen on left side of view. Near Hartville, Wyoming. (Photo by J. H. Rathbone.)

this time span, only to be eroded away before Middle Devonian time. We do know that no mountain building took place, as the strata below the disconformity are not in any way deformed.

### Angular Unconformity

Often the rocks below an unconformity are not parallel with those above it; the relation (Figures 14-26 B, 14-28) makes the term **angular unconformity** appropriate. The reason for the relation is clearly the fact that the lower set of strata was folded before being beveled by the regional erosion surface. Faulting, regional tilting, or even metamorphism of sedimentary layers, are other kinds of rock deformation that may take place before burial by a younger set of strata. The strata below the uncomformity may dip so gently that the relation at any individual exposure seems to be that of a disconformity, and only regional study shows that there is in fact a small angle between the rocks above and below the unconformity. At the opposite extreme are those unconformities below which the strata stand vertically, that is, have a dip of 90 degrees, showing that before truncation and burial they were caught in severe crustal disturbance, probably mountain building (Figure 14-29). The time significance of angular unconformities does not depend on the degree of angularity, however, but as with disconformities, on the age of the rocks above and below as determined by fossils.

**Figure 14-28**

Cretaceous limestone resting with angular unconformity on Triassic siltstone, shale, and limestone. The light-colored material is loose sand. There has been some post-Triassic folding in the region, as shown by the anticline at the left. Southeastern Iran. (Photo by J. C. Van Wagner.)

Figure 14-29
Horizontal Devonian beds lying on vertical Precambrian metasedimentary rocks, Ouray, Colorado. (Photo by J. H. Rathbone.)

## Nonconformity

Where sedimentary layers lie upon massive rocks, such as granite or gabbro, the surface separating them is termed a **nonconformity** (Figure 14-26 C). Erosion surfaces cut on massive rocks are especially numerous where sedimentary rocks are exposed overlying the ancient surface that bevels Precambrian rocks (Figure 21-3).

## "Layers of Geology"

Unconformities—in particular, angular unconformities—have been aptly said to separate "layers of geology." In composition, sedimentary features, and fossils, the strata below an angular unconformity record an early period of sedimentation; their structures show that they have been subjected to earth stresses; their aquifers and mineral deposits are dependent on this history. The flat-lying strata above the unconformity, on the other hand, reveal a totally different history. Often this younger layer completely masks the complex older one, and the latter can be probed for its geologic story and economic resources only by the drill or by geophysical devices. The layers of geology above and below a widespread unconformity are indeed often so different that they could as well be considered to lie in different regions as to be superimposed in the same region.

## 14-9 SUMMARY

At many places, sedimentary rocks have been tilted from their original horizontal position by earth forces. We describe the attitude of such rocks by determining their dip, or direction and amount of inclination, and their strike, or direction at right angles to that of dip. Rocks fold into anticlines and synclines as a result of lateral compression, under weight of thick overlying rocks. Folds range greatly in size; some are so large that they can be recognized only by obtaining the dip and strike of strata over large areas and assembling this information on a map. Equidimensional folds, known as domes, are probably the result of vertically directed stress. Large vertical uplifts may produce dome mountains.

Faults are fractures along which the rocks on one side have been displaced with respect to those on the other. In strike-slip faults, the dominant movement is horizontal, parallel to the strike of the fault. A famous large-scale example is the San Andreas fault of California. In normal faults, the upper or hanging-wall block has moved down with respect to the lower or footwall block. This type of faulting suggests crustal stretching or tension. A reverse fault is one on which the hanging-wall block has moved up with respect to the footwall block. The evidence here is of compression. Low-angle reverse faults, or thrust faults, are common in major mountain belts. Fault-block mountains are bounded on one or more sides by normal faults. Examples are the Sierra Nevada, the Tetons, and the Slate Range and many other ranges in Nevada and southeastern California.

An earthquake is a series of vibrations passing through the crust, as a result of sudden slippage on a fault. The point at which the slippage occurred is the focus; the point directly above it at the surface is the epicenter. Earthquake damage is produced by shaking, downslope movements, fault displacement, and uplift or sinking of sizable areas. The magnitude, or amount of energy released, is indicated by a number on the Richter scale; the intensity, or effect on man and his structures, may be indicated on the modified Mercalli scale. Earthquakes are common along our Pacific Coast and other parts of the great belt of volcanism and mountain-building that nearly encircles the Pacific Ocean.

Unconformities are buried surfaces that indicate interruptions in the history of accumulation of sediments. The strata above and below a disconformity are parallel; erosion or nondeposition is indicated. The strata below an angular unconformity were deformed and eroded before burial by younger rocks. The time value of both of these types of unconformity is shown by the ages of the rocks above and below, as determined from fossils. A nonconformity bevels massive, nonbedded rocks such as granite, and is overlain by sedimentary rocks. Layers of geology separated by regional unconformities record quite different geologic histories.

## SUGGESTED READINGS

Clark, S. P. 1971. *Structure of the Earth,* chaps. 1, 2, and 5, pp. 1–25 and 67–91. Englewood Cliffs, N.J.: Prentice-Hall. (Paperback.)
These chapters deal with earthquakes and geologic structures. The different types of earthquake waves and their behavior are covered in considerable detail.

Crowell, J. C. 1962. Displacement along the San Andreas Fault, California. *Geological Society of America Special Paper 71.* 58 pp.
Discussion of the nature, extent, and geologic history of the San Andreas fault system.

Grantz, A., and others. 1964. Alaska's Good Friday earthquake, March 27, 1964. *U.S. Geological Survey Circular 491.* 35 pp.
A report that includes basic information on the Alaskan earthquake and its effects.

Harbaugh, J. W. 1968. *Stratigraphy and Geologic Time,* chap. 3, pp. 34–39. Dubuque, Iowa: W. C. Brown. (Paperback.)
Chapter 3 presents the story of unconformities.

Iacopi, R. 1969. *Earthquake Country.* Menlo Park, Cal.: Lane Books. 160 pp. (Paperback.)
An authoritative and highly readable account of the where and why of earthquakes in California.

Summer, J. S. 1969. *Geophysics, Geologic Structures and Tectonics,* chaps. 6 and 9, pp. 47–59 and 79–88. Dubuque, Iowa: W. C. Brown. (Paperback.)
Chapter 6 presents good coverage of the basics of earthquakes and seismology. Chapter 9 deals with faults and folds.

U. S. Geological Survey and National Oceanic and Atmospheric Administration. 1971. The San Fernando, California, Earthquake of February 9, 1971. *Geological Survey Professional Paper 733.* 254 pp.
A comprehensive report in which over 50 separate papers by many different investigators and organizations have been assembled for use by specialists, general readers, and public officials.

# 15 METAMORPHIC ROCKS AND PROCESSES

15-1   Rocks Formed from Other Rocks

15-2   Contact Metamorphism

15-3   Regional Metamorphism

15-4   The Common Metamorphic Rocks

15-5   The Rock Cycle

15-6   Summary

The entrance lobby of a certain public building is panelled in a highly polished stone that is mostly gleaming white but in places is streaked and mottled with dark gray. Close examination shows that the white mineral is calcite and the dark one graphite. Grains of these minerals form a closely interlocking fabric. What class of rock does the stone belong to?

Such a rock cannot be igneous, if for no other reason than its mineral composition: neither calcite nor graphite is of any significance in rocks known to be of igneous origin. The rock can hardly be sedimentary, as layers of sediment would not have the observed streaked and kneaded structure. Furthermore, graphite is not a sedimentary mineral. It seems clear that we are concerned here with a different class of rock altogether.

This third large group of rocks constitutes the metamorphic rocks. They are formed through the alteration of preexisting rocks by one or more of the agents of heat, pressure, and hot gases and solutions. The processes of alteration, collectively termed **metamorphism,** take place within the crust. The resulting rocks are exposed where we can see them, or quarry them for building stone, only after uplift and long erosion.

Some metamorphism is so intense that all distinctive features of the preexisting rock are erased and we cannot tell whether it was igneous or sedimentary. More commonly, however, the nature of the parent rock can be deduced with some assurance. Although the parent rock of the building stone described above could not have been igneous, because of its bulk composition, it could readily have been sedimentary. A common variety of sedimentary rock, namely limestone, consists chiefly of calcite, and many limestones contain carbon-rich organic matter that could be altered to graphite. Thus we would classify this building stone, which is the kind called marble, as a **metasedimentary** rock. A rock clearly derived from basalt or granite, on the other hand, we would term **meta-igneous.**

Any body of intrusive magma brings heat with it, and this heat is lost to the surrounding rock as the intrusive mass cools. In the process of cooling, furthermore, a magma may give off hot gases and solutions, and these agents may transfer much material from the intrusive to the country rock. The changes that intrusive masses produce in the rocks that they invade, at and near the contact between them, result from processes inclusively termed **contact metamorphism.**

The width of the zone affected, and the intensity of metamorphism, depend on the size and composition of the intrusive mass and on the nature of the rock invaded.

15-1
ROCKS FORMED
FROM
OTHER ROCKS

15-2
CONTACT
METAMORPHISM

Where a small dike cuts through shale, contact-metamorphic effects may be confined to zones a few inches or feet wide on either side of the dike, in which the shale is baked (recrystallized) to a hard flinty rock, with little or no evidence of transfer of material from dike into wall rock (Figure 15-1). A larger intrusive body of course releases more heat and affects a wider zone. The Treasure Mountain granite, a small batholith in western Colorado, has metamorphosed the surrounding sedimentary rocks for distances as great as 3,000 feet outward from the contact. The dominant agent has been heat. One of the rocks affected is an ordinary limestone, which has been metamorphosed to a sparkling white marble near the intrusive. The dominant process has been recrystallization of the constituent calcite, with little addition or loss of mineral material.

Around many large batholiths, which released chemically active fluids as well as heat, the rocks are not only recrystallized but altered in composition. Some belts of contact metamorphism are several miles wide. They are characterized by systems of veins (Chapter 4), where solutions have deposited mineral matter in cracks in the country rock or in the solidified outer part of the batholith itself; and by replacement deposits, especially in limestone. Valuable ore minerals may be found in both types of deposit. Thus a considerable variety of contact-metamorphic effects may be ascribed to "the drainage systems of freezing batholiths."

As the name indicates, **regional metamorphism** involves rocks over large areas. Its effects are especially noticeable where erosion has cut deeply into ancient mountain systems, exposing in their "roots" a series of distinctive rocks that are clearly the products of alteration of preexisting rocks.

15-3
REGIONAL
METAMORPHISM

### Agents

As in contact metamorphism, heat and circulating solutions are important agents. Both probably come chiefly from slowly cooling batholithic magmas. Some heat may be contributed by friction generated in rock deformation, and a part of the solutions by ordinary ground water whose temperature has been raised by circulating near the magmatic furnace.

Besides producing changes in mineral composition, regional metamorphism imposes on most rocks a characteristic structure. We find that newly formed minerals tend to be platy or flaky, like muscovite, graphite, and chlorite; or elongate, like hornblende. As these minerals take shape in the metamorphic rock, they tend to align themselves with their flat faces or long dimensions at right angles to the applied pressure (as in a pile of compacted autumn leaves). The result is **foliation,** a more or less pronounced parallelism of mineral grains in a metamorphic rock (Figure 2-8). Naturally, foliation is most strongly developed in rocks made largely of flaky minerals, and is much less pronounced in rocks in which these minerals are minor. Marble, for example, generally shows little foliation, because the grains of its chief constituent, calcite, are equidimensional rather than flat-sided.

Laboratory studies have shown that certain metamorphic minerals form or persist at relatively low temperatures and pressures, others at higher ranges, and still others only under the most severe conditions. Thus the kind of environment under which a given metamorphic rock formed may be read in its constituent minerals. As might be expected, metamorphic rocks that are along the margins of mountain belts show the least effects of regional metamorphism, and rocks at and near the centers of such belts show the most profound effects. The extreme degree of alteration shown by rocks in the roots of some mountains is the result of regional metamorphism reinforced by large-scale contact metamorphism from batholithic intrusion.

The physical aspects of regionally metamorphosed rocks show that they have been more or less intensely deformed in the solid state. The agent of deformation is strong lateral compression, or **directed pressure,** of the same nature that produces folds and thrust faults (Chapter 14). Concurrently with directed pressure there must be intense **confining pressure,** from the weight of overlying rocks. The mechanism by which rocks yield to directed pressure, undergoing changes in shape, is like the internal flow deep in a glacier (Section 12-2)—an infinite number of microscopically small slippages within and between mineral grains.

### Changes Produced

Rocks affected by the agents just described undergo significant changes. Let us take as an example a clastic fossil-hash limestone. The most pronounced change here would undoubtedly be recrystallization of the fossil fragments and other grains of calcite to a mosaic of larger, interlocked calcite grains: in other words, a physical change to a form better adapted to metamorphic conditions. Other minerals of the parent rock may be unable to exist at all in the new environment, and will therefore alter to different minerals. If there was a little clay in the parent limestone, for example, the clay minerals are likely to alter to muscovite, a mineral with essentially the same composition as clay but stable under heat and pressure. Finally, we have seen that hot circulating fluids may remove some minerals and substitute others. Solutions rich in magnesia (MgO) might attack the calcite and convert some of it to dolomite. In this example, then, the various agents of regional metamorphism convert a slightly clayey shell-fragment limestone to a calcite-dolomite marble containing a little muscovite.

## 15-4 THE COMMON METAMORPHIC ROCKS

Metamorphism may affect any of the igneous rocks and any of the sedimentary rocks. Thus there is a wide variety of metamorphic rocks, both meta-igneous and metasedimentary. Nevertheless, the great majority of metamorphic rocks belong in only six general types. Three of these types show the foliation typically produced in regional metamorphism, and the remaining three show it either faintly or not at all.

### Slate

When beds of shale are caught in a mountain-making vise and tightly folded, they may be converted to an extremely fine-grained rock called **slate.** Under high magnification, slate is found to consist of shreds, scales, and fibers of mica and micalike minerals, among which grains of quartz and other minerals may be distributed.

The minute flaky grains are in parallel orientation, and this fact produces a highly developed foliation termed **rock cleavage** (Figure 15-2). Slate is different from all other rocks in its capacity to split into thin sheets along cleavage surfaces.

The character of the folds in slate districts shows that the parent shale underwent intense squeezing. This involved movement of rock material in the solid state from the limbs of the folds toward the crests of anticlines and the troughs of synclines. Rock cleavage, developed during this deformation, is oriented parallel to imaginary planes that bisect the folds (Figure 15-3). Thus cleavage is superimposed on bedding; the two are entirely different and should not be confused.

Although slate is produced by tight folding, it is formed at relatively low temperatures, for in slate we do not find minerals indicative of great heat. Nor do the other rocks in slate regions show evidence of strong metamorphism.

**Figure 15-2**
Ledge of slate. Surfaces dipping prominently to the right are cleavage. Bedding dips to the left, as may be seen by traces of beds on the cleavage and on the dark fracture surface above the hammer handle. Ordovician, eastern Pennsylvania. (Photo by A. A. Drake, Jr., U. S. Geological Survey.)

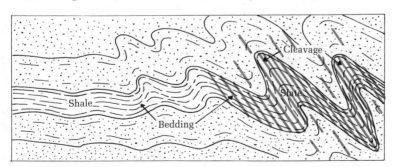

**Figure 15-3**
Development of slate from shale through close folding. Cleavage, which forms parallel to imaginary planes bisecting the folds, is superimposed on bedding.

Thus we conclude that slate is a result of low-intensity, or low-rank, regional metamorphism.

### Schist

**Schist** is a metamorphic rock made up largely of platy or flakelike minerals such as the micas, chlorite, talc, and graphite. Splinter-shaped minerals like hornblende, and nonplaty minerals, especially quartz and garnet, are also common in schist. Grain size is generally large enough so that the chief minerals may be identified with the unaided eye. The preponderance of flaky minerals gives schist a pronounced foliation (Figures 2-9, 15-4).

Schists are commonly named for the most conspicuous or abundant mineral, as chlorite schist or quartz-muscovite schist. The wide variety of minerals in these rocks suggests a considerable variety of parent rocks. Schist may be produced from shaly sandstone, sandy shale, felsite, tuff, basalt, and numerous other rocks of either sedimentary or igneous origin. Because schist is produced by more intense regional metamorphism than slate, it contains little evidence of bedding or other features inherited from the parent rock.

**Figure 15-4**
Schist, with typically well-developed foliation. (Photo by W. B. Hamilton, U. S. Geological Survey.)

**Phyllite** (Figure 15-5) is a rock transitional between slate and schist, with a texture intermediate between them and a fine wavy foliation.

### Gneiss

Rock of medium to coarse grain and crude foliation, with the minerals in bands and streaks, is termed **gneiss** ("nice") (Figure 15-6). Feldspar, quartz, biotite, and hornblende are common constituents, and the majority of gneisses have a composition near that of granite. Indeed, the two rocks can often be shown to be related, as where thin sheets of granitic magma were injected into stratified or foliated rock, altering the intruded rock and preserving in a crude way its layered structure. Some granite gneisses are the result of "granitization" (Section 4-3). The mineral composition and field relations of other gneisses point to gabbro, conglomerate, or schist as parent rocks.

### Quartzite

**Quartzite** is a rock consisting of quartz grains cemented tightly by quartz. Ground water may introduce silica into a sand, and if this cement fills all the spaces between grains, a quartzite of sedimentary origin is produced. Quartzite may also be produced from sandstone in metamorphism, either contact or regional. Sedimentary

**Figure 15-5**
Metasedimentary rocks of Precambrian age near Gatlinburg, Tennessee. Bedding extends from the vein at left toward the upper right. The dark layers were formerly shale but are now phyllite; they show a faint foliation that crosses the bedding, from upper left to lower right. The light layers are contorted beds of quartzite.

quartzite may be distinguished from metamorphic by microscopic examination of the rock, and by field study to determine whether or not the associated rocks are metamorphosed.

All the pore spaces in quartzite are filled with silica, and the grains are held so tightly that when the rock is broken the fracture surfaces pass through the grains instead of around them. Most quartzites are white or light shades of brown or red. They are typically not foliated, although there are gradations between some metamorphic quartzites and foliated rocks like quartz-mica schist.

### Marble

**Marble** is a nonfoliated rock consisting mostly or entirely of calcite or dolomite. It is derived from limestone or dolostone by either contact or regional metamorphism. Many marbles have a vaguely streaked or mottled pattern (Figure 15-7). The calcite grains are medium to coarse in texture and uniform in size, and possess the tightly interlocking fabric produced by recrystallization.

The subsidiary minerals in marble, if any, owe their composition to minor constituents in the parent limestone, or to materials introduced during metamorphism. Graphite means that organic material was present; mica suggests reconstituted clay; quartz may originate from sand grains

**Figure 15-7**
Marble. The rock is mostly coarse-grained calcite; gray streaks are produced by very small amounts of graphite. The streaked pattern represents a vague foliation. The specimen, which has been cut and polished, is 6 inches square.

in the limestone or from silica in invading solutions; dolomite may be inherited or introduced.

Since marble is dominantly calcite, it is a moderately soft rock that can be easily cut and worked; and its granular fabric allows the rock to take a high polish. These properties are also possessed by some recrystallized limestones, which owe their texture and fabric to the work of ground water rather than to metamorphism. A few such limestones are quarried and polished for use in buildings, and are known in the trade as "marble." Their sedimentary nature is generally revealed by traces of stratification and by recognizable fossil shells. In true marble, all evidence of fossils and bedding has generally been obliterated in the metamorphic processes that produced it.

## Anthracite

Bituminous coal is a sedimentary rock made up of carbonified plant remains. Its dominant constituents are not minerals, but carbon and compounds containing carbon, hydrogen, and oxygen. When bituminous coal and the strata with which it occurs are deeply buried and folded the coal loses much of its hydrogen and oxygen, and its relative content of carbon increases proportionately. This converts it to **anthracite,** a black shiny rock without foliation. The carbon content of anthracite lies in the 85- to 95-percent range.

The folding that produces anthracite from bituminous coal is not intense enough to affect the sandstones and shales with which it occurs, and these are not metamorphosed. Thus it is clear that coal is exceptionally sensitive to crustal heat and stress. In the central parts of mountain belts, where deformation has been profound and all the rocks are metamorphosed, anthracite has been altered to graphite, all constituents except carbon having been driven off by metamorphic processes.

**15-5**
**THE ROCK**
**CYCLE**

From Chapter 4 to the present point we have discussed all three main classes of rock, together with the processes that produce and modify them. Our progress among these subjects may be summarized on a chart like that of Figure 15-8.

Figure 15-8
The rock cycle.

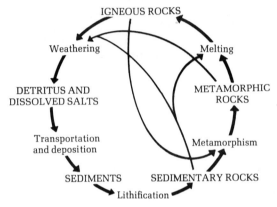

**15-6**
**SUMMARY**

Metamorphism is the alteration of rocks by one or more of the agents of heat, pressure, and hot gases and solutions. It is usually possible to determine whether a metamorphic rock is meta-igneous or metasedimentary. In contact metamorphism, heat and vapors from cooling magma alter the adjacent country rock. Small intrusions produce only minor effects, whereas batholiths may release heat and chemically active fluids that are effective for thousands of feet away from the contact. Regional metamorphism involves not only heat and fluids but also strong lateral compression, or directed pressure, under intense confining load of overlying rocks. Minerals of preexisting rocks tend to alter to minerals that are stable under heat and pressure. These include especially the platy minerals like the micas, chlorite, and talc, and splintery minerals such as hornblende. Many regionally metamorphosed rocks show a marked parallelism of mineral grains, termed foliation, which develops at right angles to the directed pressure.

Slate, formed in the close folding of shale beds, shows an exceptionally well-developed foliation, called rock cleavage. Slate is a product of low-intensity regional metamorphism. Schist is a foliated rock characterized by mica, chlorite, talc, and other platy minerals; it may be produced from a variety of igneous and sedimentary rocks. Phyllite is intermediate in texture and foliation between slate and schist. Gneiss is a crudely foliated rock, in which the minerals are in bands and streaks. Many gneisses are of granitic composition. Quartzite, an all-quartz rock derived from sandstone, may be of either sedimentary or metamorphic origin. It is not foliated, and is very hard and tough. Marble, produced by metamorphism of limestone or dolostone, is a nonfoliated rock consisting of tightly interlocking calcite grains produced by recrystallization. Anthracite is a nonfoliated rock resulting from very mild metamorphism of bituminous coal.

## SUGGESTED READINGS

Ernst, W. G. 1969. *Earth Materials,* chap. 7, pp. 126–42. Englewood Cliffs, N.J.: Prentice-Hall. (Paperback.)
  A good presentation concerning metamorphic rocks and contact and regional metamorphism.
Spock, L. E. 1962. *Guide to the Study of Rocks,* 2nd ed., chap. 9, pp. 228–72. New York: Harper & Row.
  An authoritative and relatively complete coverage of the many different types of metamorphic rocks.

# 16 OCEAN BASINS, CRUSTAL PLATES, AND CONTINENTAL DRIFT

16-1 Nature of the Ocean Basins

16-2 Sea-Floor Spreading

16-3 The Plate-Tectonic Theory

16-4 Continental Drift

16-5 Stability versus Drift

16-6 Summary

Patches of thick, relatively lightweight crust termed *continents* are separated from one another by much larger areas of thinner but denser crust, the *ocean basins*. The continents stand high with broad areas of their surface exposed to view; ocean basins, on the other hand, are low areas in the crust, which are filled to overflowing with water. Study of the ocean floor, and of the geologic processes that shape and modify it, is consequently much more difficult than study of continents from land-based observations. Indeed our knowledge of the ocean basins was very sketchy until after World War II, when a variety of coring, sampling, and electronic-sounding devices was developed, and governments became interested in financing expensive programs of both direct and remote exploration of the ocean floor. The data retrieved from these programs have been of immense geologic importance. It is now clear that the ocean basins hold the key to many geologic riddles, whose answers could only be guessed at from the seashore.

## Size and Distribution

Approximately 70 percent of the lithosphere's surface is covered by water. Of the water-covered area, 10 percent represents the shallowly submerged continental shelves. Thus the ocean basins occupy about 60 percent of the earth's surface and are crustal features of primary importance. Although all the earth's great seawater bodies form a single interconnected "world ocean," it is customary to divide this ocean into seven subordinate parts, each of which may be regarded somewhat artificially as occupying an ocean basin. Average depths and approximate areas of the seven major ocean basins are given in Table 16-1, and their location in Figure 16-1.

From Table 16-1 we see that the Arctic Ocean basin (shown in Figure 16-1 to be almost landlocked) is not only the smallest but also the shallowest; that the South Pacific

## 16-1
## NATURE OF THE OCEAN BASINS

**Figure 16-1**
The ocean basins. (Modified from J. Mainwaring, *An Introduction to the Study of Map Projection,* by permission of Macmillan Education.)

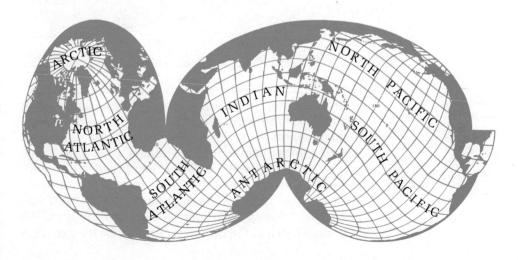

285

| Table 16-1 Size and Depth of Ocean Basins | | |
|---|---|---|
| Ocean Basin | Area, square miles | Average Depth, feet |
| Arctic | 5,427,000 | 5,010 |
| North Atlantic | 17,646,000 | 10,780 |
| South Atlantic | 14,098,000 | 13,420 |
| North Pacific | 31,639,000 | 14,050 |
| South Pacific | 32,361,000 | 12,660 |
| Indian | 28,400,000 | 13,002 |
| Antarctic (or Southern) | 12,451,000 | 12,240 |

is the largest; and that the North Pacific has the greatest average depth. We note also that the North and South Pacific ocean basins, with a combined area of about 64 million square miles, represent a little more than 45 percent of the "world ocean." They are so situated with respect to surrounding shelf seas and continental masses that they make up an almost complete hemisphere of water.

## Margins

Sloping gently away from the continental shorelines, at about 10 feet per mile, are the shallowly submerged outer edges of continents, the **continental shelves** (Figure 16-2). Seismic evidence, and the study of cores extracted from these shelves, indicate that they are the upper surfaces of thick masses of land-derived sediment, delivered by rivers and spread out by oceanic currents of various sorts.

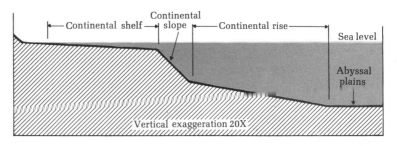

**Figure 16-2**
Principal features of a continental margin and adjacent parts of an ocean basin.

Continental shelves are relatively wide along the margins of the Atlantic and Indian ocean basins. Around the Pacific basin, however, the shelf is mostly very narrow; in many places along the west coasts of North and South America, for example, it is less than a mile wide. Long, narrow **sea-floor trenches,** rather than broad shelves, are characteristic features of the Pacific margin (Figure 16-3). Most of these trenches are depressed at least 6,000 feet below the adjacent sea floor, and in one of them, the Mariana Trench of the northwest Pacific, the floor reaches

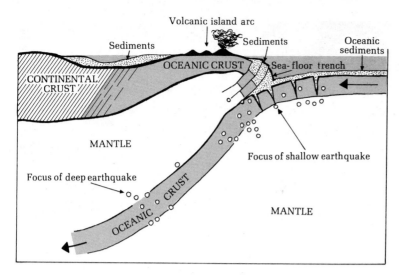

Volcanic island arc

Sediments

Sediments

Oceanic
sediments

OCEANIC CRUST

Sea-floor trench

CONTINENTAL
CRUST

MANTLE

Focus of shallow earthquake

Focus of deep earthquake

OCEANIC CRUST

MANTLE

**Figure 16-3**

Cross section of part of the lithosphere in the vicinity of a sea-floor trench, an island arc, and a continental margin.

its greatest known depth, 36,198 feet. Along the landward sides of many of these trenches are parallel festoons of active volcanoes, which form **island arcs** such as the Aleutians, the Japanese islands, and the Philippines. There are only two arcs and sea-floor trenches in the present-day Atlantic basin (Figure 16-5). The deepest point in the northern trench is 27,498 feet, reached in the so-called Milwaukee Deep north of Puerto Rico.

At the outer edges of continental shelves, below an average depth of about 600 feet, is a relatively narrow belt of abruptly steeper sea floor, the **continental slope** (Figure 16-2). Although much of the surface of the continental shelves and slopes is essentially featureless, both are cut in many parts of the world by deep **submarine canyons.** Some of these, like the Hudson Canyon off New York, are extensions of large river valleys from the land, cut in the Pleistocene Epoch when sea level was much lower than now. But other canyons begin on the outer edges of the shelves, have no landward counterparts, and seem to have been carved by great masses of water-laden sediment that slid down the continental slope toward the deep-sea floor.

**Floors**

Seaward of the continental slope is a broad area of very gentle gradient, the **continental rise** (Figure 16-2), which extends outward for hundreds of miles to merge in deeper parts of the ocean basins with smooth, sediment-veneered **abyssal plains.** Here and there, the featureless surface of the latter is broken by groups of irregular rocky **abyssal hills,** which stand from 150 to as much as 300 feet above the abyssal plains. The central third of the Atlantic Ocean Basin is occupied by a broad **mid-oceanic ridge** (Figure 16-4), the crest of which rises 10,000 feet or more above

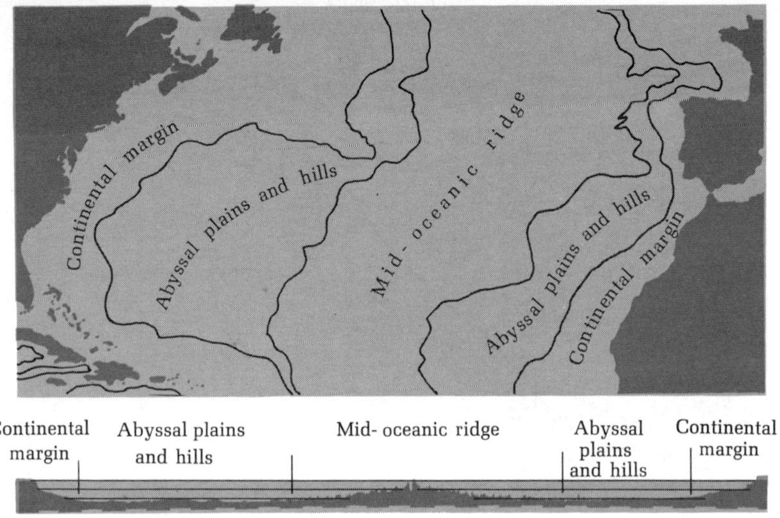


**Figure 16-4**
Principal regions of the North Atlantic Ocean Basin. (From Heezen and Tharp, 1959.)

the abyssal plains to east and west. Soundings indicate that this feature, the *Mid-Atlantic Ridge,* is extensively fractured parallel to its length, probably by normal faults, and that a strip 15 to 30 miles wide along its crest has subsided 3,000 to 9,000 feet below adjacent parts of the ridge. Earthquakes are common along the ridge, volcanism is extensive, and heat loss is six times greater than in the abyssal floor of the Atlantic. A limited surface view of some of these features can be obtained in Iceland, which is astride the ridge and is famous for volcanoes, hot springs, and earth tremors.

The Mid-Atlantic Ridge is only part of a continuous oceanic ridge system that extends the full length of the North and South Atlantic oceans and crosses the Indian Ocean and the eastern part of the Pacific (Figure 16-5). Despite the fact that it is largely below sea level, this ridge system constitutes the longest and perhaps the most important chain of mountains on earth.

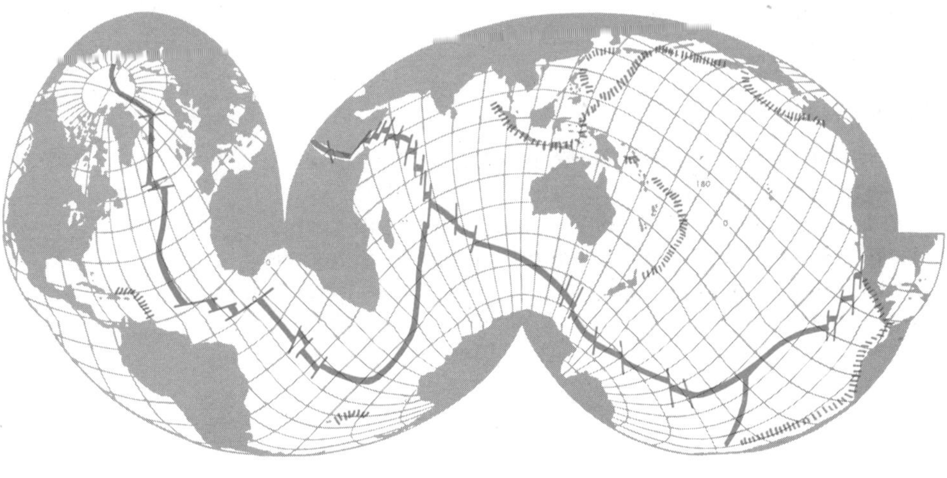

**Figure 16-5**
Trenches, island arcs, and the oceanic ridge system. The crest of the ridge system is indicated by heavy solid lines. Trenches are shown by hachures; island arcs are the chains of islands along the edges of some of the trenches. The crest of the oceanic ridge system is offset in many places by major faults, only a few of which are shown. (Base map modified from J. Mainwaring, *An Introduction to the Study of Map Projection,* by permission of Macmillan Education.)

## Sediments

The thick wedges of sediment that extend seaward from continents beneath the continental shelf and slope are composed largely of land-derived gravel, sand, and clay, transported to the shoreline by rivers and streams, the wind, and glacier ice. At some places, beds of gypsum and rock salt are present at depth, and nearly everywhere the shells of animals and the calcareous deposits of marine plants are important constituents. Loose sediments of the outer shelf and slope may break away from time to time, slump to the foot of the slope, and spread oceanward as **abyssal fans** similar to alluvial fans at the foot of terrestrial mountains. Beyond the reach of slump transport, sediments of the deep ocean floor consist largely of relatively thin calcareous and siliceous **oozes,** formed from the shells of microscopic animals that inhabit the waters above; and in deeper parts of the ocean basins, of **brown clays,** which consist of wind-blown silt and clay and volcanic matter that settles from the atmosphere over the oceans. Oozes and brown clays both contain meteoritic dust and minor quantities of material derived from land. This last includes gravel and finer material rafted out to sea by icebergs and floating trees; volcanic matter that settles into the sea from the atmosphere; and waste of various sorts dumped into the sea by man.

Oozes and brown clays accumulate very slowly; hence the 1,000 to 1,200 feet that are known to occur in deep ocean basins should provide the record for a very long period of time. It is therefore curious to find that the oldest rock yet dredged from the deep-sea floor or collected from oceanic islands is of Middle Jurassic age. This surprising fact can be interpreted in several ways. Perhaps we have been sampling only the younger layers of oceanic sediment; or present-day ocean basins were not available as catchment basins for sediment much before the Jurassic Period; or some mechanism in ocean basins serves periodically to "sweep the floor clean" of sediments deposited in it. Although the first of these possibilities is probably part of the answer, the latter two are both considered likely by many geologists.

There are two chief schools of thought concerning the origin and subsequent history of the ocean basins. The more traditional one holds that ocean basins are permanent features of the crust, which have changed little since early stages of earth history. The other regards ocean basins as impermanent in size and location with respect to continents. Because new information is brought to light

**16-2
SEA-FLOOR
SPREADING**

289

almost daily, it would be premature to accept or reject the views of either school at the present time. The discussion that follows is cast largely in terms of the newer concept, that of ocean-basin impermanence, not because we are convinced that these views are necessarily the last word, but because the scheme advocated by this school of thought integrates a large amount of information into a single general concept. Further, the older view of ocean-basin permanence has evolved largely from land-based observations, and includes little consideration of the origin or development of features that are characteristically confined to the ocean basins. Information with respect to such features has been accumulated indirectly, for the most part, from studies of the seismic, thermal, and magnetic properties of crustal rocks in the ocean basins.

In 1960, after it had been established that the system of oceanic ridges was a feature not only of the Atlantic but of all ocean basins, a new concept of ocean-basin development was proposed. This concept, a hypothesis of **sea-floor spreading,** was at first largely speculative; but it has gained such strong support through the 1960s from direct and experimental observations, as well as from theoretical calculations, that it must now be considered a fundamental part of current theory explaining the origin, nature, and behavior of the crust.

The hypothesis of sea-floor spreading postulates that hot mantle rock, rising toward the surface, lifts, stretches, and ultimately fractures the oceanic crust in long broad welts, the oceanic ridges. Upon release of confining pressure, some of this hot rock melts to form magma, and erupts through fissures as volcanoes. The rock not reaching the surface cools, becomes denser, and makes up a strip of brand new basaltic crust along the ridge crest. As more of the heated mantle rock rises, however, the new crust is stretched, fractured, and shoved aside, and its place is taken by another, younger strip, formed in the same way. If the process is viewed as a more or less continuous one, it can be seen that the crust would migrate slowly away from the highly fractured crest of an oceanic ridge. In short, crust is generated at ridge crests and later spreads away from the central crest area.

If basalt is constantly being produced at ridge crests, as the hypothesis of sea-floor spreading holds, either the earth must be expanding or crust is being destroyed somewhere else at about the same rate as it is being created in the oceanic ridges. Because theoretical evidence suggests that the earth has probably had about its present size for much of its history, the hypothesis of sea-floor spreading postulates that the crust is being destroyed about as fast as it is made. Destruction of the crust is believed to take place in sea-floor trenches, such as those that are characteristic features of the Pacific basin margin. Presumably, as crust moves farther and farther away from the sites of its formation on ridge crests, it becomes progressively cooler and its density gradually increases, until ultimately its more or less rigid leading edge slowly begins to sink. The line along which this happens defines the highly fractured seaward wall of a trench (Figure 16-3); and as the crustal plate is deflected downward into the trench, it is gradually heated to the point that it is assimilated into the mantle. Some of the heat thus formed, together with that generated by friction between sheared segments of subsiding crust, melts overlying mantle and crust, producing magma, which rises on the landward side of the trench and perhaps spills out to form volcanic island arcs. The concept of sea-floor spreading, having begun as a hypothesis, or educated guess, as to the behavior of the crust and mantle between mid-ocean ridges and sea-floor trenches, now finds support from a variety of lines of evidence.

## Evidence from Paleomagnetism

Part of the wisdom common to Boy Scouts, mariners, and geophysicists is that the earth behaves as if it contained in its interior a gigantic bar magnet, with its long dimension essentially coincident with the earth's axis of rotation and its north and south poles approximately at the north and south geographic poles. Extending from pole to pole of this magnet are invisible lines of force that make up the earth's magnetic field. Iron-bearing materials, if free to move, are particularly susceptible to this field and tend to line up like compass needles, parallel to the lines of force. Thus elongate grains of magnetite or other iron-rich minerals in cooling magmas or lavas, or in unconsolidated watery deposits of sediment, tend to become crudely aligned with the magnetic field while they are still free to move. They are "frozen" in such aligned positions when the magma or lava crystallizes, or when the sediment is lithified. Geophysicists with the proper equipment and training can read these mineral grains as though they were fossil compasses, and can determine from them the polarity of ancient magnetic fields and the orientation of those fields with respect to the present one and with respect to various places on the earth's surface.

One of the more perplexing discoveries of the last few years is that polarity of the magnetic field has been reversed a number of times in the past. That is, the intensity of the field has repeatedly diminished, then built back up again with north and south poles reversed. Knowledge that this has happened repeatedly in the past has had an important bearing on the hypothesis of sea-floor spreading. Some years ago a group of geophysicists made several flights across a part of the Mid-Atlantic Ridge System that projects southeast from Iceland. On a magnetometer in their aircraft they recorded information about the magnetic properties of the basaltic rocks that make up the ridge. A map made from this information is shown in Figure 16-6. If the magnetic properties measured had been uniform in nature and uniformly distributed through the area studied, there would have been no differences to map and the map of Figure 16-6 would be a blank. As you can see, however, the map shows that magnetic properties of the crust are not uniformly distributed over the Mid-Atlantic Ridge. On the contrary, they show substantial variations, and the map gives a pattern of narrow bands, elongated parallel to the ridge crest and arranged with rough symmetry on either side of it. Similar anomalies were soon found to be associated with other parts of the mid-ocean ridge system.

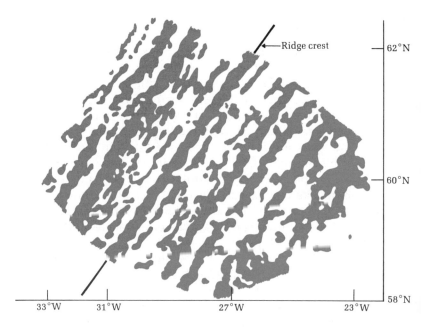

**Figure 16-6**

A map of part of the Reykjanes Ridge southeast of Iceland. The crest of the ridge is indicated by a heavy line. White and colored bands on either side of the ridge crest are strips of reversely and normally polarized basaltic crust. (Modified from Heirtzler, LePichon and Baron, 1966.)

Now if we combine our knowledge of repeated reversals of the magnetic field with the hypothesis of sea-floor spreading, magnetic anomalies like those of Figure 16-6 are readily explained. As heated mantle rock rises toward the surface at ridge crests, it loses heat and ultimately becomes rigid, to form a strip of brand new crust. When temperatures drop below a critical level, magnetically susceptible minerals in this new crust come to be oriented in accordance with lines of force in the magnetic field and are then locked into that orientation in the resultant strip of solid basalt. As more hot rock rises to the surface, the ridge crest is stretched, crust is displaced laterally on the crustal conveyor belt, and a new strip of basalt is added in the breach. If, between formation of the first basalt strip and that of the second, the magnetic field reversed its polarity, this would be faithfully recorded in the newly solidified rock. Strips of normally and reversely magnetized basalt, symmetrically arranged on either side of the ridge crest, thus form a record of repeated reversals of the earth's magnetic field, and they can be used in support of a hypothesis that advocates slow creation and lateral spreading of crust at the mid-ocean ridges.

## Evidence from Gravity and Earthquakes

If the materials of which the core, mantle, and crust of the lithosphere are made were uniformly distributed, we should expect the force of gravity to be the same everywhere at the sea-level surface of the earth. We know this is not the case. For example, the force of gravity is always less than theoretically expected in continental areas, because continents are concentrations of rock that is much lighter in weight than a uniform crust would be. However, because ocean basins seem to be floored everywhere by crust of essentially the same composition, we should expect measured values of the force of gravity to be the same throughout these basins, at least if we make slight corrections for lightweight oceanic sediments. For most of the ocean basins, we find that gravity measurements are monotonously uniform; in sea-floor trenches, however, the values are much lower than expected. The only plausible explanation thus far made is that rocks lighter than the mantle have been thrust or dragged downward in the trenches in such a way as to make these parts of the lithosphere less dense than they would otherwise be. This is just what is postulated by the hypothesis of sea-floor spreading, in which trenches are visualized as the places in which crust is finally returned to the mantle on the crustal conveyor belt.

Plunging of crust into the mantle might also produce slippage of crustal segments past one another; hence repeated tremors in the crust near the sites of sea-floor trenches should be common. Seismic records show that the large majority of earthquakes originate in narrow belts along the landward side of sea-floor trenches, and most of the others along the crest of the mid-ocean ridge system.

Earthquakes along the mid-ocean ridge originate at rather shallow depths and can be interpreted as the results of rupturing associated with spreading of the ridge. Earthquakes originate at many depths beneath sea-floor trenches and on their landward side, however, and it has been known for many years that the points of origin, or foci (Section 14-7) of these earthquakes lie along relatively thin plates that dip downward from the sea-floor trenches and away from the adjacent ocean basins. This arrangement is shown in cross section in Figure 16-3. Thus the distribution of anomalous gravity values and earthquake foci lends strong support to the hypothesis of sea-floor spreading.

## 16-3 THE PLATE-TECTONIC THEORY

If the hypothesis of sea-floor spreading is followed to its logical conclusion, it leads to a much grander concept, the **plate-tectonic theory,** which explains present and historic features of both continental and oceanic crust as the results of interaction between large mobile crustal segments, or plates. According to this theory, we may visualize the earth's crust as divided into an interlocking mosaic of rigid but independently movable plates, each of which is continuously generated along a segment of mid-oceanic ridge, moves slowly away from that ridge, and terminates by descent into the mantle along a sea-floor trench. Each plate might be regarded as a semi-independent conveyor-belt system, composed largely of thin, rigid oceanic crust, but with segments of thick, relatively lightweight continental crust here and there on its back. In this view, the oceanic crust is the active part of the conveyor-belt system; continents are incidental riders.

Outlines of the crustal plates recognized today are shown in Figure 16 7. Note that much of the Pacific Plate is without continental passengers, and forms a single

**Figure 16-7**
Crustal plates of the present. Ridges, along which crustal material is manufactured, are shown by rows of Xs. Trenches, where crust is consumed, are shown by heavy dashed lines. Plate boundaries that fall in continental areas are marked by ranges of mountains. Arrows on plates indicate general direction of motion. (Base map modified from J. Mainwaring, *An Introduction to the Study of Map Projection,* by permission of Macmillan Education.)

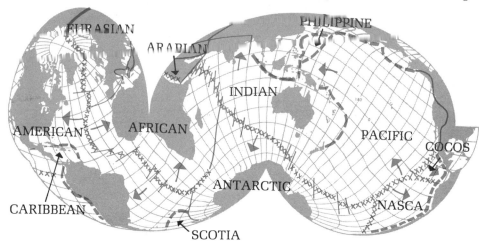

gigantic unit, which moves slowly west from the East Pacific ridge and is consumed in the extensive system of sea-floor trenches along the opposite margin of the basin. The Indian Plate, on the other hand, transports both peninsular India and Australia on its back as it moves northeast toward a partly active trench system. Most other crustal plates have at least one main continental rider.

Although parts of the Pacific and Indian plates are bounded by lines of sea-floor trenches, more complex features mark other plate boundaries. For example, the African Plate, which is moving northeastward from the Mid-Atlantic Ridge, has not only overridden an actively spreading ridge along the southwest side of the Indian Plate, but it has collided with part of the southern edge of the Eurasian Plate in the latitude of the Mediterranean. Northeastward movement of the Indian Plate has apparently brought peninsular India into contact with a segment of the Eurasian Plate, and westward migration of the American Plate seems to have caused North America to override not only a part of the East Pacific ridge system but perhaps a segment of sea-floor trench, as well. Particularly striking in Figure 16-7 is the indication that where continental crust occurs along the outer edge of a spreading plate it is greatly deformed and rides high, to form systems of mountain ranges of great internal complexity. Note also that the converse of this association is not true; that is, not all systems of mountains are at the outer edge of a spreading plate. The Appalachians, for example, which stretch along the eastern margin of the North American continent, are well within the American Plate and not at its outer edge; and the Urals are in the center of the Eurasian Plate, not at its outer edge.

You may wonder why northern Africa, southern Europe and Asia, peninsular India, and western North America have not gradually disappeared as the crustal plates on which they ride are slowly consumed in marginal sea-floor trench systems. This can be explained if we remember that continents, which ride passively on crustal plates, are thick masses of relatively lightweight materials. As such, they are able to resist being drawn down into the mantle during descent of the crustal conveyors on which they ride. In effect, continents and piles of oceanic sediment are scraped off the conveyor system at sea-floor trenches, and although they may be much deformed in the process, they continue to ride high in iceberglike floating equilibrium.

As thus briefly explained, the plate-tectonic theory provides a framework for understanding not only the origin and evolution of ocean basins but also the history and development of continents. If crust is indeed being constantly created and destroyed, it is not so surprising as it seemed at first that the oldest sedimentary rocks found thus far in any ocean basin date back only to the middle of the Jurassic Period; nor will it be so surprising to discover that pre-Jurassic history can be read only from continental rocks, which resist destruction as crustal plates come and go.

**16-4**
**CONTINENTAL**
**DRIFT**

In Chapters 19 through 24 we shall summarize evidence indicating that North America has not always had the size or shape it now has. Comparable evidence from other continents suggests a similar conclusion. To the school of thought that holds ocean basins to be permanent features of the crust, such evidence means simply that continents have grown with time, in their present position, at the expense of the basins around them. Indeed, North America and South America, Europe and Asia, and Africa and Asia are thought to have grown together with time through such gradual enlargement; this thought is held in the face of clear evidence that these continents were separated by seas, if not by ocean basins, through considerable stretches of the past.

If continents have always had the same position relative to one another, even though they expanded in area with time, they should also have had the same position through time with respect to the north and south magnetic poles. That is, the north-seeking end of a compass needle should always have pointed to the same spot on the crust, the north magnetic pole, regardless of whether the compass was used in Europe, Asia, Africa, or North America. But we find that positions of the magnetic poles, plotted from studies of mineral compass needles locked into pre-Cenozoic rocks, do not cluster around the modern rotational poles as might be expected, but diverge greatly from them. Even more puzzling, fossil compass needles in North America point to different spots on the crust from those in Europe, Africa, Australia, and other continents. It would seem from this observation that each continent had its own north and south magnetic poles until very recent times.

Students of the earth's magnetism do not know how the magnetic field is produced; hence the possibility that there was more than one north magnetic pole through much of the geologic past cannot be ruled out entirely. However, the present condition of the magnetic field is the simplest one imaginable, and most geologists assume it is the one that has obtained through the earth's long history. If this assumption is correct and our reading of fossil compasses has not been consistently in error, we must conclude that the north magnetic pole was the same for all continents in the past as it is now and that it was close to or coincident with the north rotational pole, as is now the case. The fact that fossil compass needles in Europe, for example, point to different poles from those in rocks of the same age in North America must then mean that these two continents have changed position relative to one another and relative to the magnetic and rotational poles since their respective compasses were locked into place millions of years ago.

The suggestion that continents bordering the Atlantic basin were once parts of a continuous landmass, which has subsequently been separated or fragmented, is not a new one. Until recently, however, such a concept has received serious attention from only a small number of geologists. With formulation of the hypothesis of sea-floor spreading and the ultimate expansion of this concept into the plate-tectonic theory, the notion of continental fragmentation and dispersal has come to be considered seriously by a much larger number of scientists. Thus in this section we sketch current views on this aspect of sea-floor spreading and its effect on the distribution of continents and ocean basins.

## Pangaea

If, in your imagination, you reverse present motions on the American, Eurasian, and African plates as shown in Figure 16-7 (that is, if you run these conveyor-belts back toward the Mid-Atlantic Ridge rather than away from it) the North and South Atlantic oceans will close, and segments of continental crust now separated by the Atlantic will assemble along the line of the ridge to form a single large block of continental crust. If the same imaginary procedure is applied to the other plates shown in Figure 16-7, the result is the very large area of continental crust shown in Figure 16-8, a supercontinent named **Pangaea.**

## Evidence for Pangaea

In many respects, conditions are quite different at present in the several continents thought once to have been parts of Pangaea. The animals and plants of South America, for example, differ from those in Africa; and Africa, Australia, and South America have climates markedly unlike that of Antarctica. Before Pangaea separated, present-day continents would have been much closer neighbors, and this should be reflected geologically by similar or identical rock and fossil records in continental

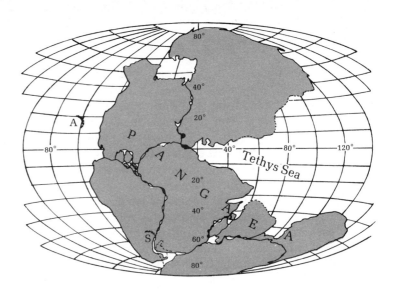

areas now separated by thousands of miles of open ocean. Such records are indeed found, as the following examples indicate.

In parts of South America, Africa, Antarctica, Australia, and peninsular India, brought close together in our reconstruction of Pangaea (Figure 16-8), a thick sequence of nonmarine rocks, largely of Pennsylvanian and Permian age, consists of coal-bearing sandstones and shales that grade upward into tillites. Some of the sandstones contain fossils of a remarkable seed plant, *Glossopteris* (Figure 16-9). *Glossopteris* has no very close living relatives, but botanists agree that its seeds could not have withstood transport over the wide bodies of water that now separate the continents on which its fossils are found. Thus, it is thought that for *Glossopteris* to have achieved such wide distribution, modern continents of the southern hemisphere would have to have been parts of a single landmass in the latest Paleozoic. The tillites tell much the same story. As evidence of widespread episodes of continental glaciation, they record similar conditions of climate and topography. Yet today they range from the equatorial tropics to polar regions. If Pangaea is reassembled as suggested, late Paleozoic tillites, and associated beds with *Glossopteris*, come to be united in areas below 40 degrees South Latitude.

Advocates of the permanence of continents and ocean basins point out that the most likely sites for the growth of continental glaciers are landmasses immediately adjacent to bodies of open water. These geologists believe that the very location of tillites in a Pangaea such as we have reconstructed in Figure 16-8 is strong evidence against the former existence of that supercontinent.

Late Paleozoic tillites, coals, and *Glossopteris*-bearing

sandstones are succeeded in the southern continents by Triassic sandstones and conglomerates with fossils of distinctive land plants and animals. Among the latter are several types of reptiles and amphibians, the same species of which are represented in Triassic rocks in both South Africa and Antarctica. Because these cold-blooded land animals were not very well equipped to navigate large bodies of water like those that now separate the two continents, their occurrence has been taken to suggest that Africa and Antarctica, at least, were joined together in the early part of the Triassic.

Other lines of evidence, including mountain ranges whose continuations are found on other continents, and continental margins that fit together like pieces in a jigsaw puzzle, contribute support for Pangaea. The supercontinent may have had about the form shown in Figure 16-8 at the beginning of the Mesozoic.

## Mesozoic and Cenozoic Drift

Continental margins on east and west sides of the Atlantic have obviously similar outlines, and the concept of Pangaea came into being many years ago, long before we knew much about the ocean basins or the structure and composition of the crust. It was held that Pangaea broke apart by some unknown mechanism, and that its constituent parts "drifted" slowly away from one another, like icebergs of granite in a sea of basalt. This view, based on a mechanism no longer thought to be sound, came to be termed **continental drift**. Because much has been written about continental drift, we retain the expression, although "continental dispersion" would probably be better. Remember that current theory regards the continents as passengers on crustal plates, and not as drifting granitic icebergs.

Evidence presented in Chapters 21 and 22 indicates the existence of ocean basins to the east, north, and west of North America through much of the Paleozoic. However, imaginary reversal of motion on present-day crustal plates, and examination of sediments that have accumulated in the Atlantic basin, indicate that North America was part of Pangaea at the beginning of the Mesozoic: it was not separated from Europe and Africa by an ocean basin. We have no clear picture as yet of the events in the Paleozoic that may have led to opening, then closing, of the Atlantic. On the other hand, we do have a fairly clear record of the fragmentation of Pangaea, and the reopening of the Atlantic basin, in the Mesozoic and Cenozoic. The main steps in that episode of continental drift are shown in Figures 16-10 through 16-13.

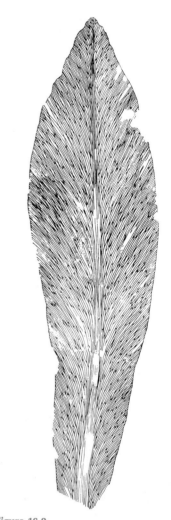

**Figure 16-9**
Leaf of *Glossopteris*. Ohio Range, Antarctica. (Courtesy of J. M. Schopf, U. S. Geological Survey.)

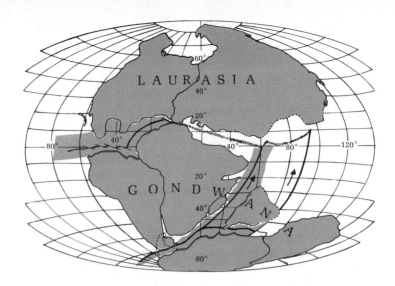

Pangaea is thought to have had a shape something like the one shown in Figure 16-8 by late in Permian time. During the Triassic Period, hot mantle rock began to rise, forming ridges in what are now the North Atlantic and southern Indian ocean basins, and an extensive trench, the **Tethyan Trench,** that stretched east from the present Alps through the Himalayas. Stretching, thermal thinning, and subsidence of the crust along normal faults was recorded in eastern North America by block-faulting and the extrusion of lavas in Middle and Late Triassic time (Palisades Disturbance; see Section 23-1). By the end of the Triassic (Figure 16-10), Pangaea had split into two main parts: a northern one, **Laurasia,** consisting of North America and Eurasia; and a southern one, **Gondwanaland,** which had already begun to separate into blocks consisting of South America–Africa, peninsular India, and Antarctica–Australia.

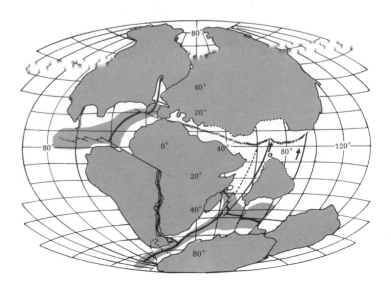

During the Jurassic (Figure 16-11), continued spreading on the northern section of the Mid-Atlantic Ridge further separated North America from the South America–Africa block; northward growth of the ridge began a separation of Greenland from North America; and northward movement brought the South America–Africa and Indian blocks closer to the Tethyan Trench. Also during the Jurassic, a new ridge beneath the South America–Africa block produced a narrow Red Sea-like rift incompletely separating the two continents.

In the Cretaceous (Figure 16-12), the spreading ridges between North America and South America, and between North America and Greenland, apparently became inactive, and ridges in the North Atlantic and South Atlantic joined forces to form the present Mid-Atlantic Ridge System. Active spreading away from the ridge between Africa and Antarctica–Australia moved Africa closer to the Tethyan Trench; India was carried northward toward the same trench on a plate of its own, bounded on the east and west by faults and on the south by a spreading oceanic ridge.

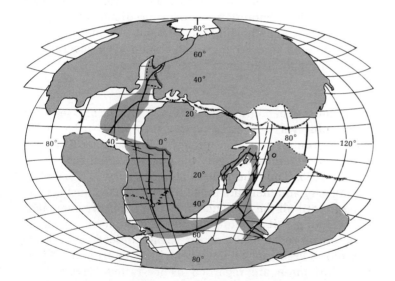

Since the Cretaceous, northward extension of the Mid-Atlantic Ridge has resulted in complete separation of North America and Europe (Figure 16-13); Africa and peninsular India have collided with Eurasia along the old Tethyan Trench to form mountains (the Alps and Himalayas among them) at the southern edge of the Eurasian block; Australia has separated from Antarctica and moved northeastward; a new ridge system has worked its way northwest beneath Africa, so that further fragmentation of that continent is now in progress along a line marked

**Figure 16-12**
By the end of Cretaceous time, the South Atlantic had widened into a major ocean and the North and South American plates had begun their westward drift. (Adapted from "The Breakup of Pangaea," by Dietz and Holden. Copyright © 1970 by Scientific American, Inc. All rights reserved.)

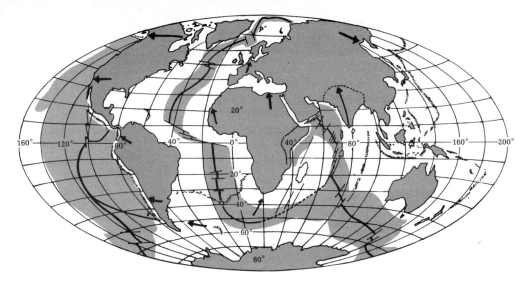

by the Red Sea; and spreading has apparently ceased along
the ridge that separates Africa and Antarctica.

Because spreading apparently continues along segments
of the mid-oceanic ridge system today, and volcanism and
earthquakes indicate that trenches are actively consuming
crust around the Pacific margins, the future should see
further rearrangements of the pieces that were once
Pangaea. New mountains can be predicted along plate
boundaries, and charting the new outlines and positions
of continents and ocean basins should keep geographers
busy through all of the foreseeable future.

**16-5
STABILITY
VERSUS DRIFT**

The concept of continental drift is an old one and sea-floor
spreading was suggested as long ago as 1931. However,
elaboration of sea-floor spreading into a general scheme
of plate tectonics is largely the result of intensive studies
since 1965. Thus the plate-tectonic theory is a scientific
youngster and there are those who caution against accept-
ing it until all the evidence has been carefully evaluated
For example, it is held by some that we have not found
the oldest rocks in the ocean basins. Indeed, some rocks
as old as Precambrian have been dredged from the Atlantic, but these are regarded as ice-rafted "erratics" by
advocates of plate tectonics. The evidence provided by
paleomagnetism is also questioned, for it can be used as
support for continental displacement only if we accept
the assumption that the earth's magnetic field has always
had just two poles that are more or less coincident with
the rotational poles. Also, the distribution of ancient
evaporites, limestones, coals, and tillites suggests to some
that these rocks have always formed in the latitudes where
we should expect them to form today if other conditions

were right. Finally, there are wide differences of opinion about the significance of *Glossopteris* and other fossils, whose distribution is cited as evidence in favor of continental drift, and thus indirectly as evidence for the plate-tectonic concept.

The question of continental stability versus continental drift is by no means settled. Much of the ocean floor and great areas of the continents have yet to be thoroughly explored. Each new map, each new dredge-haul, and each new core from the sea floor provides additional information to digest. These are exciting times for geologists, geophysicists, and oceanographers, and providing answers to the provocative questions raised by the stability-versus-drift debate will surely challenge earth scientists for years to come.

## 16-6 SUMMARY

The ocean basins, which occupy about 60 percent of the earth's surface, are important features of the crust. Studies in recent years, embraced in a hypothesis of sea-floor spreading, suggest that oceanic crust is produced at the crest of broad mid-ocean ridges and that it moves slowly away from ridge crests as new crust displaces it. Ultimately, the crust that spreads away from ridges descends and is consumed in sea-floor trenches along the edges of ocean basins. Continents are viewed as passive riders on spreading plates of oceanic crust. The history of large crustal features is given unified explanation in a general plate-tectonic theory, which includes the concept of sea-floor spreading. This theory may also provide explanation for continental drift, which deals with fragmentation of a single supercontinent, Pangaea, in the Mesozoic and Cenozoic eras.

### SUGGESTED READINGS

Bullard, E. 1969. The origin of the ocean. *Scientific American,* vol. 221, no. 3, pp. 66–75. (Offprint No. 880. San Francisco: Freeman.
Evidence that the deep ocean floor is remarkably young and is growing outward from mid-ocean ridges, pushing most of the continents apart as it does so.

Dietz, R. S., and Holden, J. C. 1970. The breakup of Pangaea. *Scientific American,* vol. 223, no. 4, pp. 30–41. (Offprint No. 892. San Francisco: Freeman.)
A report on the breakup of the supercontinent Pangaea into the present continents, with further continental disruption projected into the future.

Fisher, R. L., and Revelle, R. 1955. The trenches of the Pacific.
*Scientific American,* vol. 193, no. 5, pp. 36–41. (Offprint No.
814. San Francisco: Freeman.)
A report on the deep trenches of the Pacific, which extend
farther below sea level than the highest mountains extend
above, and the clues to the history of the earth's crust that
they provide.

Gross, M. G. 1971. *Oceanography,* 2nd ed. Columbus, O.: Merrill.
150 pp. (Paperback.)
An introduction to oceanography, which includes sections
on the ocean floor, sea-floor spreading, and exploration of
the oceans.

Heirtzler, J. R. 1968. Sea-floor spreading. *Scientific American,*
vol. 219, no. 6, pp. 60–70. (Offprint No. 875. San Francisco:
Freeman.)
An account of the hypothesis of sea-floor spreading, which
shows how it is being used to explain geophysical phenomena ranging from earthquakes to continental drift.

Hurley, P. M. 1968. The confirmation of continental drift. *Scientific American,* vol. 218, no. 4, pp. 52–64. (Offprint No. 874.
San Francisco: Freeman.)
Presents many lines of evidence favoring the idea that the
present continents were once assembled into two great
landmasses, Gondwanaland and Laurasia.

Menard, H. W. 1969. The deep-ocean floor. *Scientific American,*
vol. 221, no. 3, pp. 126–42. (Offprint No. 883. San Francisco:
Freeman.)
Reports on the discovery that the deep-ocean floor is growing outward from the mid-ocean ridges, suggesting that huge
plates act as units in the dynamic processes of the earth's
crust.

Scientific American. 1969. *The Ocean.* San Francisco: Freeman.
140 pp. (Paperback.)
A compilation of ten articles that were first published in
the September 1969 issue of *Scientific American.* Includes
readings by Bullard and Menard suggested above, in addition to others on the continental shelves, resources of the
oceans, and technology used to explore the oceans.

Takeuchi, H., Uyeda, S., and Kanamori, H. 1970. *Debate about
the Earth* rev. ed. San Francisco: Freeman, Cooper. 281pp.
(Paperback.)
Presents the theory of continental drift from the early ideas
of Wegener through modern concepts of sea floor spreading
and the new global tectonics. Includes chapters on earth
magnetism and the confirmation of drifting continents by
paleomagnetic studies.

Turekian, K. K. 1968. *Oceans.* Englewood Cliffs, N.J.: Prentice-
Hall. 120 pp. (Paperback.)
A fairly comprehensive coverage of the geology and chem-
istry of the oceans, written for the beginning student.

Wegener, A. 1966. *The Origin of Continents and Oceans.* New York: Dover. 246 pp. (Paperback.)

This book is an English translation of an original work by Wegener, which was published in German in 1929.

Wilson, J. T. 1963. Continental drift. *Scientific American,* vol. 208, no. 4, pp. 86–100. (Offprint No. 868. San Francisco: Freeman.)

An article written at the time when the continental drift hypothesis was being revived as a result of new evidence suggesting that Wegener's 1912 proposal is correct.

# 17 MOUNTAINS AND THEIR HISTORY

17-1 Features of Continental Mountain Systems

17-2 Mobile Belts

17-3 Development of Mountain Systems

17-4 The Role of Heat in Crustal Development

17-5 Summary

Mountains are parts of the lithosphere that project conspicuously above the general level of the surrounding countryside. The word may bring to mind thoughts of high adventure amid the peaks of the world's towering ranges; but neither great summit altitude nor ruggedness is necessary to our definition. The term applies as well to the rounded tree-clad ranges of the Appalachians or Ozarks as to the icy peaks of the Alps or Himalayas. Geologists even include among mountains tracts of the crust that no longer project above their surroundings but that are shown by their structure to have done so in the past. Thus we include a considerable assortment of both topographic and structural features in a discussion of mountains.

Several minor crustal features are commonly classed as mountains: volcanoes, residual mountains, dome mountains, and fault-block mountains. In addition, the system of oceanic ridges constitutes the longest belt of mountains known although it is largely below sea level. What still remain to be considered are the earth's major continental mountain chains like the Alps, the Himalayas, and the Andes, all of which are principal crustal features with a long and complicated history.

The great continental mountain systems of the world differ from each other in dimensions and internal geologic detail, but they have many features in common, and this suggests that they have had similar histories. The Alps, Appalachians, Andes, Himalayas, and Urals, and also the western mountains of North America, consist of many long, parallel **ranges** that together constitute **systems.** Each continental mountain system (for example, the Appalachian System) is composed primarily of folded sedimentary and metamorphic rock, broken by large-scale thrust faults; such mountains thus exhibit the effects of crustal compression. Granite batholiths that cut across the folds and faults are widely exposed in the central portions of many systems.

Systems of continental mountains are many times longer than wide, and their constituent folds, as well as the strike of their thrust faults, are parallel to their long dimensions. For example, the Appalachian System, a portion of which is shown in the Frontispiece, stretches for more than 1,500 miles along the east coast of North America, from Newfoundland to Alabama, but it is mostly less than two hundred miles wide. Appalachian folds also parallel the coast, and the strike of thrust faults has a similar orientation. Furthermore, the major anticlines and synclines are overturned northwestward, toward the continental interior; and the upthrust side (hanging wall) of the thrust faults also moved toward the interior. Folds

## 17-1
## FEATURES OF CONTINENTAL MOUNTAIN SYSTEMS

and faults of other continental mountain systems have much the same relationship to the continents with which they are associated. For example, many of the major folds and faults in the eastern part of the Cordilleran System, as the western mountains of North America are collectively known, show evidence of having been pushed eastward toward the interior of the continent. Folds in the Alps incline northward toward the interior of Europe.

Three additional features of continental mountain systems deserve emphasis. First, most of these systems are divisible into narrow longitudinal belts, in which rocks of different origin and thickness are brought into contact with one another. In many systems, a belt immediately adjacent to the continental interior is characterized by sedimentary rocks that are many times thicker than flat-lying rocks of the same age in the nearby interior. These thick rocks contain ripple marks, mud cracks, coal beds, and fossils of shallow-water organisms, all of which indicate that the rocks formed either in shallow water or above sea level. Other belts in mountain systems contain rocks that give evidence of accumulation in environments like that of the continental rise (Section 16-1), or around the flanks of island arcs. In some systems, certain belts appear to be composed of much-deformed oceanic sediments; others include greatly altered slivers of oceanic crust. What is particularly significant here is the fact that oceanic crust and sedimentary rocks that formed in very diverse environments are brought side by side, or are even superimposed, in adjacent belts of continental mountain systems.

Secondly, studies of earthquake waves, and measurements of heat loss and the force of gravity, indicate that mountain systems are not heaps of deformed rock piled up on the surface of an unyielding crust. On the contrary, the main mass of such mountains extends downward into the mantle beneath, in the form of a gigantic "root." In short, mountain systems form substantial parts of the crust that project into the mantle. In this respect, they are quite unlike volcanoes, dome and fault-block mountains, and maturely dissected plateaus, which are all largely surficial features.

Finally, it is undoubtedly significant that mountain systems that formed in the past, or are in the process of formation today, are all marginal to continents or are thought to have been so when they first developed. The Appalachian, Ouachita-Marathon, Cordilleran, and Innuitian systems, which are considered more fully in Chapter 18, form a highland border that nearly encircles the broad, low-lying interior of North America and separates it from the surrounding ocean basins. In like manner, the Andean System separates the interior of South America from the Pacific basin. By contrast, the Himalayas are sandwiched between the vast interior lowlands of Asia and the subcontinent of India, and can be considered marginal to both continental blocks.

In the next several sections of this chapter we look at mountain systems in the light of plate-tectonic theory (Section 16-3), which has been put together from fragmentary evidence in an attempt to explain the origin and development of a number of complexly interrelated earth features. There are, however, two main theories for the development of continental mountain systems. The *geosynclinal theory*, the principal alternative to the concept we discuss below, was developed and has long been held by the school of thought that regards continents and ocean basins as static and probably very ancient features of the earth's crust. Evidence presented in Chapter 16 suggests that this view may be inadequate as an explanation for youthful oceanic sediments and for ancient magnetic-pole positions. Plate-tectonic theory, on the other hand, is broadly synthetic in that it considers the history of continents and ocean basins within a single framework. Because the development of mountain systems is apparently tied very closely to evolution of both continental and oceanic crust, we discuss mountain building in a plate-tectonic context.

## The Concept of Mobile Belts

The folded sedimentary rocks of the Appalachian Mountains are all Paleozoic in age and are about 35,000 feet thick. Westward, in the continental interior, the average thickness of Paleozoic rocks is about one-tenth this amount. For ten times as much sedimentary rock to have accumulated in the Appalachian area as in the interior, the crust of the former must have subsided much more during the Paleozoic than did that beneath the interior. The fact that Paleozoic rocks in both areas are largely marine indicates that both were beneath sea level most of the time; greater subsidence made more space for sediment in the Appalachian area than in the interior.

Further, because the Appalachians and other mountain systems are many times longer than wide, the sinking area in which their thick sediments accumulated must also have been long and narrow. Finally, for all sediments in these sinking areas to have been deposited in shallow water or on land, the rate of sediment accumulation must have been approximately equal to the rate of sinking.

Thus the surface of sedimentation was never far beneath sea level, or was actually above it at times. In other words, the belt of subsidence developed gradually as sediments accumulated in it. Thus greatly elongate, slowly sinking strips of crust, blanketed with sediments as they sank, formed early in the history of continental mountain systems. Crustal strips with such a history may be termed **mobile belts.**[1]

## Where Do Mobile Belts Form?

There are only a few places where long linear belts of thick sediment accumulate today or could have accumulated in the past. Central parts of continents can be ruled out. Although even these regions may periodically sink beneath sea level, and accumulate thin veneers of marine sediment, their dominant tendency has been to stand higher than nonmountainous parts of their borders. Thus, in the long run, the centers of continents have tended to serve more as sources of sediment than as basins in which thick sediments accumulate. Deep ocean basins, far from continents or oceanic islands, receive few deposits and thus are also unlikely as sites of thick sediment accumulation. The search for potential sites is narrowed to two places: the edges of continents (Figure 17-2), and the vicinity of island arcs (Figure 16-3). In such areas, continents or island arcs are sources of abundant sediment, and space is available nearby for that sediment to accumulate to great thickness. Thus, since we derive our concept of mobile belts from elongate strips of thick, now deformed sedimentary and volcanic rock, we must conclude that such belts are features of the continent ocean-basin interface and of areas of oceanic crust in the vicinity of island-arc systems.

## The Nature of Mobile Belts

If mountains like the Appalachians, which formed in areas that were once mobile belts, are "unfolded," it can be seen that the mobile belt was divided lengthwise for at least part of its history into two parallel parts (Figure 17-1). The part adjacent to the continental interior is floored by continental crust and is occupied primarily by limestone and shale, but it also contains wedge-shaped masses of sandstone that thicken toward its outer side. In the outer part of the mobile belt, which has no identi-

---

[1] In geosynclinal theory, these crustal strips are termed **geosynclines.** The inner and outer parts of a mobile belt mentioned in Section 17-2 are, respectively, the **miogeosyncline** and **eugeosyncline** of geosynclinal theory.

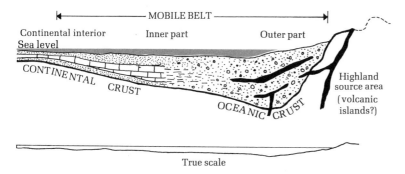

**Figure 17-1**
Restored section across a mobile belt at an early stage in its development. Remember that this is a "snapshot" at a moment in geologic time. Vertical scale greatly exaggerated; lower diagram shows the section drawn to approximately true scale.

fiable continental crust beneath it, are great thicknesses of dark-colored shale, sandy shale, sandstone, and conglomerate; chert; tuff and other pyroclastic rocks; and at many levels, extrusive igneous rocks.

Many rocks in the inner part of a mobile belt are well sorted and of fairly even texture, and accumulated in relatively shallow water. Sediments of the outwardly thickening sandstone wedges are of less uniform texture: the thinner parts of these wedges commonly contain fossils that indicate deposition in shallow water, but the thicker parts of many clearly accumulated above sea level. It is also common to find that the number of sandstone wedges increases in younger parts of the inner segments of mobile belts, and the boundary between the marine and nonmarine portions of these wedges typically migrates toward the continental interior in younger and younger parts of the section. Rocks of the outer part of a mobile belt, on the other hand, are much thicker; many of the clastic layers contain mixed particle sizes, have a "dumped-in" appearance, and bear evidence of accumulation in deep water. Here and there, masses of altered oceanic sediment and chunks of oceanic crust can also be recognized. Whereas rocks of the inner part of a mobile belt are typically more or less strongly folded and thrust-faulted, rocks of the outer part are not only intensely folded and faulted, but also metamorphosed.

These facts suggest that mobile belts like the one from which the Appalachians developed were composed of a pair of laterally adjacent parts, the outer one of which sank much more rapidly than the inner one, so that sediments deposited there were exposed briefly, if at all, to the "winnowing" action of waves and currents. The outward-thickening sandstone wedges of the inner portion of a mobile belt, and the thick clastic sediments of the outer portion, imply a sediment source in the outer segment or beyond its margins, during at least part of the history of a mobile belt. The common occurrence of volcanic rocks in the outer portion of mobile belts suggests that the source areas included or comprised volcanic highlands well above sea level. Such highlands may have been like the festoons of volcanic island arcs along the present-day coast of Asia.

### Modern Analogs

The continental shelf of eastern North America is more than 200 miles wide in places, and is underlain by a prism of sediments and sedimentary rocks more than 15,000 feet thick at its outer edge (Figure 17-2). The shelf prism, of Mesozoic and Cenozoic age, rests on continental crust and is made up of sandstone, shale, limestone, and their unconsolidated equivalents. In all respects but age, this prism resembles the inner part of the "unfolded" Appalachian System, minus the sandstone wedges, which have no counterpart on the continental shelf.

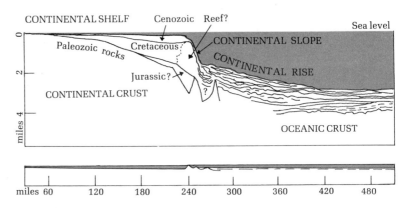

**Figure 17-2**
The margin of North America and adjacent parts of the North Atlantic basin east of Cape Hatteras, North Carolina. Vertical scale exaggerated 25 times; lower diagram is drawn to approximately true scale. (After Emery and others, 1970.)

Seaward from the shelf, oceanic sediments and material that has slumped into the deep Atlantic from the outer edge of the shelf have accumulated on oceanic crust, in the region of the continental rise, to form a chaotic deposit of sand and mud that in places is even thicker than the prism of sediment beneath the continental shelf. The sediments beneath the continental rise lack abundant volcanic material, but their other features are so like those in parts of the outer segment of the Appalachians that the continental rise might be considered a modern-day analog.

An even more attractive analog for the outer part of the Appalachian mobile belt may be found in the region of island arcs and sea-floor trenches along the east coast of Asia (Figure 16-3). In that area, an oceanward-thickening prism of sediments and volcanic rocks occupies the narrow strip of sea floor between the outer edges of the volcanic islands of the arc and the landward edge of the trench. The trench itself contains little sediment in most places, but locally thick accumulations of mixed volcanic and sedimentary materials have spread out from the mouths of submarine canyons.

In brief, the inner parts of mobile belts, minus their conspicuous sandstone wedges, have modern-day analogs in prisms of sediment like the one beneath the continental shelf of eastern North America. The outer parts of mobile belts have analogs in piles of chaotic sediment beneath continental rises, and in prisms of volcanic and sedimentary materials on the sea floor between island arcs and sea-floor trenches.

## Subsidence in Mobile Belts

During the Paleozoic Era, a pile of sedimentary rock ten times thicker than that of the continental interior accumulated in the mobile belt from which the Appalachian Mountains ultimately formed. Not all this rock was deposited beneath sea level, but a substantial part of it was, and this implies that, for much of its history, the Appalachian mobile belt subsided more than adjacent parts of the continental interior did. Why? What are the causes of subsidence in mobile belts?

In Section 2-3, we noted that the continental crust is not so dense as mantle rocks beneath it, and thus appears to "float" on them. It follows that segments of the crust made thinner by erosion will rise, whereas those parts made thicker through deposition of sediment will tend to sink. By this means, the crust tends always to be in a state of floating balance, or **isostatic equilibrium.**

Subsidence in mobile belts was at one time explained by noting that extra-heavy localized sedimentation would cause the crust to sink, thus making room for more sediment, which in turn would cause continued sinking. It was further believed that the subcrustal rock displaced by the subsiding crust would be squeezed laterally, causing areas adjacent to the mobile belt to be uplifted.

This argument might have seemed logical some years ago, but it cannot be accepted today. It is true that localized deposits of sediment thicken the crust and cause it to subside; and that some of the denser rock beneath a sinking crustal segment would have to be compressed or pushed aside—that is, displaced—to make room for the sinking material. But the rock to be displaced is denser than the load placed upon it, and only an amount equal in weight to that of the load could be squeezed from beneath the subsiding mobile belt. More specifically, addition to the crust of 100 feet of sedimentary rock with a density of 2.5 would cause subsidence of less than 100 feet into subcrustal material of density 3.0. Clearly, subsidence in a mobile belt cannot be wholly ascribed to sedimentary loading of the crust, even though the weight of such sediments would surely cause some subsidence.

If sedimentary loading is mechanically inadequate, we might conclude that mobile belts sink because they are dragged, pulled, or in some other way forced downward into the mantle; and that during long intervals of time, these elongate strips of crust are not in floating equilibrium. In Section 16-5 we mentioned evidence indicating that the crust is not in floating equilibrium in sea-floor trenches, because these are the places where slabs of cold oceanic crust are thought to begin their plunge back into the mantle. We have also noted that crustal strips between trenches and island arcs are likely modern analogs of the outer parts of mobile belts. Perhaps the same mechanism that promotes the plunge of crust into trenches also produces subsidence in the outer parts of mobile belts.

We are in trouble, however, if we extend this reasoning to the inner parts of mobile belts (minus their sandstone wedges), for all indications are that the continental shelf of eastern North America, a likely analog for the wedgeless inner part of a mobile belt, has developed in an area of considerable stability, far from any sea-floor trenches or island arcs, and is in floating equilibrium with the mantle beneath it. Subsidence of such segments of mobile belts must then be the result of some other process. We look to plate-tectonic theory for an explanation.

If you look again at Figure 16-7, you will see that the broad continental shelf of eastern North America is on the side of the continent farthest from the leading edge of the American Plate, and the same relative position is characteristic of other wide shelves. Despite their similarity to the wedgeless inner parts of mobile belts, wide continental shelves underlain by thick deposits of sediment form today on the quiescent trailing edges of continents, in the stable central parts of crustal plates, rather than along continental margins immediately adjacent to the more active edges of those plates. To explain the subsidence necessary for accumulation of a thick continental-shelf prism in a region of isostatic stability, we examine a hypothetical sequence of events.

If hot mantle rock were to rise beneath a continent (Figure 17-3 A), its crust would be up-arched, stretched, and somewhat thinned by heating, with the results depicted in Figure 17-3 B. Note in this figure that stretching of the crust has caused crustal blocks to sink along normal faults; trenches between them are rapidly filled with sediment eroded from adjacent fault-block mountains. As additional uplift is produced by upwelling of more hot mantle rock, the crust is further fractured; but more important, it is greatly thinned by stretching, in much the same way that bars of hot metal or glass are thinned, or "necked," as they are drawn out into wires or tubes. As we show in Figures 17-3 C and D, the initial chunk

314

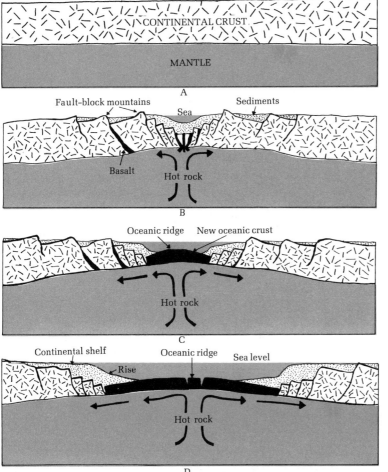

Figure 17-3
A continuous tract of
continental crust (A) is arched
upward, faulted, and stretched
by heating above an upwelling
mass of hot mantle rock (B). As
time passes (C, D) new crust is
generated at the crest of an
oceanic ridge and a slowly
widening ocean basin appears
between formerly continuous
segments of continental crust.
Sediments that accumulate
along margins of
continental-crust segments to
form gradually broadening
continental shelves and rises
are analogous to those in the
inner parts of mobile belts.

of continental crust is ultimately broken into two parts,
each with thin, highly fractured edges facing the ocean
basin that has developed between them. Subsidence of
these edges, permitting accumulation on them of thick
prisms of continental-shelf sediment, is seen to be partly
a result of slow thermal thinning of an originally thick
segment of continental crust, and partly a result of downward displacement of that crust along major normal faults.

Our discussion thus far indicates that continental mountain systems originate from mobile belts, which are linear
strips of crust where thick sediments and volcanic rocks
accumulate. Yet the sediments of mountain systems are
now folded, faulted, and metamorphosed, and form great
systems of mountains hundreds or thousands of feet above
sea level. Clearly, originally horizontal sediments are
deformed, and subsidence is replaced by uplift, at some
stage in the development of continental mountain systems.
Such a sequence of events, converting linear belts of thick
sediments and volcanic rocks into a system of folded
mountains, is termed an **orogeny,** or an **orogenic cycle.**

17-3
DEVELOPMENT OF
MOUNTAIN SYSTEMS

## Orogeny at Plate Margins

According to the theory of plate tectonics, the outer shell of the lithosphere is divided into a mosaic of crustal plates, each of which grows along one edge by the addition of basalt in oceanic-ridge systems, spreads slowly away from the ridge, and is ultimately consumed as it plunges back to the mantle in long, narrow sea-floor trenches. Because mountains are forming today, or have formed in the recent past, along plate margins where crust is being consumed, orogeny is seen to be a feature of such margins. Obviously a variety of different situations can occur there (Figure 16-7), and these differences are thought to be responsible for differences in the mountains that develop along various margins. We consider only one situation, for it may have been a common one, at least in the history of North America.

The somewhat puzzling conclusion of our discussion of mobile-belt subsidence was that modern analogs of the inner and outer parts of many mountain systems subside as a result of different processes and develop in different places on present-day crustal plates. Yet the deformed equivalents of these now commonly occur side by side, or are even partially superimposed. Clearly, we need a way of bringing these crustal provinces together, as well as a mechanism to explain their deformation and ultimate uplift into a great system of continental mountains.

We begin in Figure 17-4 A with a continental margin of the type produced by the sequence of events shown in Figure 17-3 A–D. As the ocean basin on the right side of Figure 17-4 A widens by sea-floor spreading, the continent to the left is carried farther and farther away from the ridge where new oceanic crust is generated. During this stage, the trench toward which the crustal plate and its continental rider are moving is somewhere beyond the left margin of the figure. But there are surely limits to the width of an ocean basin, imposed by crustal rigidity and temperature. Thus we might expect that when the ocean basin on the right side of Figure 17-4 A reaches these limits of width, oceanic crust would begin to plunge mantleward along the outer edge of the basin, initiating a sea-floor trench near the continental margin. Such a trench has formed in Figure 17-4 B.

As the slab of oceanic crust is thrust downward into the mantle, it comes to be more and more heavily loaded by rock above it, and is subjected to greater and greater compressive stresses, of the sort that might be visualized to occur in a piling as it is driven into regolith or soft sedimentary rock. Under such a load, and in response to such stresses, part of the descending crustal slab and

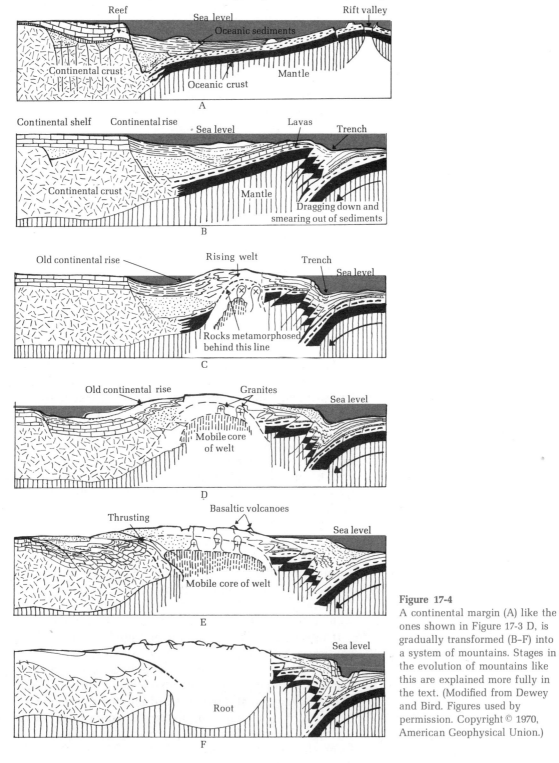

**Figure 17-4**
A continental margin (A) like the ones shown in Figure 17-3 D, is gradually transformed (B–F) into a system of mountains. Stages in the evolution of mountains like this are explained more fully in the text. (Modified from Dewey and Bird. Figures used by permission. Copyright © 1970, American Geophysical Union.)

the mantle near it melt, giving rise to magma. Upward migration of this magma brings it to the surface on the continental side of the trench, where it erupts to form submarine lava flows and volcanic island arcs (Figure 17-4

B). Wedges of volcanic rock, and sediment eroded from these islands, then spread out on the ocean floor to coalesce with sediments of the continental rise, thereby forming the suite of materials characteristic of the outer part of continental mountains.

As heat generated by further melting and rising magma increases, a linear welt is uplifted in the vicinity of the island-arc system. This welt, shown in Figure 17-4 C, continues to rise and becomes broader as its hot, mobile core expands. Heat from the expanding core invades the sediments and volcanic rocks that are buried deep in the outer belt, and metamorphism and plastic deformation result. With further uplift (Figures 17-4 C and D), sediments and volcanic rocks on the flanks of the rising welt become unstable, begin to crumple, and slide toward the inner segment of the mobile belt. Erosion of the crest of the welt, as soon as it is above sea level, also produces sediments, which build out as wedge-shaped prisms on both sides. Deposits like this are represented by sandstone wedges in the inner parts of most continental mountain systems.

Ultimately, as the hot mobile core of the welt continues to expand, its upper part begins to spread laterally, or "mushroom," and compressive forces generated by the spreading core are directed strongly against the old continental margin. A result of this is mechanical failure of the inner segment of the old mobile belt, which is thrown into folds and broken by thrust faults along which large chunks of deformed sedimentary rocks are transported in toward the continent. This stage in orogeny is shown in Figure 17-4 E.

Not only does the solidified core have a conspicuous "root," but it acts as a buttress, resisting further compression. It continues to rise, however, to maintain isostatic equilibrium (Figure 17-4 F). And this uplift, which also involves deformed and metamorphic rocks marginal to the core, produces a broad, high mountain system, firmly welded to the margin of the old continental block.

If subsidence of crust into the marginal sea-floor trench continues, as is happening along the west edge of the Andes today, for example, heat and rising magma will continue to be generated beneath the mountain core. Some of the magma might move upward into the solidified core to crystallize as small batholiths, and some might erupt through tensional fractures in rocks above the core to form chains of composite volcanoes like those in the Andes or in the Cascade Mountains along the west coast of North America.

Continued subsidence of crust into the trench promotes faulting of the oceanic crust on the landward side of the trench, and the faults extend upward into the sediment prism built out on the oceanic side of the old welt (Figure 17-4 F). A situation like this, in which a thick pile of sediments is being deformed and broken by active faulting, is to be found along parts of the California coast today. Analogous parts of older mountain systems, such as the Appalachians, are no longer to be seen, for trenches that are presumed to have been active in their formation long ago ceased to be active, extensive erosion has effaced their seaward margins, and they have been somewhat dismembered by later episodes of sea-floor spreading that involve crustal plates of different size and extent.

### Later History of Continental Mountains

As soon as mountain systems come to be above sea level, they are attacked by the various agents of erosion and are gradually lowered. By this means, the thickened crustal segment of which they are a part is slowly made thinner, and must continue to rise to maintain floating equilibrium. This adjustment commonly proceeds at a slower rate than erosion, however, for we deal here with rigid solids "floating" in plastic solids, not with a solid floating in a liquid. Only gradually, over long periods of time, does the mantle respond plastically to such unbalanced conditions.

Because of the lag in isostatic adjustment, stream erosion may proceed to a late stage in the cycle of regional reduction before the deformed mass of old sediments moves plastically upward to reestablish floating equilibrium. As a result, many old systems of continental mountains, beneath which crustal subsidence ceased long ago, show evidence in the form of rejuvenated and dissected erosion surfaces of having been reduced more than once in their history to low-lying flat or hilly country. Ultimately, the interplay of surface erosion and isostatic adjustment serves to thin the complex prism so that it no longer projects conspicuously above surrounding regions. Geologists may recognize the former existence of mountain systems in such regions from study of their exposed and deeply eroded roots (Figure 1-2); all trace of them as prominent topographic features has disappeared.

Although we have avoided specific discussion thus far of the engines that move the crustal conveyor-belt systems, the role played by heat has been alluded to several times, and the effects of heat in orogeny have been mentioned. Now it is appropriate to give consideration to the earth's heat as a source of power for these engines.

Convection currents in the mantle (Figure 17-5) are believed to be the engines that move crustal conveyor-belts, and radioactive decay of elements irregularly dis-

**17-4
THE ROLE OF HEAT
IN CRUSTAL
DEVELOPMENT**

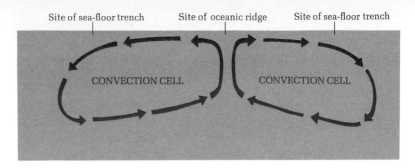

Site of sea-floor trench   Site of oceanic ridge   Site of sea-floor trench

CONVECTION CELL          CONVECTION CELL

**Figure 17-5**
Convection cells and their presumed
relation to oceanic ridges and sea-floor
trenches. Arrows indicate the general
directions of movement of heated rock.

tributed through the upper mantle is thought to be a primary source of the heat that provides the needed power. According to this view, rocks near the site of radioactive decay become hotter, and consequently expand and become less dense than rocks above them. Just as heated water at the bottom of a saucepan rises into cooler water above it, so the heated rock rises slowly toward the surface. Its place is taken by cooler rock, which becomes heated and begins a slow rise. In this way, slow overturn of rock beneath the crust is visualized to occur.

The mass of hot rock above the source of radioactive heat forms the ascending limb of a convection-current cell (Figure 17-5). As heated rock of this limb approaches the surface beneath an oceanic ridge, its upper portion cools and becomes rigid basalt. Somewhat deeper parts of the ascending limb, however, are not rigid and tend to spread out plastically, parallel with the surface. As deeper layers spread laterally, they define the upper limbs of convection-current cells that carry the rigid upper layer along with them. This spreading moves the crustal conveyor-belt system that is central to the hypothesis of sea-floor spreading and the theory of plate tectonics. As spreading continues, both the rigid upper crust and the plastically spreading layer beneath it gradually lose heat, become denser and thicker, and begin to sink back toward the mantle. Oblique subsidence of these crustal layers defines the descending limb of a convection-current cell.

Heat that melts descending crust and forms magma that rises to power the orogenic cycle deduced from plate tectonic theory is generated partly by friction, as crustal plates are sheared under high compressive forces. Thus we must add this source of heat to that provided by radioactive decay if we are to understand the full role of heat in crustal development.

**17-5
SUMMARY**

Continental mountain systems are long, linear features with a complex internal structure. They include adjacent or superimposed belts of thick sedimentary and volcanic rocks, and slivers of oceanic crust, which formed or accumulated in a variety of environments and have been deformed or metamorphosed, intruded by granite, and uplifted to form narrow highland systems. All the great mountain systems are elongated parallel to the strike of their folds and faults, and they are marginal to continents or are thought to have been so when they formed.

If continental mountain systems are "unfolded," it becomes clear that they developed in places that were once greatly elongate, slowly sinking strips of crust to which we have applied the term mobile belt. Sediments deposited in the inner parts of ancient mobile belts are like those of the present-day continental shelf of eastern North America, whereas those of the outer part of such mobile belts are similar to rocks that accumulate today in the continental rise and in areas adjacent to island arcs. Subsidence in the vicinity of sea-floor trenches and island arcs, where sediments and volcanic materials like those in the outer parts of mobile belts form, is probably a result of the same process that causes cold slabs of oceanic crust to plunge downward to the mantle. However, subsidence of continental margins, where prisms of sediment like those in at least the older parts of the interior portions of mobile belts accumulate, is apparently related to thermal necking and down-faulting of continental crust on the trailing edges of continental blocks moving away from oceanic ridges on growing crustal plates.

Because the inner and outer parts of mountain systems like the ones marginal to North America are composed of rock masses of the kind that form today in different settings on crustal plates, a concept of mountain development that relates subsidence, orogeny, and uplift to the evolution of ocean basins and plate margins is a plausible explanation. Heat, generated either by radioactive decay or by friction, powers engines that move crustal plates and drive orogeny.

## SUGGESTED READINGS

Cailleux, A. 1968. *Anatomy of the Earth,* chaps. 3 and 7, pp. 103–39 and 217–32. New York: McGraw-Hill. (Paperback.) Presents at an elementary level the story of mountain building and orogenic belts.

King, P. B. 1959. *The Evolution of North America,* chap. 4, pp. 41–75. Princeton, N.J.: Princeton University Press. A good reference on the development and evolution of the Appalachian Mountain system, which serves as an example of the development of a complex mountain system.

Milne, L. J., and others. 1962. *The Mountains.* Life Nature Library, New York: Time Inc. 192 pp. The story of mountains is presented for the layman at an elementary level. Good color photographs.

Summer, J. S. 1969. *Geophysics, Geologic Structures and Tectonics,* chaps. 10 and 11, pp. 89–108. Dubuque, Iowa: W. C. Brown. (Paperback.) These chapters deal with mountain building and tectonic hypotheses.

# 18 GEOLOGIC FRAMEWORK OF NORTH AMERICA

18-1  Geographic and Geologic Provinces of North America

18-2  Main Events in North American Geologic History

18-3  Summary

Information about the shape, size, and topography of North America has accumulated impressively during the more than 450 years since Columbus landed in the West Indies. Knowledge has been gained in scores of exploratory and mapping expeditions, and the outlines and topographic features of North America's main geographic provinces have been well known for at least a hundred years.

Systematic large-scale geologic studies began in the United States and Canada in the early decades of the nineteenth century. The object of these studies, made primarily by state and national governments, was to accumulate information on the location and distribution of mineral resources. From the geologist's point of view, however, the most important result of these surveys, and of similar studies by universities and industries, has been an ever increasing store of facts concerning the distribution, structure, and relative ages of the rocks that make up the North American continent. Such studies continue, but a new dimension has been added recently: we now have the capacity to take photographs, or to produce other types of images, of vast regions from space.

Although there are still parts of our continent that are almost unknown geologically, and other areas about which we have much to learn, the distribution of at least the principal rock and structural elements is known for all of North America (and for most of the world), and many conclusions as to the historical development of our continent and others are now widely accepted.

Fortunately, North America is an ideal continent for geologic study. It is not crowded close to any others, and seems to be relatively complete. Furthermore, its development has probably been fairly regular, and in outline at least, its history appears straightforward. For these reasons, and because we are already familiar with its geography, we shall deal with North America as an ideal continent.

With respect to the distribution of its topographic and structural features, North America is very nearly symmetrical. In Figure 18-1, which is an outline map of our continent showing the location of its principal geographic and geologic parts, this symmetry is clearly displayed. The broadly arched continental nucleus, the Canadian Shield, is nearly everywhere separated by the Interior Lowlands from a highland border that includes several systems of complex mountains. On its eastern and southern margins, as well as in the Arctic, the continent is fringed by coastal plains that continue beneath sea level as broad continental shelves.

## 18-1
## GEOGRAPHIC AND GEOLOGIC PROVINCES OF NORTH AMERICA

**Figure 18-1**
Principal geologic and geographic provinces of North America. (Base map, Goode Base Map Series, Dept. of Geography, University of Chicago. Copyright by University of Chicago.)

In the figure, the labels read: Innuitian System, East Greenland System, Coastal Plains, Canadian Shield, Cordilleran System, Appalachian System, Colorado Plateaus, Interior Lowlands, Coastal Plains, Ouachita-Marathon System, Shelf, Approximate edge of continental crust.

## The Canadian Shield

The great central part of North America, the **Canadian Shield,** includes most of northeastern Canada, the greater part of Greenland, and some 75,000 square miles of the United States along the west and south sides of Lake Superior and in the Adirondack Mountains of northern New York. In the shield, which has an area of nearly 3 million square miles, the surface is sculptured into low hills and rough highlands, only a few of which are more than 1,500 feet above sea level. This subdued-to-gently-rolling terrain includes thousands of lakes; much of the surface is swampy, and bedrock is widely exposed because ice sheets in the recent past scraped off most of the soil and disarranged the drainage (Figure 18-2).

For the most part, the bedrock of the Canadian Shield consists of granite, lavas, and various kinds of deformed and metamorphosed sedimentary rock. Patches of the bedrock are concealed by glacial deposits. Only a few

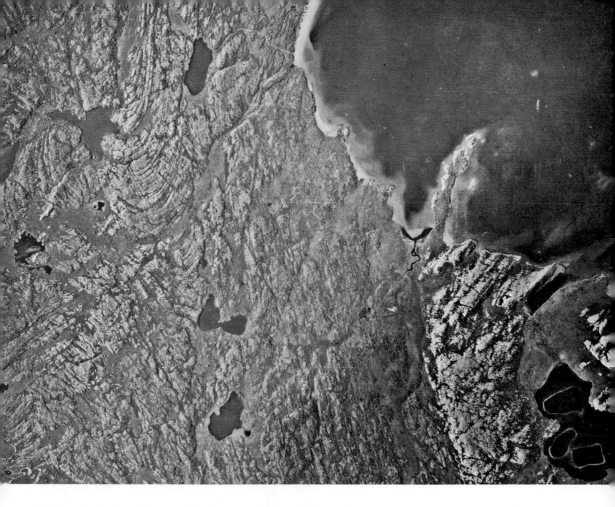

of the strata contain fossils. These facts make geologic work in the shield at once difficult and challenging.

## The Interior Lowlands

Except in Greenland, the Canadian Shield is bounded on the east by the Atlantic Ocean basin. In all other directions, the bedrock of the shield passes beneath a relatively thin cover of Paleozoic, Mesozoic, and Cenozoic sedimentary strata that dip very gently away from its margins. These nearly flat rocks underlie the **Interior Lowlands.**

Topographically, the Interior Lowlands are largely plains country, for the most part only a few hundred feet above sea level. Southeast across the Allegheny and Cumberland plateaus, however, and southward across the Ozark Plateau of Missouri, the lowlands rise into deeply dissected uplands. Westward through the High Plains (Figure 18-3), the surface climbs gradually; it is more than a mile above sea level along the front of the Rocky Mountains (Figure 18-4).

Here and there in the Interior Lowlands are dome mountains, like the Black Hills and the Ozarks, where

**Figure 18-2**
Air view of the Canadian Shield in the Ross Lake area, Northwest Territories. Folded Precambrian metasedimentary rocks (left and center) are cut by a granite batholith, which makes the light-colored area south of the lake on the right. The surface is without soil, and the lake basins in bedrock suggest glacial scour. (Courtesy Department of Mines and Technical Surveys, Canada.)

the thin sedimentary cover has been breached by erosion. In such mountains, igneous and metamorphic rocks like those of the Canadian Shield are exposed at the surface. Similar rocks have also been encountered at many places below the surface of the lowlands, where drill holes have penetrated the sedimentary cover and continued into older rocks. Precambrian rocks that are buried beneath a veneer of sedimentary strata are referred to as the **basement complex,** or simply as the **basement.** The buried extension of the Canadian Shield constitutes the basement complex that everywhere underlies the Interior Lowlands.

**Figure 18-3**
The Great Plains: westward extension of the Interior Lowlands. This farmstead, in the wheat fields of Montana, is well protected from winter wind and snow by planted trees. (Photo courtesy of U. S. Department of Agriculture.)

**Figure 18-4**
Air view looking north along the Front Range of Colorado. The ridges of dipping sedimentary rocks mark the boundary between the Rocky Mountains on the west (left) and the Interior Lowlands on the east. The city of Denver is just east of the area shown. Mesa on the right side of the view in the middle distance is formed by a Paleogene lava flow, which has protected the soft Cretaceous shales from erosion. The prominent ridge in the center is made by the Dakota sandstone (Cretaceous). Note the water gap. (Photo by T. S. Lovering, U. S. Geological Survey.)

## The Appalachian System

Forming an almost complete rim around the Canadian Shield and Interior Lowlands is a narrow belt that includes five major mountain systems, each of which is composed of one or several ranges. The **Appalachian System** to the east extends for more than 1,500 miles along the southeastern margin of North America, from Newfoundland to Alabama. This system consists of two distinct parts. Parallel to the outer edge of the Allegheny and Cumberland plateaus, which are underlain by horizontal Paleozoic rocks, is the Valley and Ridge Province. In this narrow segment of the Appalachians, which extends from central Alabama to northern New Jersey, stream erosion of folded and faulted Paleozoic strata has produced a series of nearly parallel valleys (in weak rocks) and intervening sharp ridges (of resistant rock) (Frontispiece; Figures 11-19, 18-5). East of the Valley and Ridge Province, for the full length of the mountain system, is an elongate area of greatly deformed and variously sculptured igneous, sedimentary, and metamorphic rocks. New England and the Maritime Provinces of Canada, which form the northern segment of this belt, are mountainous

**Figure 18-5**
Looking eastward across the Valley and Ridge Province of the Appalachian System in central Pennsylvania. Note the even-crested ridges. The Juniata River has cut a water gap through one of the mountains in the middle distance to left of center. (Photo by John S. Shelton.)

327

country underlain mostly by igneous and metamorphic rocks of Precambrian and Paleozoic ages. The surface bears the obvious imprint of extensive scouring by widespread ice sheets. From New Jersey southward, the outer belt of the Appalachian System consists of two elongate provinces side by side. On the northwest, next to the Valley and Ridge Province, is a range of high mountains that includes the Blue Ridge of Virginia, Maryland, and Pennsylvania, and the Great Smokies of North Carolina and Tennessee. Parallel to this on the southeast is the Piedmont Upland, a hilly to almost flat region blanketed by thick residual regolith. Both the Great Smoky–Blue Ridge mountains and the Piedmont Upland are underlain by greatly deformed Precambrian and Paleozoic sedimentary rocks, and by igneous intrusives of Precambrian and Paleozoic age.

## The Cordilleran System

Between Alaska and Panama, the west side of North America is occupied by the **Cordilleran System,** or the North American Cordillera, as it is sometimes termed. The highest summit, Mount McKinley in Alaska, is 20,300 feet above sea level, and peaks with elevations of more than 14,000 feet are numerous throughout its great extent. Oddly enough, the Cordilleran System also includes the lowest elevation in North America, 282 feet below sea level, in Death Valley, California.

For much of its length, the Cordilleran System consists of three longitudinal belts, the Western, Eastern, and Central Cordillera. There are considerable differences in topography and geologic structure from place to place within each belt and there are many subdivisions of each.

The Western Cordillera includes the high mountains of the Pacific Coast from Alaska to the southern tip of Baja California. The many ranges of this belt are composed largely of Mesozoic granite and of all types of deformed and metamorphosed Mesozoic and Cenozoic rocks. The Cascade Range of Washington, Oregon, and northern California includes Lassen Peak and Mounts Shasta, Rainier, Hood, and Baker, a series of famous late Cenozoic volcanoes. In eastern California is the Sierra Nevada, a great batholithic block of Mesozoic granite, uplifted along a fault on its east side (Figure 18-6), and deeply sculptured by valley glaciers (Figure 12-5). The Sierra is separated from contorted late Mesozoic and Cenozoic strata of the Coast Ranges by the Central Valley of California.

The Eastern Cordillera is a narrow chain of disconnected mountainous areas, separated from each other by

alluvium-floored basins, and fronting eastward on the
High Plains (Figure 18-4). From eastern British Columbia
southward to New Mexico, the Eastern Cordillera is
composed of the Rocky Mountains. The Rockies are joined
structurally to the ranges of northeastern Mexico through
a belt of low, isolated fault-block mountains in southern
New Mexico and adjacent Texas; to the north the Rockies
continue through the Yukon into Alaska via the Richardson, Franklin, and Mackenzie mountains. Throughout its
extent, the Eastern Cordillera consists of folded Paleozoic,
Mesozoic, and early Cenozoic sedimentary rocks. Volcanic rocks are present in some segments, but are not particularly important. In those segments of this belt that have
been uplifted the most, the sedimentary cover has been
removed by erosion and a core of Precambrian igneous
and metamorphic rocks is exposed (Figure 18-7). The
eastern edge of the northern Rockies is grandly shown
in the scenery at Glacier National Park, Montana; the high
southern Rockies of Colorado can be studied to advantage
in Rocky Mountain National Park, northwest of Denver.

Between the Eastern and Western cordilleras is a broad
belt of high country that includes some of North America's
most varied scenery. This belt is the Central Cordillera.
In Canada and the northwestern United States, the Central
Cordillera includes subdued mountain ranges and high
plateaus underlain largely by complexly folded Paleozoic

**Figure 18-6**
East front of the Sierra Nevada, viewed
across Owens Valley. The face of the
Sierra is a fault scarp, much modified
by erosion, and the rock exposed in it
is largely granite. The general structure
is like that shown in Figure 14-21 A.
(Photo by Mary R. Hill.)

**Figure 18-7**
The Gore Range, north central Colorado, from the east. An upfaulted block of Precambrian rocks forming one of the ranges of the Colorado Rockies. Cretaceous shale makes the slope across the valley in the middle distance. (Photo by J. H. Rathbone.)

and Mesozoic sediments, and Mesozoic granites. In southern and western Idaho, and in eastern Washington and Oregon, are the Snake River and Columbia plateaus, held up by thousands of feet of basaltic lava extruded in fairly recent Cenozoic time from vents and crustal fissures (Figure 4-4). South of these lava fields, the Central Cordillera includes two strikingly different provinces, the Basin and Range and the Colorado Plateaus.

The Basin and Range Province, which includes the Great Basin of Nevada and western Utah and stretches in modified form southward through Arizona into Mexico, is rough arid country that is generally lower than the Sierra on the west or the Colorado Plateaus on the east. It includes many isolated fault-block mountain ranges separated by desert basins filled with alluvium (Figures 3-2, 7-18, 18-8). The entire province is underlain by rocks of Precambrian through Cenozoic age that were first complexly deformed by compressional forces and then differentially uplifted along normal faults (Figure 14-21 B). In many places notably in the Sierra Madre Occidental and the Plateau Central of Mexico, the Basin and Range Province includes great areas of Cenozoic lavas and pyroclastic rock.

The Colorado Plateaus of Utah, Colorado, New Mexico, and Arizona are magnificent high desert country underlain by brightly colored horizontal sedimentary rocks of Paleozoic through Cenozoic age. Stream erosion has carved these rocks into spectacular cliffs and canyons, among which are those in Grand Canyon (Figure 2-11), Bryce Canyon (Figure 5-7), and Zion Canyon (Figure 23-2) national parks, and Monument Valley (Figure 18-9). At the Grand Canyon, the sedimentary blanket of the plateau has been entirely breached by the Colorado River, and

Precambrian rocks like those in the core of the Rockies are exposed at the bottom of the canyon.

The higher summits of the Cordilleran System bear evidence of extensive valley glaciation, and in that part of the system north of the Basin and Range and Colorado Plateaus, all three belts have been broadly modified in surface form by glaciation in the very recent past. The Cascades and the Pacific Ranges in British Columbia and Alaska support large valley-glacier systems today, and much of what we know about glacial processes has been gained from study of active glaciers in these districts of the North American Cordillera.

## The Ouachita-Marathon System

In east-central Alabama, the folded rocks of the Appalachian System disappear from view beneath a thick cover of horizontal Mesozoic and Cenozoic rock, and the Coastal Plain projects northward along the Mississippi Valley to merge with the Interior Lowlands. Beyond this break in the highland border are the **Ouachita Mountains,** which

**Figure 18-8**
View southward across the Basin and Range Province in southern Nevada. Fault-block mountains in the foreground are the Arrow Canyon Range. Note the coalescing alluvial fans along the mountain front. (Photo by John S. Shelton.)

occupy a belt 50 to 60 miles wide and some 200 miles long in southeastern Oklahoma and west-central Arkansas. The Ouachitas are generally lower than the folded Appalachians, but they resemble the Valley and Ridge Province of the latter system in topography and structure. They are formed of folded Paleozoic sedimentary rocks broken into thin slices by thrust faults. As in the Appalachians, the folds and faults record compression directed from the southeast toward the northwest.

In the Marathon region of trans-Pecos Texas, uplift associated with formation of the Cordilleran System produced a broad dome in Mesozoic rocks. Subsequent removal of the Mesozoic strata from the dome through erosion has exposed another belt of folded and faulted Paleozoic rocks, the **Marathon Mountains.**

There seems to be little doubt that the Ouachita and Marathon mountains are parts of the same system, now mostly buried beneath younger sediments. Because exposed segments of this system are of deformed Paleozoic rocks similar to those of the Appalachians, many geologists believe them to represent a southwestern extension of the Appalachian System. We refer to the largely buried belt of mountains as the **Ouachita-Marathon System.**

### The Innuitian and East Greenland Systems

From the northern tip of Greenland south and westward across islands of the Canadian Arctic Archipelago stretch mountains of the complex **Innuitian System.** Geologic exploration of this system has barely begun, but its major provinces have been delineated by geologists of the Geological Survey of Canada. As a whole, the Innuitian System is similar in age and type of rock, and in topographic aspect, to the northern segment of the Appalachians. An outer belt of metamorphosed volcanic rock and sediments is represented in northern Ellesmere Island; an inner belt of deformed but little metamorphosed Paleozoic sedimentary rocks can be traced from Greenland through Ellesmere Island (Figure 18-10) to the western end of the archipelago, where it disappears beneath flat coastal-plain sediments. Rocks of the folded belt are bounded on the south by horizontal or broadly arched Paleozoic sediments of the Interior Lowlands.

The highland border that nearly encircles North America is completed by the **East Greenland System,** which occupies Greenland's east coast north of 70° North Latitude. These mountains, like the Appalachians, the Ouachita-Marathon System, and the Innuitian ranges, are of folded and extensively faulted Paleozoic rocks.

**Figure 18-10**
Air view of the Innuitian System on the east coast of Ellesmere Island, Canadian Arctic. The folded sedimentary rocks exposed in the walls of the glaciated valleys are mostly limestone and shale of Early Paleozoic age. (Photo courtesy of Department of Mines and Technical Surveys, Canada.)

## Coastal Plains

Along the eastern, southeastern, and far northern sides of the continent are **coastal plains,** which are low, almost flat areas developed on nearly horizontal beds of sand, clay, and limestone (Figure 18-11). These strata contain Mesozoic and Cenozoic fossils, and in most places they overlap the beveled outer edges of rocks in the highland border. Indeed, the Ouachita-Marathon System is largely buried beneath sedimentary rocks of the Gulf Coastal Plain.

Rocks of the coastal plain are continuous seaward with sedimentary blankets on the submerged edges of the continents. This drowned edge of the continent, sloping gently seaward, is the continental shelf (Section 16-1; Figure 16-2).

**Figure 18-11**
Coastal plain along the Gulf of Mexico. Louisiana in the foreground, Texas in the distance beyond smoke. (Photo by John S. Shelton.)

The distribution of geologic structures and topographic features in the chief provinces of North America suggests a pattern in the geologic evolution of our continent. A knowledge of the principal steps in continental development will help to focus our attention more profitably on the details to be studied later.

As a group, the granites, lavas, and metamorphic rocks of the Canadian Shield are continuous with the basement complex beneath the Interior Lowlands and below the mountains of the highland border. Thus, according to the law of sequence, these rocks are older than the strata of the lowlands and the mountain systems. Because we know of no rocks in North America older than the bedrock of the Canadian Shield and the basement complex, we may safely infer that the legible history of the continent began with their formation.

The oldest rocks in the Interior Lowlands and mountain systems are unfossiliferous; these strata are overlain by several thousand feet of fossiliferous sedimentary rocks, the oldest of which are confidently correlated with rocks in the Cambrian System of Wales. The youngest bedrock identified in the lowlands is of Neogene age. Presumably then, the ancient rocks of the Canadian Shield and the basement complex are North American records of events in the Precambrian Eras, whereas Paleozoic, Mesozoic, and Cenozoic history is at least partly documented by strata in the Interior Lowlands.

The Appalachian, Ouachita-Marathon, Innuitian, and East Greenland systems are composed primarily of deformed sediments containing Paleozoic fossils, into which in some places have been intruded batholiths of Paleozoic age. Neither Mesozoic nor Cenozoic rocks seem to be involved in the structures of these mountain belts, but instead they overlap the beveled edges of the older rocks and extend seaward to form the coastal plains. Clearly then, mountain systems were formed along the eastern, southern, and northern borders of the continent before the Mesozoic, but after passage of at least a considerable part of Paleozoic time. The Cordilleran System, on the other hand, includes folded and faulted Paleozoic, Mesozoic, and Cenozoic rocks. Thus Cordilleran ranges are younger than those in other mountain systems. Finally, the blanket of glacial drift that indiscriminately mantles parts of nearly all the important geographic provinces of the continent records a very recent episode of continental glaciation in North America.

From this brief résumé, the following chronology of events seems evident. Legible geologic history began in North America in the Precambrian Eras with the formation of the rocks of the Canadian Shield and its buried

## 18-2
## MAIN EVENTS IN NORTH AMERICAN GEOLOGIC HISTORY

extensions. Formation of these rocks was followed in the Paleozoic by deposition of sedimentary rocks in the Interior Lowlands and in areas now occupied by the rim of mountainous highlands. After passage of at least some of Paleozoic time, the Appalachian, Ouachita-Marathon, Innuitian, and East Greenland mountain systems were formed. During the Mesozoic and Cenozoic eras, sediments continued to accumulate in the lowlands, on the site of the present Cordilleran System, and along the southern, eastern, and northern coastal plains. After great thicknesses of these rocks had been deposited, folding and faulting produced the Cordilleran System. Finally, within very recent times, great masses of glacial ice overspread the shield and adjacent parts of the lowlands and highland border, disrupting drainage and leaving moraines, deposits of outwash, and lake sediments.

**18-3 SUMMARY**

North America is divisible into four principal geologic and geographic provinces. In the Canadian Shield, which occupies much of the center of the continent, topography is subdued and bedrock is largely deformed igneous, metamorphic, and sedimentary rock of Precambrian age. Except on its east side, the shield is bounded by Interior Lowlands, the bedrock of which consists of a thin veneer of flat-lying Paleozoic, Mesozoic, and Cenozoic sedimentary rock that lies on a basement complex of Precambrian rocks like those exposed in the Canadian Shield. A highland border, including five extensive mountain systems, forms an almost complete rim around the shield and Interior Lowlands. The Innuitian, East Greenland, Appalachian, and Ouachita-Marathon systems, on the north, northeast, east and southeast sides of the continent, consist of deformed Precambrian and Paleozoic sedimentary rocks; whereas the Cordilleran System, which stretches along the western side of the continent from Alaska to Panama, is composed of greatly folded and faulted Paleozoic, Mesozoic, and Cenozoic strata. Along the eastern, southeastern, and far northern sides of the continent are coastal plains, flat areas developed on nearly horizontal Mesozoic and Cenozoic sedimentary rocks that overlap the beveled edges of rocks in the highland border and are continuous seaward with sedimentary blankets of the continental shelf.

## SUGGESTED READINGS

Hunt, C. B. 1967. *Physiography of the United States.* San Francisco: Freeman. 480 pp.

A college text written for students not majoring in geology or the earth sciences. The book is divided into two sections, one dealing with general physiography, the other with descriptions of each of the physiographic provinces.

King, P. B. 1959. *The Evolution of North America.* Princeton, N.J.: Princeton University Press. 190 pp.

A reference that provides the story of the geologic evolution of North America on a regional basis.

Murray, G. E. 1961. *Geology of the Atlantic and Gulf Coastal Province of North America,* chap. 8, pp. 493–564. New York: Harper & Row.

A reference that gives detailed information on the geology and physiography of both the submerged and emerged coastal segments of North America.

Thornbury, W. D. 1965. *Regional Geomorphology of the United States.* New York: Wiley. 609 pp.

An excellent reference within the scope of many introductory students, which provides detailed information on the major geomorphic provinces of the United States.

# 19 THE PRECAMBRIAN ERAS

19-1  Geologic Features and History of the Moon

19-2  Precambrian Rocks

19-3  Interpreting the Precambrian Rock Record

19-4  Precambrian Mineral-Age Provinces

19-5  Precambrian Rocks and History of the Lake Superior Region

19-6  Inferences as to Precambrian History of North America

19-7  Precambrian Ice Ages

19-8  Precambrian Mineral Resources

19-9  Summary

The geologic record tells us that the point farthest back in earth history is 4,600 million (4.6 billion) years ago in the scale of absolute time (Table 3-1). The Precambrian Eras constitute the interval from that incredibly remote date to the beginning of the Paleozoic Era about 570 million (0.57 billion) years ago. Thus Precambrian time had a duration of roughly four billion years.

The figure 4.6 billion years is believed to represent approximately the age of the earth. It is based on evidence that comes from outside the earth. History that can be reconstructed from direct geologic evidence collected on the earth begins with rocks only about four billion years old. No one has so far found a rock record on earth for the first 0.6 billion years of its history. Indeed, this earliest record has probably been obliterated by erosion and sedimentation, and by sea-floor spreading, which concurrently renews and destroys the crustal record in ocean basins.

The earth, however, has a large companion in space, a satellite that we call the moon (Figure 19-1). Since both the earth and the moon are parts of the solar system,

**Figure 19-1**
Full-face view of the moon, taken by astronauts from about 10,000 miles away. Maria (dark areas), highlands (light areas), and craters are distinguishable. (Photo courtesy of NASA.)

**Figure 19-2**

Photo from *Apollo 12* mission, showing imprint made more than 2½ years previously by *Surveyor* unmanned spacecraft (left). Freshness of print indicates negligible rate of erosion on the moon's surface. (Photo courtesy of NASA.)

their early histories were probably affected by similar processes acting under similar conditions. Study of the moon might thus provide important information on the earliest stages in the history of our own planet. This is particularly likely because the moon has no atmosphere or hydrosphere, and thus no wind, clouds, rain, rivers, or oceans. Rocks exposed at its surface should therefore not be greatly altered chemically or much disturbed by those processes of erosion, redeposition, and regeneration that have obliterated the record of the earth's earliest history (Figure 19-2). The moon might turn out to be a museum, in which a record of the very earliest part of earth-moon history is prominently on display. A good beginning for our study of earth history is therefore a brief description of the moon's "geologic" features and a short account of its history.

## 19-1 GEOLOGIC FEATURES AND HISTORY OF THE MOON

### Features

The moon is an approximately spherical body, 2,160 miles in diameter, which makes a complete revolution about the earth in a little less than a month. On the average, earth and moon are separated by 238,857 miles, a distance that can be crossed in about four days by astronauts on exploratory missions. The moon has a relative density of 3.33, or only about 0.6 of that of the earth.

Present evidence indicates that the moon's outer layers are cold and over wide areas may consist of unconsolidated material several kilometers thick. At greater depths, magma may be generated, and some may rise to the surface periodically, for sporadic volcanic activity has been observed at a few sites. Compared with the earth, the moon is quiet; only a few minor "earthquakes" have been recorded, and these are probably triggered by tidal forces generated by the earth's attraction.

Major topographic provinces of the moon's surface are *highlands* and *maria* (singular: *mare*, Latin "sea"; so named because they were thought to be seas by seventeenth-century astronomers) (Figure 19-1). The highlands, which cover most of the moon's surface, are rugged mountainous areas, pockmarked by craters. We have little reliable information on composition of the lunar highlands in most places, but isostatic models suggest that they include crustal rock that is less dense than that in the body of the moon. The highland crust was sampled directly by *Apollo 15* astronauts, and some of the rock fragments brought back from the Mare Tranquillitatis (or Sea of Tranquillity) by *Apollo 11* astronauts are thought to have been derived from adjacent highlands. Samples returned by both missions consist of an igneous rock termed *anorthosite*, which is very similar to gabbro in composition.

Lunar maria are irregularly shaped dark basinlike areas of various sizes. From earth they appear to be smooth, featureless plains; through a telescope or from a spacecraft in orbit around the moon, however, their surfaces are seen to be extensively cratered. Far from being filled with water, the lunar maria are apparently seas of basalt, blanketed here and there by regolith (Figures 19-2, 19-3) composed of material ejected from lunar craters when they were formed by meteorite impact.

Working from greatly enlarged telescopic photographs, or from photographs taken from spacecraft in orbit about the moon, geologists have been able to make geologic maps of the lunar surface, and have assigned relative ages to events in the moon's history. They have followed two general procedures. In the first method, craters within certain areas are counted and a figure representing crater density as a function of diameter is calculated. If it is assumed that craters are produced by meteorite impact, then crater density should be greatest in areas that have been exposed the longest to meteorite bombardment and least in areas whose surficial deposits are youngest. Thus a rough idea of the relative age of surficial features on the moon can be determined from relative crater density. Also, because lunar craters were not all formed at the

same time, the loose material ejected from them, or from the lunar maria when they were formed, accumulated to form blankets of regolith, or *ejecta,* that are also of different ages. A second means of determining the relative age of events in the moon's history involves study of these ejecta blankets. For example, if the blanket that accumulated around one crater overlaps the edges of another crater, the latter must be older than the former.

Using these procedures, a chronology of lunar events has been worked out. The highly cratered highlands turn out to be the oldest features of the lunar landscape. Ejecta blankets derived from the lunar maria overlap the highlands and indicate that the next event was excavation of the maria, presumably by infall of large meteorites. However, ejecta blankets from the maria are more extensively cratered than the material that now fills the maria. Thus a third important event in lunar history must have been formation of the basalts that now fill the maria. Later events include formation of some of the larger craters in the maria, and development of several volcanic regions. Absolute ages can now be assigned to a few of these events in moon history from samples brought back by astronauts, and information from these samples and from lunar geologic maps has been assembled with theoretical inferences into a still incomplete account of the moon's early history.

## History

Hypotheses about the origin of the solar system, an event that involved the earth and moon, are numerous and varied, and there is little agreement as to which of them is the best. Many of these hypotheses, however, have in common the idea that the earth, moon, and other rocky planets all originated at about the same time, by rapid accretion or "sweeping up" of small solid dust-like particles that were once widely dispersed in the orbits of these planets. According to one version of the accretion hypothesis, the earth and moon grew independently in the orbits they now occupy; the earth simply grew to be larger than the moon. Another hypothesis holds that although the moon was formed by accretion of dustlike particles, it grew somewhere else and was snared by the earth's gravitational field early in the history of the solar system.

Studies of stars, and of gas and dust clouds that have not yet condensed into stars, suggest that the process of accretion of earth and moon did not take very long by geologic standards. (Stars have apparently been born in the gaseous Orion Nebula in the incredibly short time of 8 years.) Some hypotheses of the earth's thermal history suggest that accretion may have taken less than half a million years.

However the earth and moon originated, and whatever the time it took for them to grow from their ancestral dust clouds, the time at which they formed can be fixed as about 4.6 billion years ago. This date has not been determined directly from lunar or terrestrial rocks, but from meteorites and from various models that have been evolved to explain the evolution of matter and its eventual fractionation into chemical elements in our galaxy and solar system. Meteorites, thought to represent fragments of a now disrupted rocky planet that formed in the same part of the solar system as the earth and moon, and at about the same time, yield maximum ages of about 4.6 billion years. Model ages, calculated from samples of lunar regolith in which there are rock fragments more than 4.4 billion years old, suggest an age of at least 4.5 billion years for some of the parent material of the regolith. Finally, studies of the ratio between radiogenic and nonradiogenic isotopes of strontium in lunar samples and in meteorites indicate that they both started out with strontium of the same isotopic composition; and this suggests that both began condensation from their respective dust clouds at about the same time. That time was apparently 4.6 billion years ago.

Rapid accretion of dust to form the moon and earth

would have produced much heat, and this would have been increased by radioactive decay and by compression and contraction. Thus, even though these rocky planets might have been cold to begin with, it is likely that their internal temperatures rose rapidly as they grew larger. This would have produced at least partial melting.

We should expect that as parts of the moon and earth melted, there would have been some gravitational rearrangement of their originally randomly distributed components. Materials of lighter density should have risen toward the surface as those of greater density sank toward the center. This process, which may have been partially responsible for initial segregation of the earth's core, mantle, and crust, is one of the ways in which geologists explain formation of the lunar crust, now exposed in the moon's extensive highland areas. In any event, samples of highland crust brought back by *Apollo 15* astronauts indicate that a lunar crust of anorthositic composition was formed between 4 and 4.6 billion years ago.

During formation of the lunar crust, meteorite infall took place at a rapid rate, sculpturing what are now the moon's highland areas into rugged, extensively cratered mountains with thick blankets of ejecta. Especially large meteorites, striking the moon's surface obliquely, are thought to have excavated the mare basins. Because ejecta blankets from some of the basins are superimposed on parts of the highlands, the basins were undoubtedly produced after the highland crust had formed.

Samples of bedrock from the lunar maria indicate that the mare basins were formed between 3.9 and 3.2 billion years ago, and that impact by the giant meteorites that formed them triggered repeated eruptions of lava; the lava then hardened to form the basalt that now fills the basins. Subsequent infall of meteorites, apparently at a much lower rate than in early moon history, produced craters and ejecta blankets in the maria and highlands. It is likely that these, and a few features produced by downslope movement and local volcanic activity, are the only changes made in the lunar landscape in the last three billion years or so.

Very possibly, the earth's first 0.6 billion years were marked by events similar to those in the moon's early history. The earth's envelopes of air and water, however, and especially the constant internal stirrings associated with greater size and heat, have acted to alter, cover, or erase the record.

On the other hand, we must credit air and water for their parts in accumulating what record we do have of Precambrian and later time. Specifically, we would probably have a record for no more than post-Paleozoic time were it not for the activities of the agents of weathering and erosion, for they have worked to form even lighter-weight masses of sedimentary rock from the relatively lightweight slag that probably formed the earth's original crust. The process of sea-floor spreading may serve both to renew and to destroy the oceanic crust in relatively short intervals of geologic time. Only continental crust survives, for the low-density rocks of which it is composed resist destruction in sea-floor trenches. We can then profitably turn for an account of the remainder of the earth's Precambrian history to the continents, specifically to North America.

## Division of the Record

It may seem strange that the rock record of the latest 570 million years of earth history can be divided with little trouble into three great divisions, each representing an era of geologic time, but that the great mass of rock beneath it, recording more than six times as great a stretch of time, has not yet been divided into units that find universal favor among geologists. To be sure, this thick and significant division of very old rocks has at one time or another been divided into many units, but no one has yet been very successful in tracing them from one continent to another or, indeed, in following them from one place to another in the same continent.

As we shall presently see, however, the great thickness of rock below Cambrian strata includes the record of at least four important episodes in the history of North America. Most probably each of these lasted for an interval of time equal to, or longer than, the Paleozoic and later eras. Because there are no widely accepted names for the several eras of time represented, we shall refer to all time before the Cambrian Period as the Precambrian Eras, and refer to the rocks formed during all this time as Precambrian rocks.

## Distribution of Precambrian Rocks

In North America, Cambrian fossils occur in the oldest sedimentary rocks of the Interior Lowlands and low in the sedimentary sections exposed in the highland border. According to the law of sequence, then, a record of the Precambrian Eras must be contained in the igneous and metamorphosed basement rocks (Section 18-1) beneath the lowlands and the marginal mountains, as well as in the Canadian Shield. By and large, the basement is not exposed to direct observation, although patches of it can be studied at the surface in such places as the Ozarks, the Black Hills, and the cores of both the Appalachian and Cordilleran mountain systems. The Canadian Shield,

**19-2**
**PRECAMBRIAN ROCKS**

on the other hand, is the largest exposed tract of Precambrian rock in the world.

**19-3**
INTERPRETING THE
PRECAMBRIAN
ROCK RECORD

## The Difficulties

As with other parts of geologic history, what we can deduce of Precambrian events must come from study of the rock record, which is well displayed and widely distributed. Consequently, the Precambrian history of North America should be rather well known. Unfortunately, however, it is quite incompletely known, for its record is very difficult to decipher.

### Deformation and Erosion

There are several reasons for our presently limited knowledge of Precambrian history. First, the Precambrian record is contained in igneous and sedimentary rocks, most of which have been complexly folded and faulted, and, in the process, more or less metamorphosed. Furthermore, they have been exposed to erosion for such an inconceivably long time that sequences are no longer complete and the record is full of gaps. In combination, these facts make it very difficult, or altogether impossible, to identify even closely adjacent sequences by lithologic equivalence.

### Scarcity of Fossils

Another attribute of Precambrian rocks that makes them difficult to interpret historically is their almost complete lack of fossils. To be sure, even very old rocks contain a few fossils, but the fossil record is so sketchy, and the fossils themselves so small and primitive, that the usual procedures of biologic correlation have not yet been applied successfully.

### Exposure and Accessibility

The basement beneath the Interior Lowlands and the mountain systems includes an area of Precambrian rock at least as great as that exposed in the Canadian Shield and elsewhere. Because the basement is almost entirely concealed, what we know about most of it has been learned largely through study of chips and cores of rock brought up from the bottom of wells that have been drilled through the sedimentary cover. The widely scattered distribution of these wells, and the small volume of rock that they make available, sharply limit our knowledge of the buried Precambrian rocks.

Finally, great expanses of the Canadian Shield are in places that have been difficult of access until very recently. For this reason, they either have not been studied at all or have been mapped and described in only a reconnaissance fashion.

## Radioactive Minerals: Aids in Correlation

Even though faced with the difficulties we have just mentioned, geologists have spent more than a century in making detailed maps and descriptions of Precambrian rocks in many widely separated parts of North America. Although this work is constantly revised and periodically refined, enough has been completed to provide a general account of Precambrian history in each area studied. Furthermore, within recent years, Precambrian rocks in several areas have been dated by analysis of radioactive minerals in them.

Absolute ages derived from analysis of radioactive minerals must be treated with caution, and geologists have avoided attaching too great importance to scattered determinations in areas for which there is little other geologic information. Nevertheless, the number of reliable determinations has increased steadily in the last decade or so, and the absolute age of major events in geologically well-known Precambrian areas is now pretty firmly established. Thus, the absolute time scale provides a framework within which to make correlations between widely separated Precambrian sequences.

Figure 19-4 shows North America with Interior Lowlands and Coastal Plains sediments removed. The basement rocks thus exposed, together with the Canadian Shield, are divided into provinces on the basis of rock structure and the distribution of radioactive minerals for which reliable age determinations have been obtained. Several interesting features stand out. Note first that the oldest minerals dated thus far occur in the Superior Province near the center of the continent, and that the structural "grain" of the rocks in that province is distinctly different from that in adjacent ones. Note also that mineral ages are younger outward from the Superior Province.

Other features of more than passing interest are shown in the figure. For example, at least some (generally all) of the rocks in each mineral-age province have been deformed. If this were not so, it would not be possible to map the structural trends as we have done. In addition, the mineral-age provinces are all elongated parallel to the structural trends within them. Although it is not apparent on Figure 19-4, we should note that some (probably all) provinces overlap others containing older rocks.

Figure 19-4 suggests that there is some inherent order, or pattern, in the *distribution* of the Precambrian rock

**19-4
PRECAMBRIAN
MINERAL-AGE
PROVINCES**

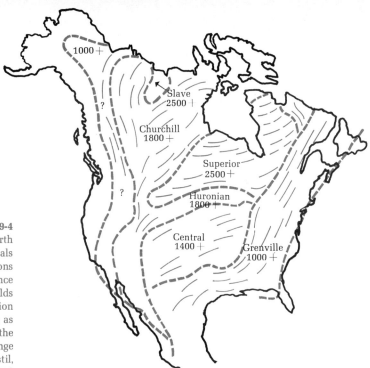

**Figure 19-4**
Chief mineral-age provinces of North America. Ages of radioactive minerals in each province are shown in millions of years. Patterns within each province show general trend of principal folds and faults. This map is a generalization based on widely scattered evidence; as additional observations are made, the patterns shown will probably change substantially. (Data largely from Gastil, 1960, and Engel, 1963.)

record, but it tells us very little about the record itself. For these facts, we must turn to a great mass of reports and maps in which the nature and distribution of Precambrian rocks are described. It would be an exhausting task, and likely a frustrating one as well, to attempt a comprehensive summary of all the information now at hand about North American Precambrian rocks. Not all areas of these rocks have been studied with the same care; each area has its own special features and problems; and such a wide variety of names has been applied to divisions of Precambrian rocks and time in different areas that it is often confusing to try to make sense from them.

The large tract of Precambrian rock exposed in the Lake Superior region of southern Ontario and northern Michigan, Wisconsin, and Minnesota merits close inspection, however, for it includes parts of the Superior and Huronian mineral-age provinces, and its geology is probably better known than that of any comparable area of Precambrian rocks in North America. Furthermore, the history deduced from sequences of Precambrian rock in the Lake Superior region can probably be extended with many local modifications through much of the Superior and Huronian provinces.

**19-5**
**PRECAMBRIAN ROCKS AND HISTORY OF THE LAKE SUPERIOR REGION**

### The Oldest Rocks

In Figure 19-5, which is a very generalized geologic map, the oldest rocks in the Lake Superior region form the large, irregular patch northwest of the lake that bears the number 1. Rocks in this area represent two geologic systems, the older, termed the **Ontarian,** and the younger, the **Timiskamian.** The Ontarian sequence begins with gneiss that represents an unknown original thickness of muddy sandstone and shale. These ancient rocks are succeeded by basaltic lava, tuff, and a limited thickness of cherty rock rich in iron. Timiskamian rocks, separated from strata of the Ontarian System by an unconformity, begin with a conglomerate that is locally more than 4,000 feet thick. Above this is a complex mass of metamorphosed shale, sandstone, tuff, volcanic breccia, and iron-bearing beds, with an aggregate thickness of about 15,000 feet. Both Ontarian and Timiskamian rocks have been intensely folded and metamorphosed since they were formed; however, from sedimentary features in sandstones it is possible to determine that they were deposited in water.

**The Algoman Orogeny**

Rocks numbered 2 in Figure 19-5 are granite, or its metamorphic derivative, gneiss. A study of the field relations between these rocks and deformed Ontarian and Timiskamian sediments indicates clearly that the granites intruded the sediments, hence are younger. Radioactive minerals in the granite, and biotite that came into being in the deformed sediments when they were metamorphosed, give similar ages of about 2.5 billion years. This suggests that a prolonged interval of sedimentation and volcanism, during which Ontarian and Timiskamian rocks were formed, was brought to a close about 2.5 billion

**Figure 19-5**
Geologic map of the Lake Superior region. All rocks shown are Precambrian in age. Numbers refer to rocks described in the text. (Adapted from Goldich and others, 1961.)

years ago by intrusion of great granite batholiths and by folding and metamorphism of the intruded rocks. Almost certainly, this sequence of events produced a range of mountains much like the Appalachians of later times. This range, only the roots of which now exist, has been termed the **Algoman Mountains,** and the sequence of events that produced it, the **Algoman Orogeny.**

### Post-Algoman Strata

Rocks in area 3 in Figure 19-5 are included in the **Huronian System.** They are separated from the Algoman granite and from contorted Ontarian and Timiskamian rocks by an angular unconformity. This fact implies that, before Huronian rocks began to form, the Algoman Mountains had been reduced by erosion to an essentially featureless surface at low elevation. It has been estimated that reduction of the Algoman Mountains to such a surface took more than 200 million years. Thus slowly do the mighty perish.

Northwest of Duluth, Huronian rocks are made up of thousands of feet of conglomerate, sandstone, and shale. A considerable thickness of iron-bearing chert, deeply weathered and intricately folded, is included in this mass of contorted strata, and forms the Mesabi Range, shown in Figure 19-5 by the heavy dark strip separating areas 2 and 3 northwest of Duluth. As they are traced southeastward from the Mesabi Range, Huronian rocks become thicker and grade into dark gray muddy sandstone and black shale that include both tuff and basaltic lava. All these rocks have been folded, and although they are little altered where they exist northwest of Duluth, they become increasingly metamorphosed southeastward of there.

### The Penokean Orogeny

Area 4 in Figure 19-5 includes several varieties of granite and their metamorphosed derivatives. Although these rocks formed in several different batholiths they have two features in common. All intrude Huronian rocks, and all are about 1.8 billion years old. Radioactive minerals formed during the metamorphism and deformation of sediments in area 3 have a similar age. These facts together indicate that the cycle of sedimentation that produced the Huronian rocks was brought to a close about 1.8 billion years ago by folding, metamorphism, and the intrusion of granite batholiths. As in the cycle that terminated with intrusion of the Algoman granite, this sequence of events undoubtedly produced a range of high mountains. This one, whose roots now trend diagonally northeast-southwest from southern Ontario across Michigan, Wisconsin, and Minnesota, is the **Penokee Range;** the processes that produced it are known collectively as the **Penokean Orogeny.**

## Keweenawan Rocks

Rocks of area 5 in Figure 19-5 are collectively termed the **Keweenawan System.** They rest on the beveled edges of the sedimentary and igneous rocks of the Algoman and Penokee mountains. Clearly, before the great mass of Keweenawan rock could accumulate, the highlands of the Penokee Mountains must have been largely removed by erosion, and the entire area planed down to an essentially featureless surface. This surface, which now separates Keweenawan rocks from older ones, records an interval of time that may have lasted half a billion years.

Sometime after the Penokee Mountains had been worn down, a great elongate trough developed in the Lake Superior region, and in it were deposited the Keweenawan rocks that make up area 5 in Figure 19-5. The sedimentary part of this sequence is mostly conglomerate and sandstone, derived primarily from remnant Penokee highlands southeast of the trough. These rocks are interbedded with thick flows of both felsitic and basaltic lava, some of which contain extensive deposits of metallic copper. Great masses of gabbro and related dark igneous rock invaded the sediments and lavas a little more than a billion years ago. Unlike previous intrusive masses, however, these were more or less parallel to the bedding, and so formed sills and sill-like bodies. Intrusion was not accompanied by great folding, deformation, or metamorphism. Keweenawan strata have not been greatly disturbed by earth movements since their formation, although they have been broadly folded and broken by long faults. There is no evidence that they were ever invaded by batholiths or elevated to mountains.

## Post-Keweenawan Sediments

Although we do not show them on the map, Figure 19-5, more than 17,000 feet of conglomerate, sandstone, and shale were deposited in the southern part of the Lake Superior region after the mild deformation of Keweenawan strata. These rocks are separated from Huronian sediments and Penokean granites by an unconformity, and a similar surface separates them from fossiliferous Upper Cambrian sediments above. Many geologists hold that these sediments are latest Precambrian in age; others believe them to be Lower Cambrian. At present, there seems to be no way of resolving the argument, other than to say that they are older than Late Cambrian and younger than one billion years in the absolute-time scale. Similar rocks, also of indeterminate age, occur beneath Lower Cambrian rocks in northern Europe and Australia. In both the Cordilleran and Appalachian regions of North America, Lower Cambrian strata are underlain by thousands of feet of unfossiliferous sedimentary rock, showing that deposition in these mobile belts began well before the beginning of the Cambrian.

## 19-6
## INFERENCES AS TO PRECAMBRIAN HISTORY OF NORTH AMERICA

With the preceding sketch of the Precambrian geology and history of the Lake Superior region in mind, let us derive certain generalizations about Precambrian history. First, we should remember that in both the Superior and Huronian provinces, tens of thousands of feet of water-laid sedimentary rock, lava, and tuff accumulated in areas that were apparently much longer than wide. Presumably, these areas sank rapidly, so as to permit the accumulation of such great thicknesses of rock. Second, after a prolonged interval of sedimentation, these elongate masses of rock were considerably deformed, to produce chains of folds whose axes trend essentially parallel to province boundaries. Finally, during each episode of folding, batholiths were intruded into the pile of deformed sediments, and uplift followed. Analysis of minerals formed in the granites and sediments at the time of intrusion and deformation helps to fix the time of these episodes in the absolute-time scale.

There is a striking similarity between the events inferred from observations made in the Lake Superior region and those reconstructed from geologic features observed in younger and more completely preserved systems of folded mountains. Each of these younger mountain systems seems to have risen from a mobile belt along a continental margin, through folding and metamorphism of thick piles of sediments and invasion by batholiths. It is certainly not unreasonable, then, to conclude that sedimentary rocks of the Lake Superior region were also deposited in ancient mobile belts that were ultimately deformed by orogeny.

In general, the Slave, Churchill, Central, and Grenville provinces of Figure 19-4 exhibit the same types of rock that characterize those parts of the Superior and Huronian provinces in the Lake Superior region we have just studied. There are unique aspects to the rock record in parts of each of these provinces that bear significantly on details of their history: for instance, the now deformed sedimentary mass in the northeast part of the Grenville Province contains a vast thickness of limestone and dolostone, rocks not encountered in quantity in either the Superior or Huronian provinces. Even so, the history recorded by rocks and structures in each principal province seems to be much the same as that reconstructed in segments of the Superior and Huronian provinces in the Lake Superior region.

All these facts and inferences taken together suggest that the Precambrian history of North America included at least four complete orogenic cycles, during each of which there came into being a long, broad mountain system that was situated somewhat farther from the continental center than the ones that preceded it. In addition, the record in the Lake Superior region and elsewhere indicates that the culmination of each orogenic cycle was followed by a long interval of erosion, during which the mountains were reduced to low-relief surfaces near sea level. The fact that the beveled edges of old mountain ranges are found beneath sediments deposited in younger mobile belts also indicates that newer mobile belts somewhat overlapped the sites of old ones. Indeed, the one now represented in the Churchill Province may have been entirely floored by the beveled folds of the Superior and Slave provinces, which apparently were formed at about the same time.

The trough that received the lavas and sediments of the Keweenawan System had a different history. Rocks in this trough accumulated and were mildly deformed at about the same time that rocks in the Grenville Province formed and were folded. Almost certainly, a mobile belt at this time lay in the Grenville Province and another in western North America: Keweenawan rocks probably accumulated in an elongate basin well within the continent. These rocks, then, provide us with a notion of history in the continental interior during the same interval of time in which rocks were accumulating in the mobile belts along the eastern and western continental margins.

In summary, we may conclude from the history and distribution of North American Precambrian provinces that the continent grew during Precambrian time by the somewhat irregular addition to its margins of progressively younger belts of deformed and intruded sedimentary rock. From Figure 19-4 and our discussion thus far, it appears that growth was the result of four discernible orogenic cycles, terminating about 2.5, 1.8, 1.4, and 1.0 billion years ago.

The general conclusion that North America has grown with time may seem, at first sight, to be contrary to the plate-tectonic view of crustal development. To be sure, a conclusion that North America grew in Precambrian times by marginal accretion of successively younger belts of folded mountains implies that the consuming margins of crustal plates have not always had the same positions they occupy today. However, the hypothesis of continental drift suggests that the outlines of crustal plates have

changed considerably since early Mesozoic time, so it is
not unreasonable to suppose that these outlines may have
changed repeatedly in an interval of time as long as the
Precambrian Eras. Furthermore, because relatively lightweight masses of continental crust resist destruction at
the consuming margins of crustal plates, a picture of
slowly enlarging continents is just the one we might
expect if the crust behaved in the Precambrian as plate-
tectonic theorists hold it behaves today.

The oldest rock in the Superior Province is metamorphosed sediment, which must have been derived from
a source in even older crustal highlands. Recently, 3.6-
billion-year-old crystals of the mineral zircon ($ZrSiO_4$)
have been found in the old metamorphosed sediments
of the Superior Province. These crystals undoubtedly
formed during an episode of intrusion and metamorphism,
so their great age is almost certainly that of the metamorphic and igneous event that produced them. Thus, they
are evidence of a probable orogenic cycle in North
America preceding that in which rocks of the Superior
Province were formed. When rocks of this very old cycle
were eroded to produce sediment for the Superior mobile
belt, the zircon crystals, carrying their ages with them,
were deposited in the younger sediments. Apparently they
were resistant enough to escape the effects of later metamorphism, and have survived as relics of a pre-Superior
orogenic cycle. Sedimentary rocks, now gneisses, in
southwestern Greenland, may represent some of the strata
deposited during that pre-Superior cycle. These rocks,
very recently determined to be 3.95 billion years old, are
the oldest ones yet discovered in the Western Hemisphere.

There is also good evidence for an orogenic cycle
culminating 3 billion years ago or more in both Africa
and Europe, and rocks about 4 billion years old were
found in the former continent in 1971. Thus it appears
that continental growth began in Europe, Africa, and
North America well over 3 billion years ago. Most probably, it began soon after the earth came into being.

**19-7
PRECAMBRIAN
ICE AGES**

North of Lake Huron, Huronian sedimentary rocks are
quite thick and the midportion of the sequence includes
500 feet or more of boulder conglomerate and associated
water-laid sediments. The conglomerate is of special
interest, for not only is it made up of large boulders of
several different kinds of rock, but many of these are
flattened and striated on one or several sides. These facts,
together with other features of the deposit, indicate that
it formed at the margins of a mass of glacier ice. Because
of its thickness and the fact that it is traceable for some
hundreds of miles in the country north of Lake Huron,
this deposit is regarded as the record of a continental

glacier, the oldest one for which we have evidence in North America.

Similar glacial deposits also occur in the very latest Precambrian (or very earliest Cambrian) of South Australia, China, Norway, and France. These and the Huronian deposits indicate at least two episodes of continental glaciation during the Precambrian Eras, an early one between 1.8 and 2.5 billion years ago in North America, and a much later one not much more than 600 million years ago in Australia, Asia, and Europe.

## 19-8 PRECAMBRIAN MINERAL RESOURCES

Because the Precambrian Eras constitute the major portion of geologic time, it is not surprising that Precambrian rocks figure prominently as producers of the mineral resources on which our industrial civilization is based. It will suffice to cite a few examples. Iron has long been the most significant of the Precambrian metallic resources of North America. Iron-rich metasedimentary rocks are mined in the solid-color areas on Figure 19-5. Notably productive among these areas are the Mesabi Range northwest of Duluth, and the Marquette Range far to the east in Michigan. The ore that gave these iron ranges early and lasting fame is mainly hematite, which contains 54 to 61 percent iron as mined. Many millions of tons have been produced. The more recently developed deposits of iron-bearing chert, or **taconite**, are more widespread, but average only 30 to 40 percent iron and must be upgraded before shipping. First the ore is ground very fine, and the iron-bearing minerals (mainly hematite and magnetite) are magnetically separated from the waste. The concentrate from this process is then formed into pellets, which contain 60 to 67 percent iron. More than half the production shipped from the Lake Superior region to the steel furnaces is now derived in this way from taconite ore.

The Lake Superior region has long been an important producer of copper. Once-open spaces in both the lavas and sedimentary rocks of the Keweenawan System are now filled by masses of native copper; and copper sulfide minerals, associated with nickel, cobalt, and platinum, occur in adjacent parts of both the Superior and Huronian provinces. In the latter areas, metallic minerals are found near igneous intrusives that invaded these provinces at about the same time the copper-bearing Keweenawan lavas and sediments accumulated. More than 11 million tons of copper, and about 4 million tons of nickel, have thus far been extracted from Precambrian rocks in the Lake Superior and adjacent districts.

One of the world's largest concentrations of uranium occurs in Precambrian rocks north of the town of Blind River, Ontario, on the north side of Lake Huron. The ore minerals are found filling pore spaces in quartz-pebble conglomerates, which lie at or near the base of the Huronian System. The quartz pebbles were derived from older rocks of the Superior Province, and were probably deposited along the shoreline of a northward-advancing sea. The ore minerals are oxides of uranium, thorium, and rarer elements; pyrite and other sulfides are also present. The uranium-thorium minerals are believed to be of Precambrian age, but whether they were laid down as sediments at the same time as the conglomerate, or were introduced later as precipitates from solution, is a subject of debate among geologists in the district.

Several nonmetallic minerals and rocks of Precambrian age are also used commercially. These include a feldspar-rich igneous rock, a metamorphic rock high in talc, and a quartzite that averages more than 99 percent silica. All these are used in the manufacture of glass, and in ceramic products such as wall tile and porcelain. In addition, each has individual uses, for example, talc in paint and quartzite as a source of elemental silicon.

**19-9
SUMMARY**

Ages of meteorites, and of materials in lunar regolith, suggest that earth and moon originated some 4.6 billion years ago. Their origin probably involved rapid accretion of dustlike particles, followed by compaction, heating, and gravitational rearrangement of original constituents into core, mantle, and crust. Ages of samples from the lunar highlands indicate that the moon had a crust of anorthosite about 4 to 4.6 billion years ago; and rocks about 4 billion years old from Greenland and Africa suggest the earth also acquired a crust in this time interval. Following crust formation, between 3.9 and 3.2 billion years ago, the moon's surface was struck obliquely by giant meteorites, which excavated the mare basins and triggered extrusion of the basaltic lavas that now fill those basins. If similar events marked the earth's early Precambrian history, their records have been largely or completely erased by the action of water, wind, and sea-floor spreading.

The earth's earliest decipherable history, recorded in parts of the continental crust, begins 3.95 to 4 billion years ago. In North America, this history seems to have involved growth of the continent to about its present size through marginal accretion of successively younger belts of folded mountains. Mineral ages in the roots of these old mountain ranges suggest that the Precambrian history of North America involved orogenic cycles that terminated about 3.6, 2.5, 1.8, 1.4, and 1.0 billion years ago. Study of the rock record of two of these cycles in the Lake Superior region indicates they were broadly similar: a probably long interval of sediment deposition and volcanic extrusion was followed by a relatively short period of folding and metamorphism, during which granitic rocks were intruded in the form of large batholiths and long systems of continental mountains were uplifted. That continental glaciation was also a feature of Precambrian history is indicated by glacial deposits more than 1.8 billion years old in North America and by similar rocks only a bit more than 600 million years old in Europe, Asia, and Australia.

Precambrian rocks are prominent producers of iron, copper, uranium and other metallic elements, and also yield important nonmetallic materials such as talc, which are useful in the manufacture of ceramic products.

## SUGGESTED READINGS

Ebbinghausen, E. G. 1971. *Astronomy,* 2nd ed., chap. 2, pp. 14–30. Columbus, O.: Merrill. (Paperback.)
This chapter provides general information on the moon, including recent lunar exploration.

Engel, A. E. J. 1963. Geologic evolution of North America. *Science,* vol. 140, no. 3563, pp. 143–52.
Geologic evidence that the North American continent has grown and differentiated through time.

Gastil, G. 1960. The distribution of mineral dates in time and space. *American Journal of Science,* vol. 258, pp. 1–35.
This paper demonstrates the validity of using igneous and metamorphic mineral dates to test concepts relating to orogenic activity and continental accretion. Major mineral-age provinces have been delineated for the Precambrian using these dates.

Goldich, S. S., and others. 1961. The Precambrian geology and geochronology of Minnesota. *Minnesota Geological Survey Bulletin 41.* 214 pp., maps.
A reference to the Precambrian rocks and history of the Lake Superior region.

Kay, Marshall. 1955. The origin of continents. *Scientific American,* vol. 193, no. 3, pp. 62–66. (Offprint No. 816. San Francisco: Freeman.)
Presents evidence suggesting that land masses grow by cycles in which chains of volcanoes rise from the sea and sediment is washed into the deep troughs along their flanks.

Lowman, P. D. 1970. *The Geologic Evolution of the Moon.* Pub. no. X-644-70-381, Goddard Space Flight Center, Greenbelt, Maryland. 44 pp.
Synthesizes information on the geologic evolution of the moon, through Apollo 11, from three lines of evidence: relative ages of important lunar landforms and rock types; absolute ages of lunar samples; and lunar petrography.

Wood, J. A. 1970. The lunar soil. *Scientific American,* vol. 223, no. 2, pp. 14–23.
Reports on the structure and early history of the moon as revealed by the analysis of a tablespoonful of soil from the sample gathered by *Apollo 11* astronauts at Tranquillity Base.

# 20 ROCKS, FOSSILS, MAPS, AND HISTORY

20-1 Stratigraphic Procedures

20-2 Principal Features of the Cambrian and Later Rock Record

20-3 Summary

In North America, the flat or gently inclined rocks of the Interior Lowlands and Coastal Plains, as well as the deformed strata of the highland border, contain sequences of fossils like those that define the principal divisions of geologic time in Europe. These rocks and their structures are thus inferred to be the record of Paleozoic, Mesozoic, and Cenozoic events in North America; they are clearly younger than the rocks and structures from which we reconstructed Precambrian history, because they overlie or are impressed upon them. Fortunately, they provide us with a history of Cambrian and later time far more comprehensive than we were able to deduce for the Precambrian Eras, because large tracts of these sedimentary rocks are virtually undeformed and contain abundant fossils.

The fact that the rock record of Cambrian and later time is fossiliferous permits us to correlate local histories into continentwide and worldwide accounts. Fossils also permit us to determine the conditions under which the rocks that contain them were deposited. Furthermore, because the Paleozoic, Mesozoic, and Cenozoic eras together represent only the last 570 million years of geologic time, their rock record has been less affected by erosion than that of the Precambrian Eras and is therefore more nearly complete. We might note, by way of comparison, that the preserved Precambrian record is largely one of mountain roots representing successive orogenic cycles, whereas the more widely preserved younger deposits preserve an account of history in both the orogenic and interior regions of the continent.

For these reasons, the rock record of Cambrian and later time can be interpreted by geologic procedures not generally applicable to Precambrian rocks. Because much of what we know of Paleozoic, Mesozoic, and Cenozoic history is read from stratified rocks, the business of interpreting them defines an important subdivision of geology, namely **stratigraphy.**

The methods of reading history from stratified sedimentary rocks are both descriptive and interpretive. Descriptive (or analytic) procedures produce detailed accounts of rocks and their distribution laterally and vertically; interpretive (or synthetic) procedures include identification and correlation, as well as all the methods that enable us to "read" in rocks the sites and conditions of their formation.

## 20-1 STRATIGRAPHIC PROCEDURES

### Descriptive Procedures

In the field, the geologist divides the exposed rock record into formations. He measures the thickness of each formation, and describes the constituent rocks as to color, mineralogy, sedimentary features, and other distinguishing characters. Concurrently with such description (the results of which go into his field notebook for a later report), the geologist constructs a **geologic map.** Using a topographic map or a set of air photographs as a base, he indicates by means of a distinctive color or pattern the area in which each formation forms the bedrock. He shows the attitude of the strata, by means of dip-and-strike symbols; the axes of anticlines and synclines; and any faults that intersect the surface. Geologic maps are usually accompanied by cross sections, on which the arrangement and attitude of rocks beneath the surface are shown graphically. A geologic map and section are shown on Figure 20-1. Study this figure and its caption with care.

Geologic maps and cross sections thus bring together information with respect to the lateral and vertical distribution of rock masses: they constitute models of parts of the earth's crust. As such, they are valuable to the geologist in the interpretation of earth history. They are also useful to engineers in determining the location of tunnels, reservoirs, and highways; to prospectors in searching for mineral deposits; and to urban planners in designating areas for such features as airports, waste-disposal sites, and tracts suitable for heavy construction.

### Interpretive Procedures

The distribution of fossils in sedimentary rocks provides the basis for subdividing the rocks into zones, and into larger units that include several zones. We regard these as the records of intervals of time. The interpretive phase of stratigraphy begins with correlation of rock units from one mapped area to another. The object is to piece together a picture of the distribution of chronologically equivalent rocks and structures for states, regions, and ultimately for the continent as a whole. Much of this work still needs to be done in North America and elsewhere.

A second interpretive procedure involves correlation of at least the major bodies of chronologically equivalent strata with rock sequences that serve as reference standards (or "type sections") for the international geologic

**Figure 20-1**

Geologic map of part of the Greenville quadrangle, Virginia. The bedrock formations are sedimentary rocks of Cambrian and Ordovician ages. The cross section shows the altitude of these rocks along the line A–B on the map. Note the North Mountain fault on map and section. This is a thrust fault, on which Cambrian rocks have been pushed to the left (northwest) over younger rocks. Note also that the Long Glade syncline forms low mountains, whereas the Middlebrook anticline underlies a broad valley. (Modified from U. S. Geological Survey.)

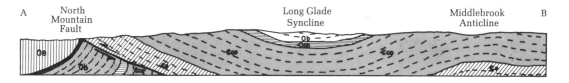

A — North Mountain Fault · Long Glade Syncline · Middlebrook Anticline — B

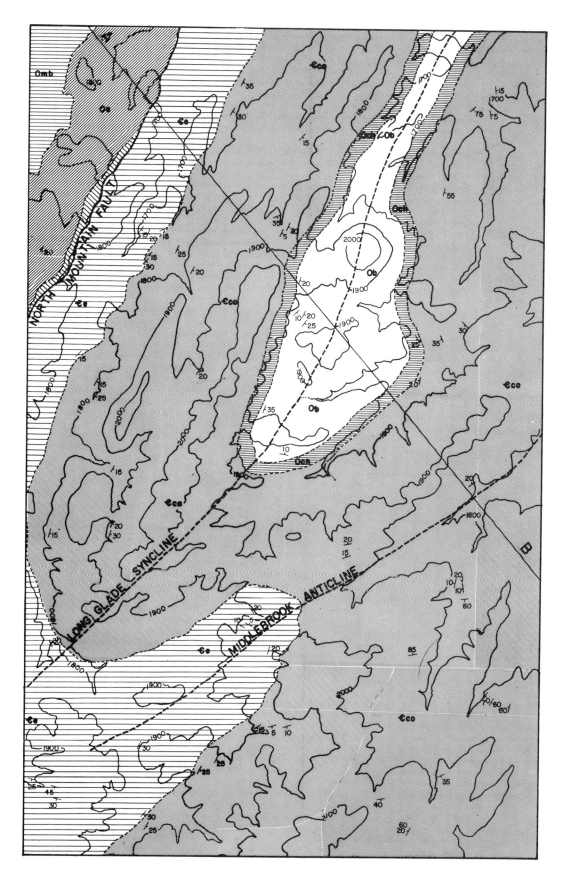

time scale. For example, we recognize that certain North American zones (or sequences of zones) are Cambrian, because they contain fossils like those in the reference standard for the Cambrian System in Wales.

Application of interpretive procedures has had two significant results. The first is that the North American rock record of Cambrian and later time can be divided into **time-rock units.** A time-rock unit is a rock body that was deposited during a specified interval of past time. All rocks, of whatever type, that were formed during a geologic period constitute a time-rock unit called a system; rocks formed during the next shorter interval, the epoch, constitute a series; and those of a still shorter interval, the age, constitute a stage (Table 20-1). Thus all those rocks that formed during the Devonian Period constitute the Devonian System (recall our remarks on dual nomenclature in connection with the geologic time scale, Section 3-5). The time-rock unit with which we will be most concerned is the system.

A second result of applying interpretive procedures has been the compilation of a geologic map of North America. Because of the vast area that must be portrayed, and the tremendous variety of rock types that are present on the continent, it is possible to show on this map only time-rock units. Though the map gives us the distribution and something of the structure of the major time-rock units, of all ages from early Precambrian to the present, it can provide no information on lithology or thickness of these rocks. Data with respect to these aspects of the sedimentary record are contained on geologic maps of hundreds of smaller areas, and in the field notes made by geologists who have measured and mapped the rocks in various parts of the continent.

A geologic map, by its very nature, shows the cumulative rock record of an area as it is today, after a long and perhaps complex geologic history. To reconstruct the step-by-step development of a region, or of the continent, we need to organize the available information in a more systematic form. A convenient scheme for unscrambling the record and presenting it progressively, one part at a time, is embodied in combined lithofacies and isopach maps.

A **lithofacies map** (literally a "rock-aspect" map) of a time-rock unit (a system, for example) shows the areal distribution of the kinds of rock deposited in the time interval chosen (in this case a geologic period). Figure 21-1 is such a map. It shows the distribution of the various types of rock deposited in North America during the Cambrian Period. Note that limestone was the principal rock formed in much of what is now the Interior Low-

| Table 20-1 | |
| --- | --- |
| Divisions of Geologic Time | Rocks Deposited during Time Divisions |
| Period | System |
| Epoch | Series |
| Age | Stage |

lands, but that there is a discontinuous fringe of sandstone and shale around the margins of the Canadian Shield and in the area of the present Rocky Mountains.

Because lithofacies maps show us the present distribution of chronologically equivalent rocks, we can interpret them as showing the past distribution of depositional environments (Section 8-6). If information derived from a study of the fossils is added to lithofacies maps, it can also be determined whether the environments were marine or nonmarine, whether the water was warm and shallow or cold and deep, and whether the lands were dry or humid.

**Isopach maps,** like topographic maps, use contour lines. Instead of joining points of equal surface elevation, however, isopach contours join points at which the rock record is of the same thickness (isopach means "equal thickness"). The heavy numbered lines on Figure 21-1 are isopach contours. They show, for example, that some 14,000 feet of rock accumulated in southern Nevada during the Cambrian Period, whereas 1,500 feet or less was deposited in Ohio during the same interval of time. Because differences in the thickness of sediment deposited from place to place are commonly the results of differences in the amount of subsidence in basins of deposition, we may safely infer that southern Nevada subsided far more during the Cambrian Period than did Ohio and neighboring states of the present Interior Lowlands. In addition, isopach maps show us the shapes of rapidly subsiding areas. The fact that the one in southern Nevada during the Cambrian was elongated in a northeast-southwest direction and received thick deposits of limestone certainly suggests that it was part of a broad continental shelf at that time.

A lithofacies and isopach map such as the one in Figure 21-1 can also be used as the basis for constructing a **paleogeographic map.** Such a map shows, as the name suggests, the distribution of important geographic features at some time in the past. Although Figure 21-1 shows only the present distribution and thickness of Cambrian rocks in North America, we have already suggested that the areas of occurrence of different rock types can be regarded as different depositional environments of the past, and that the distribution of isopachs sheds some light on the amount of subsidence in these places. If we add to this map the environmental information gained from a study of fossils, it is possible to draw another map, like the one in Figure 21-4 A, in which we show our interpretation of these features in terms of land and sea areas. For example, the distribution of Cambrian rocks away from the southern margin of the Canadian Shield certainly suggests the distribution of sediments away from the shores of modern seas; and the relatively slight amount of subsidence of areas adjacent to the sandstone belts shown in Figure 21-1 is like conditions in shallow seas on the continents, such as the Baltic Sea or Hudson Bay. Because all these Cambrian rocks contain fossils of salt-water animals, we may safely infer (as we have done in Figure 21-4 A) that the present Interior Lowlands were part of a sea floor in the Cambrian, and that the shoreline was approximately coincident with the inner edge of the fringe of Cambrian sandstone that surrounds the present Canadian Shield.

## 20-2 PRINCIPAL FEATURES OF THE CAMBRIAN AND LATER ROCK RECORD

In a standard atlas of paleogeographic maps, the rock record of Cambrian and later time in North America is divided into 84 time-rock units, for each of which a map has been prepared in the ways outlined above. Each of these maps is, in effect, a snapshot of the continent as it was at some time in the geologic past. Indeed, from the geologic viewpoint a map of present-day North America, such as we find in an atlas, is merely a snapshot. The continent was different in past time and it will be different in the future.

Careful study of paleogeographic, lithofacies, and isopach maps makes it clear that in Cambrian and later time North America was composed of several persistent structural elements. We show the outlines of these features and the names applied to them in Figure 20-2.

Throughout much of Cambrian and later time, the old Precambrian continent was surrounded by mobile belts, each at some time in its history a complex of continental shelves, island arcs, and sea-floor trenches. From each of these, long, linear systems of folded mountains ultimately developed. The **Appalachian-Ouachita mobile belt,** which extended along the eastern and southern margins of the continent from Nova Scotia into Mexico, was apparently divided through much of its length into an outer volcanic belt and an inner nonvolcanic belt; the latter included the narrow **Wichita-Arbuckle Trough.** The **Cordilleran mobile belt** is recognizable from earliest Cambrian time through the Mesozoic Era. It too was divided into an outer volcanic and an inner nonvolcanic belt. The **Franklin mobile belt,** similarly divided, stretched across the northern margin of the continent through what is now the Canadian Arctic Archipelago.

The central part of the old Precambrian continent, along with its southern extension into Texas, seems to have been above sea level during much of Cambrian and later time. However, a part of the old continent, more or less

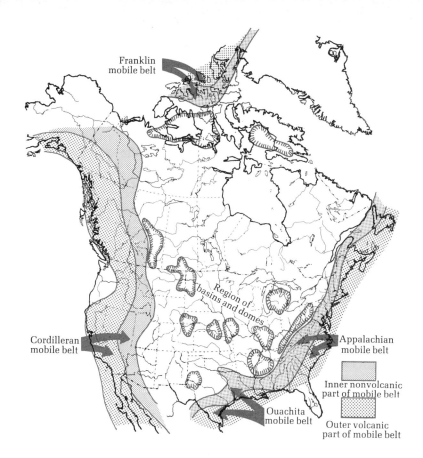

Labels on map:
Franklin mobile belt
Cordilleran mobile belt
Region of basins and domes
Appalachian mobile belt
Inner nonvolcanic part of mobile belt
Outer volcanic part of mobile belt
Ouachita mobile belt

**Figure 20-2**

Principal structural elements of North America in Cambrian and later time. (Base map, Goode Base Map Series, Dept. of Geography, University of Chicago. Copyright by University of Chicago.)

coextensive with the present Interior Lowlands, appears to have subsided slowly and to have been dimpled by broad saucerlike basins. Separating the basins were areas that were higher but still of very subdued relief, commonly termed *domes* or *arches* for want of better words. This gradually subsiding area was flooded repeatedly by shallow seas, in which were deposited the thin sedimentary cover of the Interior Lowlands.

Studied in sequence, from oldest to youngest, the 84 maps in the paleogeographic atlas previously mentioned show another striking feature of North American history in Cambrian and later time. That is, the area of the present Interior Lowlands, and adjacent parts of the Canadian Shield, were repeatedly inundated by shallow seas that moved out onto the continent from the bordering mobile belts. At the end of each inundation, the continent seems to have been broadly uplifted, and the newly deposited rocks exposed to weathering and erosion. Consequently, the rock record of each marine transgression is separated from those of earlier and later transgressions by prominent unconformities, most of which can be traced from one edge of the Interior Lowlands to the other.

All stratigraphic evidence indicates that in latest Precambrian time most of the continent was above water and being eroded, that seas flooded only its margins, and that some of the products of erosion accumulated on these margins as sediments. Beginning in the Cambrian, seas spread slowly toward the center of the old continent; by the end of the Cambrian Period they had flooded some 30 percent of it (see Figure 20-3). After an interval of stability in the Early Ordovician, the continent was broadly uplifted and the sea retreated from the region of the present Interior Lowlands. Rocks that had been deposited in this sea were exposed to weathering and erosion, and a regional erosion surface was developed on them. This surface is indicated by the lowest of the wavy lines on Figure 20-3.

Later intervals of marine inundation and regression gave rise to similar sequences of rock, each of which was laid down on the eroded surface of earlier ones. Subsequent episodes of marine transgression and regression in the continental interior mark the intervals from Middle Ordovician through Silurian; from Devonian through Mississippian; from Pennsylvanian through Early Jurassic; and from Middle Jurassic to the present.

During nearly all the time represented in Figure 20-3, marine sedimentary rocks accumulated in some part of

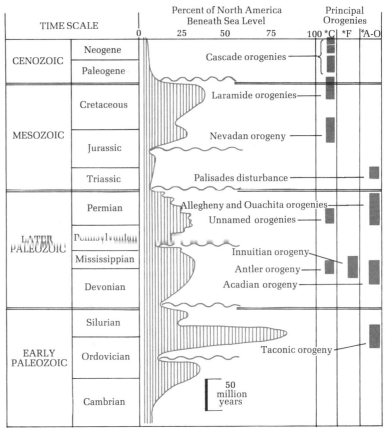

**Figure 20-3**
Chart summarizing chief aspects of the geologic history of North America since the Precambrian. Shaded areas show extent of marine waters on the continent, as indicated in scale across top. Wavy lines are regional unconformities. Orogenies in the continent's three major mobile belts are shown at right.

*C= Cordilleran mobile belt *F=Franklin mobile belt *A–O=Appalachian–Ouachita mobile belt the mobile belts bordering the present Interior Lowlands. From time to time, orogeny and isostatic adjustment produced ranges of folded mountains. These various phases of mountain building along the continental border did not follow the same cyclical development that produced the sedimentary blanket of the Interior Lowlands. Because mountains formed in mobile belts supplied sediment to seas in the present Interior Lowlands, however, sedimentary evidence of major mountain-forming events along the continental border is included in the rock record of the lowlands. Thus the histories of the two areas can be discussed in the same framework with little difficulty. Main events in the mobile belts around North America are summarized along the right side of Figure 20-3, where we also name the various orogenies that contributed to formation of the present highland border. You will want to refer to Figure 20-3 often as you read the next four chapters.

Much of Cambrian and later geologic history is reconstructed from fossiliferous sedimentary rocks. Reading history from these rocks involves both descriptive and interpretive procedures and defines an important subdivision of geology named stratigraphy. Descriptive procedures include the subdivision and description of the rock record in the field and the compilation of geologic maps, which show the distribution and attitude of various types of bedrock and the location of structures such as folds and faults. Interpretive stratigraphic procedures begin with local or regional correlation of zones and of larger units that include several zones. Next, these bodies of chronologically equivalent strata are correlated with rock sequences that constitute reference standards (or type sections) for divisions of the international geologic time scale.

**20-3 SUMMARY**

For time-rock units, which are bodies of rock deposited during specified time intervals, we can construct lithofacies maps and isopach maps. The former show the areal distribution of the kinds of rock deposited in the time interval chosen; the latter show by means of contours the thickness of the time-rock unit from place to place. If information gained from a study of fossils is added to combined lithofacies and isopach maps, we can then prepare paleogeographic maps, which show the distribution of lands, seas, and other geographic features during the time interval chosen.

Study of paleogeographic maps prepared for a sequence of 84 time-rock units indicates that the Paleozoic through Cenozoic history of North America involved five main cycles of marine transgression and regression: those of the Cambrian–Early Ordovician; Middle Ordovician–Silurian; Devonian–Mississippian; Pennsylvanian–Early Jurassic; and Middle Jurassic to present. These events are summarized in Figure 20-3, which also names and indicates the duration of mountain-building episodes in the mobile belts that bordered the continental interior in Cambrian and later time.

## SUGGESTED READINGS

American Association of Petroleum Geologists, *Geological Highway Maps,* available from the American Association of Petroleum Geologists, Tulsa, Oklahoma.
These maps present the geology of various regions in a sufficiently generalized manner to be of value to the traveler with only basic geologic knowledge. Maps currently available, in color, at a scale of 1 inch = 30 miles, include (1) Mid-Continent Region, Kansas-Missouri-Oklahoma-Arkansas (1966); (2) Southern Rocky Mountain Region, Arizona-Colorado-New Mexico-Utah (1967); (3) Pacific Southwest Region, California-Nevada (1968); (4) Mid-Atlantic Region, Kentucky-West Virginia-Virginia-Maryland-Delaware-Tennessee-North Carolina-South Carolina (1970); (5) Northern Rocky Mountain Region, Idaho-Montana-Wyoming (1972). Eleven regional maps are planned for this series.

Eardley, A. J. 1962. *Structural Geology of North America,* 2nd ed., chap. 3, pp. 12–21. New York: Harper & Row.
A résumé of the structural geology of North America, with color maps showing the major tectonic features for most of the periods of geologic time.

Eicher, D. L. 1968. *Geologic Time,* chaps. 3, 4, and 5, pp. 49–116. Englewood Cliffs, N.J.: Prentice-Hall. (Paperback.)
A reference for the beginning student of geology. Includes information on geologic maps, lithofacies maps, paleogeographic maps, and determination of stratigraphic units.

Harbaugh, J. W. 1968. *Stratigraphy and Geologic Time,* Chaps. 4 and 5, pp. 40–75. Dubuque, Iowa: W. C. Brown. (Paperback)
Chapters 4 and 5 deal with isopach maps, paleogeographic maps, and sedimentary facies.

Schuchert, Charles. 1955. *Atlas of Paleogeographic Maps of North America.* New York: Wiley.
An atlas of paleogeographic maps, one for each of 84 time-rock units of Cambrian and later ages in North America.

Sloss, L. L. 1963. Sequences in the cratonic interior of North America. *Geological Society of America Bulletin,* vol. 74, pp. 93–113.
Discusses the concept of major rock-stratigraphic sequences as delimited by major interregional unconformities in the continental interior of North America.

Sloss, L. L., Dapples, E. C., and Krumbein, W. C. 1960. *Lithofacies Maps, An Atlas of the United States and Southern Canada.* New York: Wiley. 108 pp.

An atlas in which each of the Paleozoic and Mesozoic systems is represented by a continentwide isopach-lithofacies map, in addition to regional series and formation maps and tables of formations keyed to the maps.

# 21 ROCKS AND PHYSICAL HISTORY OF EARLY PALEOZOIC TIME

21-1 Cambrian and Early Ordovician

21-2 Middle Ordovician Through Silurian

21-3 Summary

Sedimentary rocks containing fossils of Cambrian, Or-dovician, and Silurian ages are widespread in North America. In the Interior Lowlands and in parts of the adjacent highland borders, these strata are readily divisi-ble into two thick sequences, each the record of an episode of marine transgression over the old Precambrian conti-nent. Because each of these two sequences is set off from the rocks below and above by regional unconformities, and thus represents an identifiable interval of geologic time, it is convenient to discuss them separately. Thus we will first look at Cambrian and Lower Ordovician rocks and the history that we can derive from them, and then at Middle and Late Ordovician and Silurian rocks and history.

In the Interior Lowlands and the ranges of the highland border, a thick sequence of sedimentary rocks, ranging in age from late Precambrian to Early Ordovician, is sharply set off from older Precambrian rocks beneath and from Mid-Ordovician and younger rocks above by uncon-formities that can be traced for almost the complete breadth of the continent. The lithologic character and thickness of the Cambrian part of this rock body are shown in Figure 21-1. It is from a study of these rocks that we reconstruct the earliest Paleozoic events in North America.

## 21-1
## CAMBRIAN AND
## EARLY ORDOVICIAN

**The Rock Record**

Basal Cambrian rocks are an extensive blanket of con-glomerate and quartz sandstone. In the southern parts of the Cordilleran and Appalachian systems, where this blanket is thickest, it is unfossiliferous. However, fossils diagnostic of successively younger Cambrian zones are found in these basal strata as they are traced toward the present margin of the Canadian Shield. In Wisconsin, Minnesota, and Iowa, for example, they yield fossils of Late Cambrian age. This situation is shown graphically in Figure 21-2, which indicates the boundaries of both the formations and the principal time-rock units (series) into which the Cambrian and Lower Ordovician rocks are divided in the eastern United States.

Above the basal blanket of sandstone and conglomerate, the rock record of the Mid-Cambrian to Early Ordovician is largely shale, limestone, and dolostone. All these rocks contain fossils of marine animals, and as is clear from Figure 21-2, all of them become younger, and the sequence as a whole becomes thinner, from the continental border toward the Interior Lowlands. A similar situation exists in the western United States, as we noted in Chapter 10 (Figure 10-5).

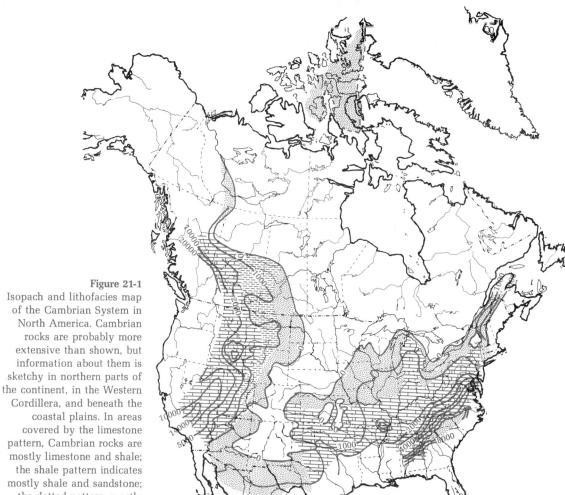

The unconformity beneath Cambrian rocks is remarkably featureless in most places in which it is exposed (Figure 21-3); in general, it separates basal Paleozoic sedimentary rocks from igneous and metamorphic rocks of Precambrian age. Lower Ordovician rocks are set off sharply from the rock records of later depositional cycles by a disconformity that is recognizable throughout the Interior Lowlands and in much of the inner, nonvolcanic

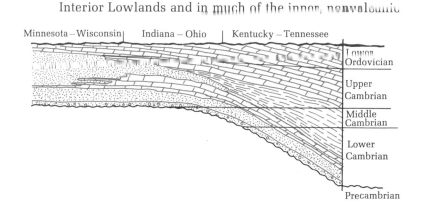

**Figure 21-3**
Base of the Cambrian System in central
Colorado. Light-colored Upper
Cambrian sandstone overlies
coarse-grained Precambrian granite on
a nonconformity that is remarkably
smooth and even. Width of view is
about 10 feet. (Photo by L. Ogden.)

parts of the highland border. In the north central United
States, there is as much as 150 feet of relief on this old
erosion surface; in other places, it is much more subdued.

## History from Rocks

### Late Precambrian Erosion

The generally smooth surface of the unconformity
separating Cambrian from older rocks suggests that prolonged Precambrian erosion had reduced much of North
America to low-lying plains before the beginning of the
Paleozoic. An idea of the length of time involved in
formation of this old erosion surface may be gained by
recalling that Cambrian and Lower Ordovician sedimentary strata rest on the truncated roots of several ancient
Precambrian mountain systems. Presumably these were
largely removed before Paleozoic sedimentation began.

### Marine Transgression

Because the formations into which we divide Cambrian
and Lower Ordovician marine rocks become younger toward the center of the continent, it is clear that the sea-
floor environments in which they accumulated migrated gradually inward onto the old Precambrian continent.

As the seas slowly flooded toward the continental center from Cambrian to Early Ordovician time, the detritus delivered to them by streams was sifted, winnowed, and spread out on the nearshore sea floor as sand and mud. As broader areas of the sea bottom came to be farther and farther from the advancing shore, limestones and dolostones, formed largely of the fragmented shells of animals and the calcareous products of plants, came to dominate the sedimentary record. Because coarse clastic rocks formed near the continental margins during only the earliest phases of this history, when the coastline was nearby, it seems probable that all the clastic material was derived from the continental center. Little, if any, came from sources outside the continental margins.

Figure 21-4 is a series of paleogeographic maps showing the inferred distribution of lands and seas at three stages in the Cambrian and in the Early Ordovician. These maps show a gradual submergence of North America, so that more than 30 percent of the present land area of the continent eventually lay beneath the sea. An even wider distribution of the seas is suggested, however, by the fact that Upper Cambrian and Lower Ordovician rocks are mostly limestone and dolostone that are practically de-

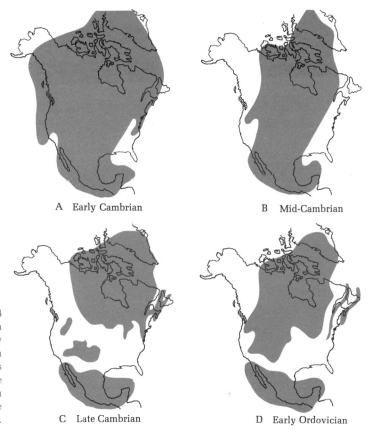

A  Early Cambrian                    B  Mid-Cambrian

**Figure 21-4**
Paleogeographic maps of North America in the Cambrian and Early Ordovician. The rock record indicates a progressive inundation of land areas (colored) by the sea during the Cambrian Period. Larger lands shown in map D suggest up-arching of the continent in the Early Ordovician.

C  Late Cambrian                    D  Early Ordovician

void of land-derived clastic materials. This may indicate that the continental center, heretofore the chief supplier of clastic sediment, was actually submerged entirely at the climax of inundation. Our maps, based on the present distribution of Cambrian and Lower Ordovician rocks, may be too conservative, for much of the record has probably been removed by erosion since the Early Ordovician.

### Uplift and Marine Regression

After a long interval of limestone deposition in widespread warm seas, the continent was broadly up-arched and great areas of former sea bottom were brought above sea level. The presence of thin layers of sandstone intercalated in the limestones and dolostones that form the youngest of the Lower Ordovician rocks suggests that uplift was gradual and that the shoreline migrated slowly outward toward the continental borders. Newly exposed strata were then subjected to erosion, large areas were undoubtedly denuded by stream erosion, and in many places the surface of this oldest sequence of Paleozoic strata was deeply dissected by streams.

### The Continental Margins.

As indicated in Figure 21-1, Cambrian rocks are nearly ten times thicker in elongate areas now coincident with parts of the Cordilleran and Appalachian systems than in the Interior Lowlands. From this we infer that the eastern and western borders subsided much more rapidly during the Cambrian than the interior areas of the continent did. Figure 21-1 also shows that sediments deposited along these mobile margins are now largely limestone and sandstone, and all the clastic material in them came from the continental center. Thus we infer that through the Cambrian and Lower Ordovician the Appalachian and Cordilleran mobile belts were similar to the present-day continental shelf along the eastern margin of North America, although the seas that covered them extended much farther inland than is now the case. The evidence is less clear in the present sites of the Ouachita-Marathon and Innuitian systems. Extensive subsidence probably began later in both of those areas.

Fossils of Mid-Ordovician through latest Silurian or earliest Devonian age are found in a widespread blanket of sedimentary rock that overlies the dissected upper surface of Lower Ordovician or older rocks and is separated from

**21-2
MIDDLE ORDOVICIAN
THROUGH SILURIAN**

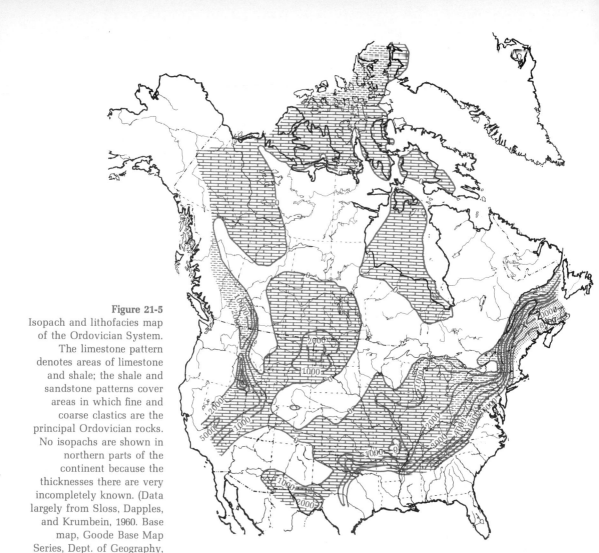

**Figure 21-5**
Isopach and lithofacies map
of the Ordovician System.
The limestone pattern
denotes areas of limestone
and shale; the shale and
sandstone patterns cover
areas in which fine and
coarse clastics are the
principal Ordovician rocks.
No isopachs are shown in
northern parts of the
continent because the
thicknesses there are very
incompletely known. (Data
largely from Sloss, Dapples,
and Krumbein, 1960. Base
map, Goode Base Map
Series, Dept. of Geography,
University of Chicago.
Copyright by University of
Chicago.)

Devonian and younger strata by an extensive unconformity. Collectively, these rocks form a sequence from which we reconstruct the second important episode in the Paleozoic history of North America. The thickness and lithologic character of that sequence can be determined from the maps in Figures 21-5 and 21-6, which are isopach and lithofacies maps of North American Ordovician and Silurian rocks, respectively. The extent and duration of this part of Paleozoic history are shown in Figure 20-3.

### The Rock Record

In many respects, the Mid-Ordovician to Silurian rock record is similar to that of the Cambrian and Lower Ordovician. Over wide areas of the Interior Lowlands, basal strata of the younger sequence are clean, loosely cemented quartz sandstones, which grade upward and laterally into thick shales, limestones, and dolostones.

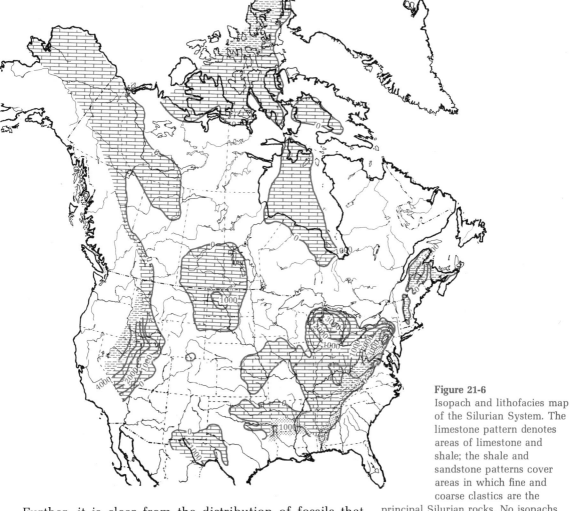

**Figure 21-6**
Isopach and lithofacies map
of the Silurian System. The
limestone pattern denotes
areas of limestone and
shale; the shale and
sandstone patterns cover
areas in which fine and
coarse clastics are the
principal Silurian rocks. No isopachs
are shown in northern parts of the
continent because the thicknesses there
are very incompletely known. (Data
largely from Sloss, Dapples, and
Krumbein, 1960. Base map, Goode Base
Map Series, Dept. of Geography,
University of Chicago. Copyright by
University of Chicago.)

Further, it is clear from the distribution of fossils that
all these rocks are somewhat younger toward the continental interior than in marginal areas of the continent.

Three features, however, distinguish the Mid-Ordovician to Silurian record and indicate that it represents a
somewhat more complex history than that which had gone
before. First, the isopachs of Figures 21-5 and 21-6 form
an irregular pattern, showing that Ordovician and Silurian
rocks of the continental interior were deposited in an
environment in which subsidence rates were far less
uniform from place to place than during the Cambrian
Period. Second, a great eastward-thickening mass of clastic sediments, the Queenston Delta, is part of the rock
record in eastern North America. It indicates the emergence of a mountainous source of sediments along the
eastern margin of the continent. Finally, Upper Silurian
rocks include considerable thicknesses of anhydrite and
salt. These rocks, which are not common in the earlier
record, suggest a widespread hot, dry climate.

## Basins, Domes, and Arches

Two nearly circular areas are especially well outlined by isopach lines in Figures 21-5 and 21-6. One of these, centering in northwest North Dakota, is the **Williston Basin;** the other, occupying the area between Lakes Michigan and Huron, is the **Michigan Basin.** Apparently both were areas of accelerated subsidence during the Ordovician and Silurian.

The Michigan Basin was separated from the more rapidly subsiding Appalachian mobile belt to the east by a broad strip of less rapidly sinking sea bottom that extended far to the northeast and southwest from the present site of Cincinnati, Ohio. This shoal, which shows especially well in Figure 21-6, is the **Cincinnati Arch.** The Williston Basin was apparently separated from the Cordilleran mobile belt by a strip of more slowly subsiding crust in northwestern Wyoming and adjacent Montana.

The areas occupied by the modern Ozark and Adirondack mountains apparently subsided much less rapidly than surrounding parts of the continental interior. Indeed the thinness or absence of sedimentary rocks over them, and the presence of sand and many unconformities in rocks adjacent to them, indicate that they were actually exposed to erosion through much of Ordovician and Silurian time. Because of their more or less circular shape, we speak of the **Ozark dome** and the **Adirondack dome.**

The basins, arches, and domes we have mentioned, together with others of less importance, greatly affected the environment in which the sediments of the second important sequence accumulated. West of the Cincinnati Arch, however, Ordovician and Silurian rocks above the basal sandstones of the sequence are largely limestones and dolostones, and include only minor thicknesses of shale. For the most part, these rocks are abundantly fossiliferous, and Silurian parts of the sequence contain broad mound-shaped reefs formed on the ancient sea floor by algae and corals.

## The Queenston Delta

East of the Cincinnati Arch, the lower and upper parts of the Mid-Ordovician through Silurian sequence are calcareous rocks similar to equivalent strata to the west but considerably thicker. The midportion of the sequence includes thick beds of gray, green, red, and black shale that grade eastward into even thicker deposits of reddish sandstone and conglomerate. As a whole, these rocks form a wedge-shaped mass of clastic material that becomes both thicker and coarser toward the eastern border of the continent. In map view (Figure 21-7 C) this wedge of coarse-grained sedimentary rock resembles a large complex of river-mouth deltas. It is termed the **Queenston Delta,** because some of the beds are exposed near Queenston, Ontario.

Because Queenston sediments become thicker and coarser toward the east, and because fossiliferous marine sediments along the western margin of the delta grade eastward into unfossiliferous red rocks typical of terrestrial deposits, they furnish good evidence that all the rocks of the delta were derived from source areas along or beyond the eastern margin of the continent. It is also evident from the distribution of coarse-grained rocks, as shown in Figures 21-5 and 21-6, that at one time or another this high source area stretched from Nova Scotia south to Georgia, a distance of at least 1,500 miles.

### Evaporites

Thick, persistent beds of rock salt characterize the Upper Silurian section in the Michigan Basin, and extend eastward across Ontario and Ohio to Pennsylvania and New York. Thin beds of anhydrite are found in Middle Silurian rocks of the Williston Basin. Evaporites like these are commonly interpreted as precipitates from an arm of the sea that became cut off from open-ocean circulation behind a barrier reef or bar, and slowly evaporated in a desert climate. The occurrence of thick beds of rock salt in the Michigan Basin indicates that in Late Silurian time the basin was alternately flooded by the sea and evaporated nearly to dryness. Almost certainly, the climate was hot and arid.

### Boundaries and Extent of the Sequence

Over much of its extent, the unconformity at the base of the Middle Ordovician separates rocks of that age from Lower Ordovician strata of the earlier sequence. In extensive areas around the southern and northern margins of the Canadian Shield, however, and in a large tract beneath and southwest of Hudson Bay, Middle Ordovician through Silurian strata rest directly on Precambrian igneous and metamorphic rocks.

Throughout the Interior Lowlands, and in areas marginal to them, Silurian rocks are separated from those of younger age by a prominent unconformity. Presumably this indicates that in the Late Silurian or earliest Devonian there was broad continental uplift, restriction of the seas, and widespread erosion.

The maps in Figures 21-5 and 21-6 suggest that both Ordovician and Silurian rocks were at one time more extensive than now. Limestones of both ages overlap Precambrian rocks of the Canadian Shield or terminate abruptly against them; eastern and western segments are separated by a broad southwestern extension of the shield; Silurian marine strata of the Williston Basin are no longer connected with marine rocks of the same age elsewhere; and isolated patches of both Ordovician and Silurian marine rocks occur near the very center of the continent. These facts and others suggest that much of the sequence has been removed by erosion since the close of the Silurian; the continent may once have been almost completely blanketed by rocks of Ordovician and Silurian ages.

## History from Rocks

### Marine Transgression and Regression

Paleogeographic maps showing the probable distribution of land and sea in North America at four stages of Ordovician and Silurian time are included in Figure 21-7. Like the four maps depicting a similar but earlier history (Figure 21-4), these show an inundation of the central part

Early Mid-Ordovician     B  Later Mid-Ordovician

**Figure 21-7**
Paleogeographic maps of North America at four times from the early Middle Ordovician to Middle Silurian. Like the maps in Figure 21-4, these show a progressive inundation of the continent, culminating in the Late Ordovician (map C). The larger areas of land in map D indicate continental uplift in the Silurian. The pattern of dots in map C shows the approximate extent of the Queenston Delta.

C  Late Ordovician     D  Mid-Silurian

of the continent by shallow seas. Figure 21-7 C indicates that, at the climax of the transgression in the Late Ordovician, at least three-fourths of the present land area of North America was submerged beneath the sea. The larger land areas and more limited seas of Figure 21-7 D indicate gradual emergence of the continent; and the presence of thick evaporites in Upper Silurian rocks suggests that during this emergence an arm of the sea in and near the Michigan Basin was repeatedly cut off from open-ocean circulation and evaporated beneath a hot sun and dry winds.

## The Continental Margins

The pattern of isopachs in Figures 21-5 and 21-6 indicates that there was prolonged subsidence in both the Appalachian and Cordilleran mobile belts in Mid-Ordovician through Silurian time. They also indicate the presence by Ordovician time of a mobile belt in the Ouachita-Marathon region of southern United States, and recent studies in the Canadian Arctic confirm the existence of the Franklin mobile belt there during the same interval.

Thick dark shales, sandstones, conglomerates, and volcanic rocks accumulated to great thickness in the outer parts of all four mobile belts, and thousands of feet of limestone and shale were deposited in their inner segments. Sometime in the Ordovician, a sea-floor trench apparently developed along the east edge of the Appalachian mobile belt. By Middle Ordovician time, a linear welt, probably surmounted by a volcanic island arc, had risen along the western border of the trench (Figure 16-3). As the welt rose, some of the Cambrian and Ordovician sediments and volcanic rocks on its flanks became unstable and crumpled, and eventually slid westward onto shelf sediments of the inner belt. In Late Ordovician and Silurian time, the welt and the mass of crumpled rocks along its western margin were prominent sources for the coarse sediments that spread westward to form the Queenston Delta and associated deposits.

Although information of a more direct nature exists in the complex geology of Newfoundland, at the northern end of the Appalachian System, the Ordovician and Silurian events just recounted are reconstructed largely from indirect evidence along most of the length of the mountain chain. The welt we infer as an Ordovician and Silurian feature of the outer part of the Appalachian mobile belt has long since disappeared as a result of erosion and burial beneath younger rocks of the present-day coastal plain and continental shelf. Sedimentary and volcanic rocks that slid westward from the welt, however,

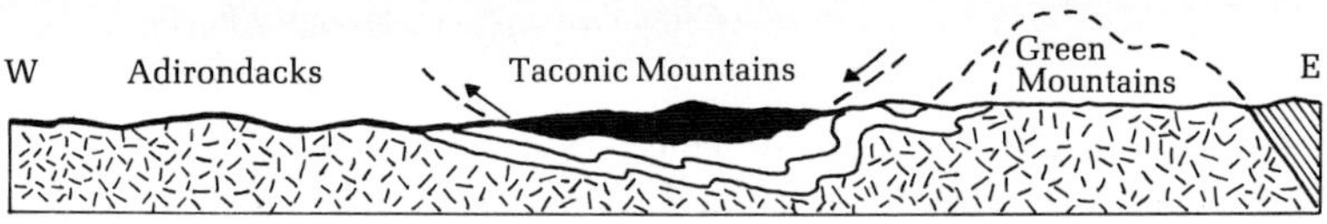

**Figure 21-8**
Diagrammatic cross section showing the relationship of the Taconic Mountains to the Adirondacks and the Green Mountains. The broken-line pattern indicates Precambrian igneous and metamorphic rocks; Cambrian and Ordovician shales and limestones are shown in white. The black mass, separated from the rocks below by a thrust fault, includes Cambrian and Ordovician slates and sandstones, which have been thrust westward to form the Taconics. A small patch of Cambrian and Ordovician schists and lavas is at the far east end of the section. (Adapted from Billings, Thompson, and Rodgers, 1952.)

are to be seen in the Taconic Mountains of eastern New York, western Massachusetts, and Vermont.

The present-day Taconics, shown diagrammatically in Figure 21-8, consist of greatly deformed dark slates and sandy rocks of Cambrian through Mid-Ordovician age, which rest on less distorted Cambrian through Mid-Ordovician sandstone, shale, and limestone. Because these mountains represent some of the best evidence for the initiation of Paleozoic mountain building along the eastern border of North America, this phase in the history of the Appalachian mobile belt is termed the **Taconic Orogeny.** Its approximate duration in time is shown in Figure 20-3.

## Climatic Zones

Thick limestones and dolostones of Cambrian, Ordovician, and Silurian ages occur as far north as Ellesmere Island and northwestern Greenland. In many places these rocks are crowded with coral and algal reefs, and with thick-shelled fossils of animals whose nearest modern relatives live in the shallow warm waters of the tropics. Thus it appears that during the Early Paleozoic many areas now far from the tropics were overspread by tropical seas favorable to the growth and proliferation of warm-water plants and animals and to the deposition of limestone in great abundance.

There are at least two ways to account for this apparently anomalous distribution of warm-water fossils and rocks in northern latitudes. Some geologists suggest that, during the Early Paleozoic, climatic zones like those of the modern world either had not been established or were only poorly defined. Seas were widespread and land areas were of limited extent in much of the northern hemisphere during this long interval of time, and warm currents may have extended farther north than now. If they did so, they could have removed some of the sharper distinctions that now exist among the several climatic zones.

To other geologists, the explanation just advanced seems unlikely. They point out that if areas of apparently tropical rocks and fossils (and similar areas containing fossils of temperate and cold-water animals) are arrayed in belts parallel to the equator, as in Figure 21-9, the south pole comes to lie in western Africa, about on the present equator! This explanation requires the continents in the

Early Paleozoic to have had a relationship to the poles much different from the present one.

Although the second explanation requires great rearrangement of continents with respect to the poles of rotation, it finds support from two unrelated lines of evidence. The first is the probable existence of an Ordovician ice sheet in the Sahara region of Africa (Section 12-6). The fact that the ice seems to have moved north/northwest suggests that it might have been moving outward from an ice cap on the south-polar landmass postulated in Figure 21-9.

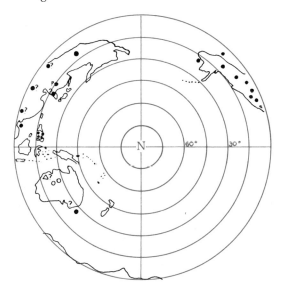 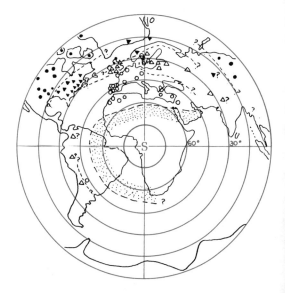

The second line of evidence is the nature of the earth's magnetic field in the Early Paleozoic. Using the iron-bearing minerals that are locked into crustal rocks as fossil compass needles, geophysicists have been able to determine the positions of the early Paleozoic magnetic poles. Interestingly enough, these poles seem to have been about where the data summarized in Figure 21-9 place the rotational poles.

Because two such different lines of reasoning lead to closely comparable results, it seems plausible that the Early Paleozoic earth was characterized by climatic belts like those of today, but that since Silurian time North America (and other continents) have in some way come to have a very different relation to the poles. The sea-floor spreading hypothesis provides an attractive mechanism for wholesale migration of continents, and helps us to understand how the frigid wastes of the Arctic may once have been tropical, and equatorial Africa may have been part of a polar continent like Antarctica.

**Figure 21-9**
The Northern (left) and Southern hemispheres during Ordovician time, based on the distribution of marine fossils in Ordovician rocks. At localities represented by black dots, fossils of warm-water animals have been collected; at black triangles, warm-temperate faunas; at open triangles, cold-temperate faunas; and at open circles, cold-water faunas. The dotted pattern on the Southern Hemisphere map indicates a land mass of Ordovician time. Note that continents of the Southern Hemisphere are shown to be closer together than they are now by some 40 degrees of longitude. Many geologists believe that they have drifted apart as a result of sea-floor spreading since the Paleozoic, and that the mid-ocean ridges (Chapter 17) represent the lines along which this separation has taken place. (From Spjeldnaes, 1961.)

**21-3 SUMMARY**

During the first third of the Paleozoic Era, accelerated subsidence of the eastern and western borders of the continent formed the Appalachian and Cordilleran mobile belts, and at least 30 percent of the present land area of North America was submerged beneath warm shallow seas that flooded slowly inward from its margins. The interval of flooding was brought to a close in the Early Ordovician by broad uplift of the continent and gradual retreat of the seas toward the continental margins. Sediments deposited along the mobile margins of the continent came from the continental center, indicating that through the Cambrian and Lower Ordovician the Appalachian and Cordilleran mobile belts were similar to the present-day continental shelf along the eastern margin of North America.

The Mid-Ordovician-to-Silurian rock record is similar to that of the Cambrian and Lower Ordovician, from which it is separated by a prominent unconformity. These rocks record a gradual inundation of the continental center by shallow seas that ultimately flooded at least three-fourths of the present land area of North America in the Late Ordovician. In the Late Silurian, seas slowly retreated and evaporites accumulated in an arm of the sea in and near the Michigan Basin, which was cut off from open-ocean circulation.

By Mid-Ordovician time, mobile belts ringed the continental center and thick deposits in them indicate prolonged subsidence from Mid-Ordovician through Silurian time. Deformation in the outer part of the Appalachian mobile belt raised a welt, from the flanks of which masses of Cambrian and Ordovician rocks slid westward. Highlands formed at this time, in the Taconic Orogeny, were the source during later Ordovician and Silurian time of great quantities of clastic sediments, which were swept westward into the continental interior to form the Queenston Delta.

In the Early Paleozoic, continents may have had quite a different relation to the poles from their present one. Paleogeographic and paleomagnetic studies suggest that Africa may have been a polar continent, and glacial deposits of Ordovician age in the Sahara tend to confirm this suggestion. North America, on the other hand, may have been submerged beneath tropical seas as far north as present-day Ellesmere Island.

## SUGGESTED READINGS

Schuchert, Charles. 1955. *Atlas of Paleogeographic Maps of North America,* maps 1–30. New York: Wiley.
   Thirty paleogeographic maps for Cambrian through Silurian time, with a list of major stratigraphic units represented on each map.

Sloss, L. L., Dapples, E. C., and Krumbein, W. C. 1960. *Lithofacies Maps, An Atlas of the United States and Southern Canada,* pp. 1–13. New York: Wiley.
   Isopach-lithofacies maps for Cambrian through Silurian time.

# 22 ROCKS AND PHYSICAL HISTORY OF LATER PALEOZOIC TIME

22-1   Devonian Through Mississippian
22-2   The Transcontinental Arch
22-3   Pennsylvanian and Permian
22-4   Summary

Rocks of Devonian through Permian age differ greatly in type and thickness from place to place, and document an increasingly restless 175 million years in the history of North America. These strata, like the Early Paleozoic rocks, consist of two distinct sequences, each recording an episode of marine transgression terminated by broad uplift and erosion of the continent.

In much of North America, rocks containing Devonian fossils are separated from older strata by a disconformity. Devonian and Mississippian rocks, on the other hand, grade into one another in most places in the continental interior; and Mississippian strata are separated nearly everywhere by a well-defined unconformity from strata bearing Pennsylvanian fossils. Clearly then, the Devonian and Mississippian periods were a time of virtually continuous deposition of sediment in much of North America, and their history is best summarized as a unit.

## 22-1 DEVONIAN THROUGH MISSISSIPPIAN

### The Rock Record

*The Continental Interior*

Comparison of the maps of Figures 22-1 and 22-2 makes it plain that limestone was the chief rock deposited in Devonian and Mississippian seas of the continental interior. In much of the eastern United States, this limestone rests on a basal unit of sandstone, and throughout most of the interior it is divided into two parts by a thin black-shale formation.

The lower limestone division, mostly less than 500 feet thick, is Devonian in age. It contains fossils in great abundance, and numerous coral reefs. Beneath the plains of Alberta, in western Canada, oil and gas are obtained from Devonian reefs that grew in a shallow sea marginal to the Canadian Shield. To the southeast, in what is now southern Saskatchewan, an unusually thick Middle Devonian section accumulated in the Williston Basin. Here these rocks consist largely of limestone, dolomite, and thick beds of salt. Some of the salt beds contain potassium-bearing minerals, which were produced by the almost complete drying up of marine waters and thus indicate a Middle Devonian climate of extreme aridity in this region. The potassium salts are of great value in fertilizers and are extensively mined.

In most places in the continental interior, the black shale that separates the lower and upper limestones of the Devonian-Mississippian sequence is less than 100 feet thick, and is rich in carbon and radioactive matter. Fossils, which indicate that the lower part of this formation is Devonian and the upper part Mississippian, are largely of floating plants and swimming marine animals; the

**Figure 22-1**
Isopach and lithofacies map of the Devonian System in North America. In areas covered by the limestone pattern, Devonian rocks are mostly limestone; the shale pattern indicates mostly shale and sandstone; the dotted pattern, mostly sandstone. (Data largely from Sloss, Dapples, and Krumbein, 1960. Base map, Goode Base Map Series, Dept. of Geography, University of Chicago. Copyright by University of Chicago.)

remains of bottom-living organisms are rare. These features suggest that this shale formed in a sea whose bottom waters were deficient in oxygen. How stagnancy of this sort could have been maintained for so long over such a wide area is a question that has not been satisfactorily answered.

The upper limestone of the sequence, which is crowded with Mississippian fossils, is similar to the lower, or Devonian, unit. It is somewhat more widespread, however (Figure 22-2), and its upper part contains thin sandstone beds shown by the attitude of their cross-stratification to have been transported into the lowlands from the northeast. Corals are numerous in these limestones, but even more conspicuous are fragmented skeletons of the plantlike echinoderms, the crinoids, and their now extinct kin, the blastoids. Indeed, remains of these animals are so numerous in places that the rocks are commonly termed *crinoidal limestone*.

**Figure 22-2**
Isopach and lithofacies map
of the Mississippian System
in North America. In areas
covered by the limestone
pattern, Mississippian rocks
are mostly limestone; the
shale pattern indicates
mostly shale and sandstone; the dotted
pattern, mostly sandstone. (Data largely
from Sloss, Dapples, and Krumbein,
1960. Base map, Goode Base Map
Series, Dept. of Geography, University
of Chicago. Copyright by University of
Chicago.)

*The Catskill Delta*

Figure 22-1 shows that east of the Cincinnati Arch Devonian rocks are largely shales and sandstones that reach a thickness of more than 12,000 feet in east-central Pennsylvania. This belt of thick, dominantly clastic sediment stretches southward from the Gaspé Peninsula of Quebec to eastern Tennessee, a distance of nearly 1,400 miles. Because stream erosion in the Catskill Mountains of southeastern New York has exposed the red sandstones and conglomerates of this belt to view, the name **Catskill Delta** is applied to the entire mass of clastic sediment.

Figure 22-2 indicates that essentially the same pattern of sedimentation was maintained through the Mississippian Period in eastern North America. However, the thickness of sedimentary rock that accumulated in the Catskill Delta region during Mississippian time was considerably less than during the Devonian.

389

The rocks along the western margin of the Catskill Delta contain fossils of marine organisms; eastward these marine strata grade into thick coarser-grained beds that are either barren of fossils or contain the remains of land plants and freshwater animals. This great eastward-thickening wedge of clastic material was clearly derived from source areas east of the present Interior Lowlands.

### Western North America

During the Devonian Period, limestone accumulated in the inner part of the Cordilleran mobile belt, and great thicknesses of chert, shale, sandstone, and volcanic rocks were deposited in the outer part. In eastern Nevada and adjacent Idaho, Mississippian limestones and shales grade into westward-thickening sandstone and conglomerate as much as 3,000 feet thick. These form a wedge-shaped deposit that greatly resembles the Catskill Delta of eastern North America. However, rock in the western wedge is largely Mississippian in age, whereas the thickest part of the Catskill Delta is of Devonian age.

### Northern North America

The Franklin mobile belt, in what is now Arctic Canada, subsided greatly during the Devonian Period. More than 17,000 feet of sedimentary rock was deposited in southern Ellesmere Island, and by the end of the Devonian at least 60,000 feet of Lower Paleozoic sediment had accumulated in the outer part of the mobile belt. All these strata are now complexly deformed, and folded rocks of Cambrian through Devonian age are overlain by nearly horizontal Pennsylvanian and younger sediments.

## History from Rocks

In most places, the disconformity that separates the Devonian-Mississippian sequence from older rocks is a subdued surface that has little relief. This, together with the fact that pre-Devonian strata were deeply eroded, or were removed completely from many areas where they must once have existed (compare Figures 21-6 and 22-1), suggests an interval of vigorous erosion before the start of Devonian deposition.

### Marine Invasion of the Interior

If rocks of this sequence are divided into four parts and these displayed on maps in terms of the environments in which they were deposited, the series of paleogeographic maps in Figure 22-3 results. Studied in sequence, these maps show that during the Devonian and Early Mississippian periods, the interior of the continent was gradually flooded by a sea that reached its maximum extent in the Early Mississippian. Because fossiliferous limestone containing many coral reefs was the chief rock deposited in this sea, we assume that it was warm and shallow, perhaps tropical or subtropical in character.

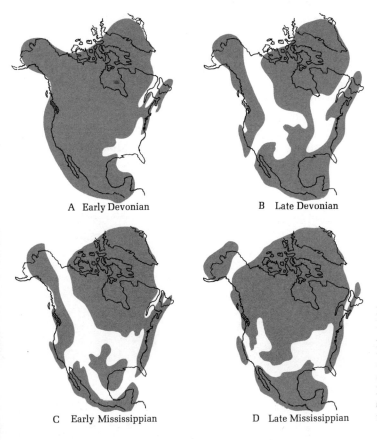

A   Early Devonian          B   Late Devonian

C   Early Mississippian     D   Late Mississippian

**Figure 22-3**
Paleogeographic maps of North America in the Devonian and Mississippian. The rock record suggests a gradual flooding of the continent during Devonian and Early Mississippian times. The more limited seas of map D indicate continental uplift in the later Mississippian.

## Mountain Making in the Mobile Belts

Rocks of the Catskill Delta are external evidence of mountain making in the northern half of the Appalachian mobile belt during the Devonian, and radioactive minerals in granites and metamorphic rocks of the Piedmont Province provide additional evidence that mountains were formed during this period as far south as the Carolinas. Because the truncated roots of the mountains produced are plainly in evidence beneath an angular unconformity in parts of Nova Scotia and New Brunswick, a district long known as Acadia, the entire chain is termed the **Acadian Mountains.** As the Catskill Delta is largely Devonian, it is clear that the major spasm of orogeny occurred during that period. Continued, but more limited, westward spread of sand and gravel during the Mississippian indicates that the Acadian Mountains were important highlands through much of the Mississippian as well.

Westward-thickening clastic rocks provide evidence of orogeny in eastern Nevada and Idaho. Furthermore, in a belt 500 miles long that stretches northeast from west-central Nevada to central Idaho, complexly deformed Ordovician, Silurian, and Devonian rocks, deposited in the outer part of the Cordilleran mobile belt, now overlie mildly deformed Cambrian through Devonian limestones of the inner part. The surface that separates these two very different types of rock is a thrust fault, from which we conclude that, at about the close of the Devonian, sediments and volcanic rocks of the outer part were folded and then thrust east over limestones deposited in the inner part of the Cordilleran mobile belt. The highlands produced during this orogeny, the **Antler Mountains,** were

**Figure 22-4**
Disconformity (traced in black) separating Mississippian limestone and shale below from Pennsylvanian sandstone and sandy shale above. Southern Indiana. (Photo courtesy of Indiana Geological Survey.)

PENNSYLVANIAN
MISSISSIPPIAN

probably similar to the Taconics, which slid west into the inner part of the Appalachian mobile belt in the Ordovician (Section 21-2).

Cambrian through Devonian sediments in both the inner and outer parts of the Franklin mobile belt were crushed and uplifted to form the Innuitian System of complex mountains some time between the end of the Devonian and the Pennsylvanian. Because Devonian sediments are the youngest rocks involved in the folds and faults of this long chain of Arctic mountains, and Pennsylvanian sediments rest on them with angular unconformity, we conclude that both the building and the principal erosional beveling of the Innuitian System took place in the Mississippian Period.

*Uplift and Erosion*

Late in the Mississippian, as shown in Figure 22-3 D, seas became less widespread and lands more extensive. By the end of the Mississippian, all the continental interior was above sea level, as indicated by the fact that, throughout the present Interior Lowlands, Mississippian and Pennsylvanian rocks are separated by a pronounced unconformity. Newly exposed sedimentary rocks were subjected to erosion, and considerable thicknesses of Devonian-Mississippian rocks were undoubtedly removed; however, the erosional interval may not have lasted very long, for high relief on the unconformity suggests that in many places only an early-mature stage in the cycle of regional reduction was reached (Figure 22-4). In much of the central and western United States, where Mississippian rocks are dominantly limestone, post-Mississippian solution by ground water produced a rough surface of sinkholes and collapse depressions, on which Pennsylvanian rocks were deposited (Figure 14-27).

**22-2 THE TRANSCONTINENTAL ARCH**

On all the lithofacies maps we have examined thus far, Paleozoic rocks are shown to be absent in an irregular area stretching southwest from the Canadian Shield into Arizona and New Mexico. This peculiarly persistent feature of the rock record suggests either that strata of Cambrian through Mississippian age were not deposited in this area or that they were removed from it by erosion after they were laid down. Each of these explanations applies to different segments of the belt. Isolated patches of pre-Pennsylvanian rock, too small to show on our maps, do occur within the belt, indicating that at least some of it was flooded by the sea; and in parts of some sequences, differences in the fossils found on either side of this belt indicate that it was a barrier to migration across the continental interior.

The strip we have just described, termed the **Transcontinental Arch,** may have been similar to the Cincinnati Arch, but it was much wider and longer. Through most of pre-Pennsylvanian time, it stretched southwest from the shield, probably as a chain of shoals or shallow "banks," dotted from time to time by islands that were nearly awash in the seas and too low to supply much coarse sediment. During uplift in the closing phases of each cycle, thin sediments deposited over the arch were among the first to be exposed, and were largely removed during the long erosional intervals that separated the Early Paleozoic cycles.

The patterns from which we deduce the former existence of the Transcontinental Arch are evident only in Mississippian and older rocks. They do not appear on maps showing Pennsylvanian and Permian strata, nor shall we encounter them later. Apparently the Transcontinental Arch was uplifted for the last time in the waning

**Figure 22-5**
Isopach and lithofacies map of the Pennsylvanian System in North America. In areas covered by the limestone pattern Pennsylvanian rocks are more than 50 percent limestone; the shale pattern indicates mostly shale and limestone; the dotted pattern shows areas dominated by sandstone and conglomerate. (Data largely from Sloss, Dapples, and Krumbein, 1960. Base map, Goode Base Map Series, Dept. of Geography, University of Chicago. Copyright by University of Chicago.)

**Figure 22-6**
Isopach and lithofacies map of the Permian System in North America. In areas covered by the limestone pattern, Permian rocks are more than 50 percent limestone; the shale pattern indicates mostly shale and sandstone; the dotted pattern, largely sandstone. The irregular broken lines in the western United States cover an area of volcanic rocks. (Data largely from Sloss, Dapples, and Krumbein, 1960. Base map, Goode Base Map Series, Dept. of Geography, University of Chicago. Copyright by University of Chicago.)

phases of the Mississippian: it had slight effect on the distribution of rocks after that.

A comparison of Figure 22-2, a lithofacies map of the Mississippian System in North America, with Figure 22-5, a lithofacies map of the Pennsylvanian, shows marked differences. Whereas the Mississippian map shows immense areas—thousands of square miles—of uninterrupted limestone, the Pennsylvanian map shows much smaller areas, in which several types of rock are present. Further, the isopach contours show that changes in thickness of Mississippian rocks are only slight to moderate, whereas those in Pennsylvanian rocks are abrupt and pronounced, with basins containing thousands of feet of sedimentary rock right beside land areas that received none. The Permian map, Figure 22-6, shows additional complication in the form of broad areas of volcanic rock. Clearly, conditions in the Late Paleozoic were very different from those of earlier times.

## 22-3
## PENNSYLVANIAN AND PERMIAN

The lateral and vertical diversity characteristic of Pennsylvanian and Permian rocks indicates great variety in depositional conditions, and rapid changes in the environment with the passage of time. A constantly changing panorama is recorded by the rocks of this sequence.

### Cyclic Deposits of the Eastern Interior

*Distribution*

As shown on the map, Figure 22-5, Pennsylvanian rocks between the Mississippi River and the Appalachian Mountains occur in three large isolated areas. The westernmost of these is in the **Illinois Basin,** the northern in the **Michigan Basin,** and the eastern in the **Appalachian Basin.** The only Permian rocks in eastern North America occur as a little remnant in the Appalachian Basin (Figure 22-6), overlying Pennsylvanian rocks, which they greatly resemble. The nature of all these strata shows that their origin is unrelated to the arches that now separate the basins in which they occur, and fossils indicate that at least the Pennsylvanian part of the section once extended in a continuous sheet from Pennsylvania to Kansas. Thus the present areas of outcrop have been isolated by erosion.

*Character*

Pennsylvanian rocks of the three eastern basins occur in thin sets of strata, or **cyclothems,** which are many times repeated in vertical succession. As indicated on Figure 22-7, the lower part of a cyclothem consists of a sandstone, conglomeratic at the base, which rests on a disconformity and grades upward into shale. Above the shale is an underclay on which lies a bed of coal. Fossils and sedimentary features of all these units indicate that they are nonmarine. Strata that overlie the coal bed, however, typically consist of shale and thin limestone that contain marine fossils. The top of this marine interval is the disconformity on which rests the basal sandstone of the next overlying cyclothem.

Most cyclothems are incomplete. One or more units may not have been deposited; or the erosion that preceded deposition of the next overlying cyclothem may have cut so deeply into the preexisting one that several beds were removed.

Not only does each cyclothem represent a dual, marine-nonmarine mode of origin, but cyclothems are monotonously repeated through hundreds of feet of Pennsylvanian rocks. (Uppermost Pennsylvanian and Permian rocks of the Appalachian Basin arc cyclical, but almost entirely

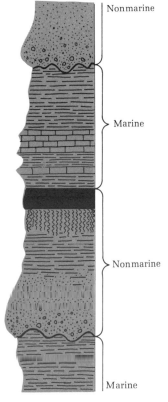

Nonmarine

Marine

Nonmarine

Marine

**Figure 22-7**

A cyclothem, typical of Pennsylvanian deposits in the eastern interior. The bed just below the coal is an underclay, thought to be the soil on which the coal forest grew.

nonmarine.) In eastern Ohio there are 44 coal beds that are thick enough to have received attention. Most of these occur in cyclothems more or less like that of Figure 22-7. These coal beds are distributed through 1,100 feet of Pennsylvanian-Permian strata; thus the section averages one coal bed per 25 feet.

The Pennsylvanian section as a whole undergoes a marked change when traced from east to west. In Pennsylvania and West Virginia, the basal sandstones are thick and conglomeratic, the shales are silty, coals and underclays are well developed, and the marine parts of the cyclothems are thin or absent. Progressively across Ohio, Kentucky, Indiana, and Illinois, the clastic rocks become finer, and the upper marine shales and limestones increase in thickness. Coals and underclays persist all across this immense area. Farther west, in Iowa and Kansas, cyclical characteristics persist but coals are thin or absent and the sections come to be dominated by marine shale and limestone.

## Origin and Significance

The nonmarine parts of the cyclothems in the Appalachian region carry within them much evidence as to their history. Conglomerates and sandstones are cross-stratified, irregularly bedded, and mostly restricted to channel-shaped troughs cut by ancient streams into the underlying sediments. Siltstones and shales above the basal sandstones contain many features of floodplain deposits, including natural levees. The coal beds are interpreted as having formed in forested swampy lowlands of wide extent; the plants were rooted in soils that have been preserved as underclays.

Systems of river channels are indicated, separated by broad floodplains on which coal swamps developed. The delta of a large present-day river such as the Mississippi has certain features analogous to those we have mentioned, though the coarse materials found in the distributary channels of the Pennsylvanian delta system imply a mountainous source of sediment close at hand.

To account for the intercalated thin marine strata is much more difficult. Even assuming that the delta systems were close to sea level, it is still hard to explain periodic inundations by shallow seas. And the difficulty increases greatly when we consider the Pennsylvanian record as a whole. How can we account for a single bed of coal like the LaSalle No. 2, which has been traced over an area that includes Oklahoma, Missouri, parts of Iowa, the Illinois Basin, and parts of Indiana and Kentucky? How can the repeated alternation of immense areas between marine and nonmarine conditions be explained? We have no modern analogs of such environments. Theories that have been proposed invoke control by crustal uplift or downsinking, by worldwide fluctuations in sea level, and by changes in climate. Each of these has weak points and none is generally accepted. The Pennsylvanian cyclothems remain an enigma.

## The Western Interior

### Distribution of Rocks

The lion's share of the Pennsylvanian-Permian sequence is in the western interior, between the Mississippi River and the east edge of the Cordilleran mobile belt in Nevada, Idaho, and western Utah. East of the Mississippi, the sequence is almost entirely Pennsylvanian in age; in the western interior, on the other hand, it contains both Pennsylvanian and Permian strata, which almost completely blanket the old Transcontinental Arch.

Pennsylvanian rocks (Figure 22-5), the most widespread, are about two-thirds marine; Permian rocks (Figure 22-6) are more limited in distribution and are less than a third marine; Triassic strata are least widespread of all (Figure 23-1) and are almost entirely nonmarine. These facts suggest that from Pennsylvanian time well into the Mesozoic, ever broader areas of the western interior were raised above sea level and subjected to erosion. As this took place, the sea floor was gradually replaced by floodplains and deserts, in which terrestrial sediments were deposited.

### Colorado-Oklahoma Mountains

In the western interior, the Pennsylvanian-Permian sequence was divided into northern and southern areas (Figure 22-8) by a belt of mountains (now subdued by erosion, buried, or broken by younger mountains), which extended from south-central Oklahoma across the Texas Panhandle and northeast New Mexico into Colorado, Utah, and Wyoming. Distribution of rocks in this belt is chaotic; in the **Colorado Mountains,** which form its northwest half, immense piles of red Pennsylvanian and Permian sediment accumulated in basins separated from one another by elongate patches of Precambrian rock. In the **Oklahoma Mountains,** which form the southeast half of the belt, ridges of Precambrian or deformed Paleozoic rock are flanked by thick Pennsylvanian deposits and are partially or completely covered by Pennsylvanian, Permian, or Triassic strata. Because Mississippian rocks are largely absent in the Colorado-Oklahoma Mountains, and

Wyoming

Nebraska

COLORADO MTNS.

Utah

Colorado

Kansas

Arizona

Oklahoma

OKLAHOMA MTNS.

New Mexico

OUACHITA MTNS.

Texas

**Figure 22-8**
Major elements of the
Colorado-Oklahoma Mountains. The
larger, more irregular patches in New
Mexico, Colorado, Arizona, Utah, and
Wyoming are the Colorado Mountains;
the narrow elongate areas stretching
from the Texas-Oklahoma Panhandle
to south central Oklahoma are the
Oklahoma Mountains. Part of the
Ouachita System, which formed in the
Late Pennsylvanian, is shown by
narrow arcuate lines in southeast
Oklahoma and adjacent states.

Pennsylvanian rocks in and adjacent to these mountains contain Mississippian fragments, we conclude that the highlands were formed late in the Mississippian with the general uplift that concluded that period.

The Colorado-Oklahoma Mountains consisted of both dome and fault-block mountains, and probably formed an archipelago across the sea of the western interior that greatly influenced the distribution of Pennsylvanian and Permian rocks. As each segment of the mountainous belt rose above sea level, it was attacked by erosion, and streams poured the resulting debris into basins between the ranges, and into the sea in nearby parts of the continental interior. In several basins within the Colorado Mountains, arms of the sea were isolated, and impressive masses of salt, gypsum, and anhydrite were deposited; in others, great thicknesses of coarse red sand and gravel were built up above sea level to form spectacular deposits of brightly colored sandstone and conglomerate that are characteristic of this sequence in the southwestern United States (Figure 22-9).

## The Rock Record North of
## Colorado-Oklahoma Mountains

In Oklahoma, Kansas, Missouri, Iowa, and Nebraska, sets of marine and coal-bearing nonmarine beds like those of the eastern interior make up the Pennsylvanian part of the rock record. Nonmarine parts of each set, which include sediment derived from the Colorado-Oklahoma Mountains and the Canadian Shield, attest to repeated

**Figure 22-9**

Garden of the Gods and Pikes Peak, Colorado. Ridges in the foreground are nonmarine red sandstone and conglomerate, and white gypsum of Pennsylvanian and Permian age, which were deposited along the edge of the northeasternmost element of the Colorado Mountains (Figure 22-8). Late in the Mesozoic, these rocks were folded to near-vertical position in the Laramide Orogeny, since then they have been sculptured by streams to form the Garden of the Gods. Snow-capped Pikes Peak in the background is part of a Precambrian batholith

uplift and erosion in both these source areas during the Pennsylvanian.

Permian rocks are mostly nonmarine red sandstone and shale. In Kansas and northern Oklahoma, however, the lowest quarter of the Permian section is marine limestone, which grades down into Pennsylvanian rocks and up into shale containing thick evaporites. These, in turn, are buried beneath a cover of nonmarine shale and sandstone, also of Permian age.

## The Rock Record South of Colorado-Oklahoma Mountains

Great thicknesses of Pennsylvanian and Permian rocks are encountered in a broad belt extending from central Texas across New Mexico into Arizona and Utah. Southeast of the blunt southern end of the old Transcontinental Arch, which was uplifted late in the Mississippian, Pennsylvanian strata are mostly marine limestone and shale. These rocks contain debris contributed by the Colorado-Oklahoma Mountains, but they also become thicker and coarser to the south and southeast, indicating the presence of highland source areas in the Ouachita-Marathon mobile belt. Lower and Middle Permian strata accumulated in the **Permian Basin,** a broad feature that today extends from central Texas some 350 miles westward beneath the sandy plains into southeastern New Mexico. In reality it consists of two distinct basins, divided by a north-south buried platform; this is known from the records of thousands of wells that have been drilled for oil and gas in the region. Along the margins of Middle Permian seas that occupied the basins, great reefs flourished. The largest, known as the **Capitan reef,** is as much as 4,000 feet thick and extends for some 225 miles in a great sweeping curve part way around the western of the two basins. On the "back-reef" or lagoonal side of the Capitan reef are well stratified dolostones and limestones; on the basinward side, thick evaporites. Upper Permian salt and gypsum blanket the whole Permian Basin. From this section, potassium-bearing minerals are mined near Carlsbad, New Mexico.

All the rocks just mentioned are buried beneath younger strata except on the western flank of the Permian Basin. Here a part of the Capitan reef has been uncovered by erosion, and forms the Guadalupe Mountains of southern New Mexico. The immense cave system known as Carlsbad Caverns is in a part of the Capitan reef.

Pennsylvanian, Permian, and Triassic rocks almost completely blanket the southern end of the Transcontinental Arch in New Mexico, Arizona, and Utah. Most of them are red sandstone and shale of the type vividly on display in Monument Valley, Utah (Figure 18-9).

## The Continental Borders

### Appalachian Mobile Belt

Pennsylvanian and Permian rocks, containing thick coal beds, accumulated in basins between ranges of the old Acadian Mountains in Nova Scotia, New Brunswick, Massachusetts, and Rhode Island. These rocks are now much deformed, and some of them have been metamorphosed and intruded by granite. South of New England, however, rocks younger than Mississippian are absent from the Piedmont and Blue Ridge provinces of the Appalachian System. This, coupled with the fact that highlands in this area must have supplied much of the coarse sediment swept into the interior during the Late Paleozoic, suggests that most of the outer part of the Appalachian mobile belt had been transformed into mountains by Pennsylvanian time.

Pennsylvanian rocks are the youngest involved in the

folds and faults of the Valley and Ridge Province of the Appalachians: Permian strata are unknown there. Triassic deposits rest on the beveled edges of Paleozoic rocks in the seaward provinces of the Appalachians, and indicate that the mountains had been eroded to a late-mature stage well before the end of the Triassic Period. All these facts imply that the remainder of the Appalachian mobile belt was converted to mountainous uplands between the Late Pennsylvanian and the Early Triassic. Paleozoic sediments of the inner part were folded into anticlines and synclines and were broken and sheared northwestward along great thrust faults. During the same episode, Late Paleozoic rocks in basins within the old Acadian Mountains were deformed, metamorphosed, and intruded by granite. We group all these events together as the **Allegheny Orogeny,** a spasm of mountain-making that resulted in final obliteration of the Appalachian mobile belt.

### Ouachita-Marathon Mobile Belt

Some of the coarse Pennsylvanian sediment in the Interior Lowlands came from sources in the Ouachita-Marathon region. This implies the existence there of mountains as early as Pennsylvanian time. In the Ouachitas of today, the youngest of the deformed sedimentary rocks are Late Pennsylvanian in age, and this is true in the Marathon Mountains as well. In the latter, however, the time of deformation is rather precisely fixed, for essentially horizontal Lower Permian rocks lie on the beveled edges of folded Late Pennsylvanian strata. Thus it appears that by the end of the Pennsylvanian Period, rocks in the Ouachita-Marathon mobile belt had been folded, faulted, and uplifted to form a complex mountain system stretching across the southern border of the continent.

### Cordilleran Mobile Belt

The Antler Orogeny divided the Cordilleran mobile belt into two parallel troughs that persisted through the Pennsylvanian and Permian and into the Mesozoic. In the eastern trough, especially in northwest Utah and adjacent parts of Nevada and Idaho, sinking was pronounced. Nearly 50,000 feet of Mississippian, Pennsylvanian, Permian, and Triassic limestone and sandstone were deposited in what had been, through Devonian time, the slowly subsiding outer edge of the interior platform.

West of the belt of mountains raised by the Antler Orogeny, subsidence must also have been great during the Late Paleozoic and Early Mesozoic, but the record has been considerably dismembered by Jurassic and later events. Most of the rocks deposited in the western trough are volcanics and clastic sediments, some of very coarse texture. In British Columbia, for example, the section is more than three miles thick, and in northwest Nevada there are 12,000 feet of Permian lavas. Only a rather thin cover was laid down over truncated roots of the Antler Mountains; indeed at times their surface may have been exposed to erosion.

Although rapid subsidence seems to have marked both the eastern and western Cordilleran troughs, there is evidence for at least four episodes of folding and thrust-faulting in the western trough during the Permian. Further, to account for the great thicknesses of volcanic rock, there must have been violently active volcanic highlands to the west during Permian time.

## Sverdrup Basin

In the north-central part of the Canadian Arctic Archipelago, rocks of Pennsylvanian and younger age rest with profound angular unconformity on the truncated roots of the Innuitian Mountains. The younger strata were apparently deposited in a sinking depression, the **Sverdrup Basin,** which extended southwest for some 400 miles from northern Ellesmere Island to Melville Island and is superimposed on the old Innuitian System.

Pennsylvanian rocks in the Sverdrup Basin are marine sandstones, conglomerates, and shales at least 2,000 feet thick. Permian strata, more widespread than Pennsylvanian, are mostly marine limestones, but include some volcanic rock and thick beds of gypsum.

## 22-4 SUMMARY

In the Devonian and Early Mississippian, the continental interior was slowly flooded by a warm, shallow sea that reached its maximum extent in the Early Mississippian. In the Devonian, the Acadian Orogeny produced mountains in the Appalachian mobile belt, and clastic sediments derived from these highlands spread westward toward the present Interior Lowlands to form the Catskill Delta and related deposits. The Antler Mountains of the Cordilleran mobile belt, and the Innuitian Mountains of the Franklin mobile belt, were produced by orogenies in the latest Devonian or earliest Mississippian.

Following retreat of the sea at the end of the Mississippian Period, the face of North America was etched to mature topography by streams, and mountains were raised from southern Oklahoma to Colorado in areas that had previously been parts of the shelflike continental interior. Early in the Pennsylvanian, erosion ended in the

interior, as it was gradually overspread by seas flooding
in again from the continental margins.

Repeated uplift and erosion of the Canadian Shield,
the Colorado-Oklahoma Mountains, and highlands in the
outer parts of the Appalachian, Ouachita-Marathon, and
Cordilleran mobile belts produced floods of coarse sediment that built out as deltas into the interior basins and
provided wide swampy environments favorable to luxu-
riant forest growth. Remains of these forests, which spread
over the continental interior time and again during the
Pennsylvanian, now form coal.

During the Late Pennsylvanian and Permian, mountains
were built in the inner parts of the Appalachian and
Ouachita-Marathon mobile belts, and repeated folding,
faulting, and extrusive volcanism affected the outer part
of the Cordilleran belt. In association with these events,
the entire continent was gradually uplifted and the seas
retreated slowly southwestward during the Permian. In
their wake came red floodplain and desert deposits,which,
together with thick evaporites in Pennsylvanian and Permian rocks, suggest that the newly exposed lowlands
baked in a hot, dry climate.

By the end of the Paleozoic, North America was much
changed. In a series of orogenies beginning in the Ordovician, the Appalachian and Ouachita-Marathon mobile
belts had been transformed into wide systems of complex
mountains. An ocean basin must have separated eastern
North America from Europe and Africa through much
of the Paleozoic. However, this basin must have disappeared before the Triassic. Not since the Early Permian
has the eastern part of the continent been flooded by the
sea. In the Arctic the central part of the Innuitian System,
built during the Mississippian Period, had foundered
beneath the sea and was again a site of thick sediment
accumulation. Deserts like those of the present-day southwest stretched from Kansas to the edge of the Cordilleran
sea in western Utah, and chains of volcanoes disgorged
lavas and ash over much of far-western North America

## SUGGESTED READINGS

Schuchert, Charles. 1955. *Atlas of Paleogeographic Maps of North America,* maps 31–59. New York: Wiley.

Twenty-nine paleogeographic maps covering the Devonian through Permian, each with a list of major stratigraphic units represented.

Sloss, L. L., Dapples, E. C., and Krumbein, W. C. 1960. *Lithofacies Maps, An Atlas of the United States and Southern Canada,* pp. 14–52. New York: Wiley.

Isopach-lithofacies maps for Devonian through Permian time.

# 23 ROCKS AND PHYSICAL HISTORY OF THE MESOZOIC ERA

23-1   Triassic and Early Jurassic

23-2   Middle and Late Jurassic and Cretaceous

23-3   Summary

Clastic rocks that are largely nonmarine dominate the Triassic and Lower Jurassic record of North America. They record a long interval during which all the continent except its western part stood above sea level (Figure 20-3), and sediments were deposited in deserts, alluvial plains, and intermontane basins. Middle and Upper Jurassic and Cretaceous rocks, on the contrary, consist largely of widespread sheets of shale, limestone, chalk, and similar rocks, which carry marine fossils. This rock record is evidence of a large inundation, at the maximum extent of which some 35 to 40 percent of the continent was beneath the sea. Extensive orogenies toward the close of the Mesozoic expelled marine waters for the last time from the continental interior, and broad uplift caused them to recede in part from the eastern and southern margins of the continent.

## Appalachian Region

The only pre-Cretaceous Mesozoic rocks in eastern North America are Triassic strata that occur in narrow strips within the Appalachian System (Figure 23-1). These strata, locally more than 20,000 feet thick, are nonmarine and include conglomerate, sandstone, red shale, and coal, as well as basalt flows, sills, and dikes. The structure of these strata and their relation to surrounding rocks show that not only were they deposited in basins that sank along normal faults, but they were broken into additional blocks by normal faulting after their deposition. These features suggest that the Atlantic basin, which was probably nonexistent at the beginning of the Mesozoic, began to open in the Mid-Triassic (Section 16-4). As it did so, the margins of continents on either side of the Mid-Atlantic Ridge were disrupted along normal faults into chains of fault-block mountains. Mud, sand, and gravel poured into basins between these mountains to form redbeds; and lava, welling up along the faults, was injected into the sediments and spread out as flows on their surface. Continued uplift broke the Triassic rocks themselves into fault-block ranges. (A sequence of events like this is shown diagrammatically in Figure 17-4.) The igneous rock that forms the Palisades along the Hudson River opposite New York City was intruded as a sill during this Triassic interval; hence we speak of the entire episode of faulting, sedimentation, and volcanism as the **Palisades Disturbance.**

## Western Interior

To the north of the Colorado-Oklahoma Mountains, Triassic rocks are of limited distribution and consist entirely of red sandstone and shale of terrestrial origin that are indistinguishable from Permian rocks beneath.

23-1
## TRIASSIC AND EARLY JURASSIC

Most of them were derived from the Colorado Mountains;
the Oklahoma Mountains had been largely removed by
erosion by Triassic time.

To the southwest, in western Texas, New Mexico, and
Arizona, Permian strata are succeeded by Triassic red
shales and sandstones that contain fossil amphibians,
reptiles, and land plants. The characteristic red color of
Triassic strata in the western interior is seen again in
the Painted Desert of northern Arizona. In large parts
of the southwest Triassic rocks are overlain by thick
cross-stratified sandstones of Early Jurassic age, which
were clearly deposited in desert areas by the wind (Figures
8-11 and 23-2).

Because only a thin portion of the Triassic and Lower
Jurassic section in this segment of the western interior
contains marine fossils, we conclude that the area was
a desert basin well above sea level. Most of the sand and
mud deposited in this basin came from the Colorado
Mountains and from the south end of the old Transcontinental Arch, both of which were buried by the middle
of the Jurassic.

## Continental Borders

The Cordilleran mobile belt, divided into parallel troughs since the time of the Antler Orogeny, received thick Triassic sections in both parts. Marine limestones and sandstones were deposited in the northern part of the eastern trough, in Utah-Nevada-Idaho. To the west, the record has been obscured by later earth movements, erosion, and volcanism; but in northwestern Nevada, for example, it is clear that Permian lavas are succeeded by 25,000 feet of interbedded volcanic and sedimentary rocks of Triassic age. We conclude that there were active volcanic highlands in westernmost North America during the Triassic Period.

In the Sverdrup Basin far to the north, more than 8,000 feet of Triassic sandstone and shale accumulated. Because the lower part of this section contains marine fossils and the top contains coal beds, we deduce that a seaway gradually filled with sediment and the region was converted to swampy lowlands in which vegetation flourished.

## Western Interior

Figures 23-3 and 23-4 indicate that in the western part of the Interior Lowlands and adjacent parts of the eastern Cordillera, Jurassic and Cretaceous rocks form a widespread deposit that thickens rapidly from east to west. A large part of the rock in this sequence is marine shale

**Figure 23-2**
Triassic and Jurassic sandstones and shales like those of Zion Canyon, Utah, shown here, are the record of widespread desert conditions in the Colorado Plateau region. Marine limestones, recording a Mid-Jurassic invasion of the sea, occur nearby just above the cliff-forming white sandstone on the skyline. (Photo by E. M. Spieker.)

## 23-2
## MIDDLE AND LATE JURASSIC AND CRETACEOUS

**Figure 23-3**
Isopach and lithofacies map of the Jurassic System. In areas covered by the shale pattern, Jurassic rocks are dominantly shale and sandstone; sandstone and limestone patterns denote areas dominantly of those types. The large areas of solid color in western North America are batholiths from intrusions during the Nevadan Orogeny. (Data largely from Sloss, Dapples, and Krumbein, 1960. Base map, Goode Base Map Series, Dept. of Geography, University of Chicago. Copyright by University of Chicago.)

and sandstone, conveniently divided into lower (Jurassic) and upper (Cretaceous) parts by a southwestward-thickening wedge of brilliantly colored nonmarine shale, sandstone, and conglomerate. We can note only the main features of each of these units.

*Jurassic Rocks*

In northern Utah and adjacent Idaho, the lower part of the sequence consists of 3,000 to 5,000 feet of marine limestone, succeeded upward by about 1,700 feet of sandstone (Figure 23-2), largely wind-deposited (Figure 8-11). These rocks, which are Middle and Late Jurassic in age, grade toward the east, south, and north into much thinner shales and sandstones. Most of the latter contain marine fossils, and from place to place they are interbedded with deposits of gypsum, anhydrite, and rock salt. Across the Colorado Plateaus, however, an appreciable part of this unit is red sandstone and shale, probably nonmarine in origin and very similar to Triassic and Lower Jurassic redbeds beneath.

Sediments forming the lower part of this unit, equiva-

**Figure 23-4**
Isopach and lithofacies map
of the Cretaceous System.
Shale, sandstone, and
limestone patterns cover
areas in which Cretaceous
rocks are dominantly of
those types. An elongate
area in which Cretaceous
rocks are absent bisects the
Cordilleran region: this is
the approximate location of
the Nevadan Mountains. East of this,
isopach contours outline the Rocky
Mountain trough; to the west, the
Pacific mobile belt. (Data largely from
Sloss, Dapples, and Krumbein, 1960.
Base map, Goode Base Map Series,
Dept. of Geography, University of
Chicago. Copyright by University of
Chicago.)

lent to the marine Middle and Upper Jurassic limestones of Utah and Idaho, were almost certainly derived from the east and south. Those equivalent to the thick overlying sandstone, however, very probably came largely from the west. Thus we infer that in the Late Jurassic, midway through deposition of this unit, a source high enough to supply a flood of coarse sand was built close to the western edge of the sea.

### The Morrison Formation

The Jurassic rocks just discussed are separated from Cretaceous rocks by the **Morrison Formation,** a famous deposit that thickens from a few feet of varicolored shale and fine sandstone in eastern Colorado to more than 800 feet of conglomerate and sandstone in eastern Utah, northwest New Mexico, and nearby Arizona. The Morrison, which is largely Late Jurassic in age, is worth special note not only for the history it records, but because it contains valuable deposits of uranium-bearing minerals, and in addition is the source of most of the large dinosaur skeletons in North American museums (Figure 9-1).

411

Dinosaurs and other fossils indicate that the Morrison Formation is entirely nonmarine, and studies in conjunction with the search for uranium show that the deposit accumulated in a maze of stream channels and on lake-dotted floodplains. Because the formation thickens and becomes coarser in texture westward and southward, we conclude that it was derived from highlands to the west and south, presumably the same ones that supplied sand for the upper part of the rocks beneath the Morrison.

## Cretaceous Rocks

Above the Morrison Formation, and spreading beyond it as far east as Minnesota and Iowa, is a thick mass of shale and sandstone that includes a few limestone beds and yields a profusion of Cretaceous marine fossils. Like the younger of the Jurassic rocks beneath, this deposit thickens rapidly and becomes coarser-grained westward. In central Utah, for example, it is represented by thousands of feet of conglomerate and sandstone, some parts of which include thick coal beds. Furthermore, along the extreme west edge of this deposit the coarser-grained beds are separated from one another by angular unconformities.

## Coastal Plains

Although Jurassic shales, sandstones, limestones, and evaporites are known to occur beneath the Gulf Coastal Plain, and Jurassic rocks occur at depth beneath the Atlantic continental shelf, the oldest part of the sequence exposed at the surface in either place is of Cretaceous age (Figure 23-4). Along the Atlantic Coast, these strata are both marine and nonmarine, dip gently seaward, and rest with angular unconformity on truncated folds and metamorphic rocks of the Appalachian System. In Georgia and Alabama, they merge into thicker, predominantly marine Cretaceous rocks of the Gulf Coastal Plain which extend northwestward beyond Jurassic rocks and almost completely bury the Ouachita-Marathon Mountains. The fact that Jurassic and Cretaceous sandstone and shale of the coastal plains grade seaward into limestones indicates that sediment was derived from the continental interior during the Jurassic and Cretaceous.

## Sverdrup Basin

Above coal-bearing Triassic strata in the Sverdrup Basin is a thick sequence of Jurassic and Cretaceous clastic rocks, shown by their fossils to have been deposited both on land and in the sea. Jurassic and Lower Cretaceous shales and sandstones, which yield marine fossils, are succeeded by an alternation of nonmarine and marine strata, the lowest of which are Cretaceous in age and the highest Paleogene. Nonmarine formations contain many thin coal beds, and in one place a series of basalt flows and pyroclastic rocks 700 feet thick occurs within the Cretaceous section. As a whole, the record indicates that the Sverdrup Basin was invaded by the sea at least four times in the Jurassic and Cretaceous, and that locally volcanism was active. The presence of coal beds suggests that forests formerly grew in an area now devoid of them.

## Cordilleran Region

To set the stage for a discussion of Cordilleran history from Mid-Jurassic through the Cretaceous, it is worth recalling that the Antler Orogeny of Early Mississippian time divided the Cordilleran mobile belt into two parallel depositional troughs. Between the Mississippian and the Late Jurassic, the eastern one sank steadily and filled with thick clastic sediments and limestone. The western trough, which was the depositional site of coarse clastic rocks and thick volcanics, underwent at least four episodes of folding and faulting in the Permian and was disturbed by several similar events in the Triassic and Jurassic.

### Nevadan Orogeny

The long-continued instability of the western Cordilleran trough built up to a Late Jurassic climax, when the rocks in it were folded, thrust-faulted, and intruded by granite batholiths in a series of closely spaced convulsions commonly grouped as the **Nevadan Orogeny.** The Nevadan Mountains, which extended from southern Alaska to the tip of Baja California, added a wide strip to the west side of the old Antler welt (Figure 23-3) and caused a westward migration of the western Cordilleran trough to form the **Pacific mobile belt.** Concurrent uplift of the Antler welt displaced the eastern Cordilleran trough eastward to form the **Rocky Mountain trough** in what had been part of the continental shelf through Jurassic time. This pattern of highlands and troughs, which strongly influenced the later history of the Cordilleran region, is plainly evident in Figure 23-4, which shows the distribution and thickness of Cretaceous rocks.

Sediments derived from the Nevadan Mountains and the rejuvenated Antler welt were carried eastward and westward by Late Jurassic and Cretaceous streams, which ultimately delivered this debris to the Rocky Mountain trough and the Pacific mobile belt. The Morrison Formation of the Rocky Mountain region, and thick Cretaceous clastic rocks both east and west of the Nevadan-Antler highlands, are records of this interval.

*Laramide Orogenies*

The Rocky Mountain trough subsided rapidly in the Cretaceous Period, and came ultimately to hold about a million cubic miles of clastic sedimentary rock. Much of this was undoubtedly derived from erosion of the Nevada-Antler highland complex to the west, but the immense volume suggests that some of the sediment was delivered to the trough by currents moving north from Mexico or south from Canada, parallel to the axis of the trough.

Near the beginning of the Late Cretaceous, rocks along the western margin of the Rocky Mountain trough were folded and faulted, while marine deposition continued farther east. In later Cretaceous and Early Cenozoic time, similar orogenies moved eastward across the trough like ponderous ripples, deformed more and more of the rocks, and displaced the sea eastward. By Mid-Paleogene time, all the strata in the Rocky Mountain trough, including nonmarine rocks deposited in basins between the earlier ranges, had been deformed and uplifted to form the Eastern Cordillera. We group all these mountain-making episodes together as the **Laramide Orogenies.**

**Appalachian Region**

The absence of Jurassic rocks in the eastern United States suggests that this region was an area of stream erosion during the Jurassic. Beneath the continental shelf, and onto the coastal plain, however, Cretaceous rocks lap westward over the beveled edges of deformed Appalachian strata. The unconformity that separates the undeformed younger rocks from the gnarled older ones exhibits little relief in the coastal plain, but it is very irregular beneath the outer part of the continental shelf (Figure 17-1). These facts imply that by Cretaceous time much of the Appalachian System had been reduced to a late-mature stage in the cycle of regional reduction, and that subsidence and block-faulting, associated with gradual opening of the Atlantic Basin, permitted the sea to invade the low-lying eastern segment of the system.

**23-3 SUMMARY**

In Triassic and Early Jurassic time, the continental interior was above sea level and exposed to erosion. During this long interval all of the Colorado-Oklahoma Mountains except their roots disappeared, unknown thicknesses of rock were removed from the continental interior, and the

Ouachita-Marathon System was largely leveled by erosion. In the Mid-Jurassic, however, the sea again spread inward toward the continental center, beginning the fifth large marine inundation since the Cambrian. Distribution of Jurassic rocks (Figure 23-3) suggests that one arm of the sea spilled southward into desert basins of the western interior, and that another marine embayment, in what is now the Gulf Coastal Plain, gradually spread northward over at least the outer portion of the Ouachita-Marathon Mountains.

In Late Jurassic time, sediments and volcanic rocks of the Western Cordilleran trough were crushed, uplifted, and intruded by granite batholiths to form the Nevadan Mountains. Debris stripped from these highlands, and from the rejuvenated Antler welt to the east of them, was carried west into the present Rocky Mountain region to form the Morrison Formation. As the Morrison sediments spread eastward from the highlands, Jurassic seas in the western interior shrank greatly in area. Uplift at this time may also have caused the shoreline in the Gulf Coast region to migrate southward: the Jurassic sediments newly deposited there may have been somewhat thinned during a brief erosional interval.

In the Cretaceous, seas spilled westward over the outer part of the Appalachians, and spread haltingly northward from the gulf and southward from the present Arctic into the Sverdrup Basin and the newly formed Rocky Mountain trough. At the climax of this inundation, a continuous seaway across western North America separated the mountainous Nevadan System from the low-lying lands of eastern North America (Figure 23-4).

In the Late Cretaceous and Early Cenozoic, sediments of the Rocky Mountain trough were compressed and uplifted in the Laramide Orogenies, and the lowlands of eastern North America were rejuvenated. As the Laramide Mountains appeared and the old continent was up-arched, seas were gradually restricted and the continent began to assume more nearly its present outline.

## SUGGESTED READINGS

Schuchert, Charles. 1955. *Atlas of Paleogeographic Maps of North America,* maps 59–76. New York: Wiley.
Eighteen paleogeographic maps covering Triassic through Cretaceous time, with a list of major stratigraphic units represented on each map.

Sloss, L. L., Dapples, E. C., and Krumbein, W. C. 1960. *Lithofacies Maps, An Atlas of the United States and Southern Canada,* pp. 53–85. New York: Wiley.
Isopach-lithofacies maps for Triassic through Cretaceous time.

# 24 ROCKS AND PHYSICAL HISTORY OF THE CENOZOIC ERA

24-1   Cenozoic Rocks

24-2   The Continental Margins

24-3   Glaciation in the Pleistocene Epoch

24-4   Geologic History in the Making

24-5   Summary

Cenozoic history, spanning the most recent 65 million years of the geologic past, is well documented in North America by widespread deposits of stratified rock and glacial drift, as well as by mountains and a variety of other landforms. Considered separately, Cenozoic rocks give evidence of a bewildering array of events; as a whole, they record the latest retreat of the sea to the continental margins, mountain building in western North America, and episodes of continental and valley glaciation. In the Cenozoic Era, North America reached its present size and shape, and may also have taken its present position with respect to other continents and the ocean basins that separate them.

**Figure 24-1**
The Badlands of South Dakota. The sands and clays exposed here are part of a sheet of brightly colored nonmarine sediment spread east from the Rockies during the Paleogene period. (Photo by H. S. Becker.)

**24-1
CENOZOIC ROCKS**

In the Rockies and much of the Colorado Plateaus, Cretaceous strata are succeeded without interruption by thick floodplain and lake deposits that contain early Cenozoic, or Paleogene, fossils. The patchy distribution of these strata shows that they accumulated in basins separated from one another by mountain ranges. Paleogene lake beds form the famous Pink Cliffs of southern Utah, into which Bryce Canyon (Figure 5-7) has been carved.

Paleogene deposits also blanket the Great Plains from Colorado north into Canada. For the most part, the rocks are nonmarine sands and clays of the type on display in the Badlands of South Dakota (Figure 24-1). In western North Dakota, fossil marine invertebrates have been collected from dark shale of earliest Paleogene age.

Later Cenozoic, or Neogene, rocks in the Rockies and Colorado Plateaus are largely volcanic ash and extrusive igneous rock of rather localized distribution. East of the Rockies, a thick sheet of nonmarine Neogene sedimentary rock covers the Great Plains from South Dakota to Texas.

Cenozoic deposits of the Atlantic Coastal Plain are marine strata similar to Cretaceous rocks beneath them, but their distribution suggests that, in general, the shoreline migrated haltingly seaward during their deposition. Beneath the Gulf Coastal Plain and adjacent continental shelf, Cenozoic rocks are more than 30,000 feet thick. They consist of a tremendous prism of sandy and clayey strata, derived from the continental interior in the same way as today's sediments in the Mississippi Delta. Intruded into these rocks are many vertical columns of rock salt, which have punched their way upward from a bed of salt, presumably Jurassic, at a depth of many thousand feet. The salt rose because it is less dense than the surrounding sediments. It moved by internal flow, much like a glacier. Some of the deformed rocks around the margins of these "salt domes" now serve as traps for petroleum; the domes also yield sulfur and salt.

**24-2
THE CONTINENTAL
MARGINS**

Paleozoic orogenies produced mountains along the north, east, and south borders of North America, and later history in those parts of the continent involved erosion of the mountains and burial of their roots beneath blankets of younger rock. Orogeny began in the Cordilleran mobile belt in the Paleozoic, continued through the Mesozoic, and affected additional segments of that belt in the Cenozoic. The Cenozoic history of all these margins now concerns us.

### Cascade Orogenies

Throughout the Cretaceous and Paleogene, volcanic rocks

and clastic sediments derived from both east and west accumulated to great thicknesses in the Pacific mobile belt. This trough, stretching from Alaska southward through California along the outer edge of the old Cordilleran mobile belt, was repeatedly disturbed by folding and faulting during the Paleogene and early Neogene; in the late Neogene it was crushed and uplifted in an intense convulsion to form the complex mountains of the Western Cordillera. Because the volcanoes of the Cascade Range were formed at about the same time (but not by folding and faulting), we speak of all these events as the **Cascade Orogenies.** Collectively, the Cascade Orogenies probably represent late stages in the evolution of a mobile belt. Such stages are shown diagrammatically in Figures 17-4 E and F.

## Neogene Events in the Cordilleran Region

By late Paleogene time, highlands of the old Nevadan-Antler welt and summits of the Rocky Mountains to the cast had been lowered by erosion to a broad low surface, with peaks of more resistant rock projecting above it. The debris derived from denudation of the Laramide Rockies spread eastward into the Interior Lowlands to form the blanket of Paleogene terrestrial deposits mentioned earlier. Early in the Neogene Period, the Rockies were rejuvenated; old streams cut deep gorges into them as they rose, and patches of the old erosion surface at high altitudes today attest to the magnitude of the uplift (Figure 24-2). The flood of clastic sediment from the rejuvenated

**Figure 24-2**
The skyline in this view of the Uinta Mountains of Utah shows evidence of a regional erosion surface, now slightly tilted toward the left (south). The surface was developed in Paleogene time, uplifted and warped during the Neogene, and has been much dissected by erosion since uplift. (Photo by W. R. Hansen, U. S. Geological Survey.)

mountains spread out to form the Neogene sediments of the Interior Lowlands.

Neogene faulting in the Central Cordillera raised the Sierra Nevada and the many fault-block ranges that characterize the Basin and Range Province today. During this same interval, basaltic lava welled upward along fissures in southeastern Washington, eastern Oregon, and southern Idaho and poured out to form the extensive Columbia and Snake River plateaus.

### The Atlantic and Gulf Coasts

The distribution of Cenozoic rocks in the Atlantic Coastal Plain suggests that during the last 65 million years the coastline has migrated slowly seaward. This implies that subsidence gave way to uplift along the eastern coast in the Cenozoic. During this uplift, the Appalachian System, or at least its westernmost part, was again rejuvenated; exposed areas of the pre-Cretaceous erosion surface were deeply trenched by invigorated streams; and new tributaries to these old rivers carved youthful valleys into belts of soft sediment.

Remnant patches of the old pre-Cretaceous erosion surface are widespread in the Appalachians, and are responsible for the remarkably accordant ridge elevations in much of the system (Figure 18-5); entrenched streams, inheriting their courses from the old surface, cut through these ridges in deep water gaps, like those shown in Figure 11-19. The Appalachian region has thus been beveled by erosion at least twice since the Paleozoic, and a third major cycle has begun. The mountains owe their relief in large part to broad Cenozoic uplift, not to the ancient orogenies that produced the folds and faults so grandly on display in the Ridge and Valley Province of today.

Thick sediments, intruded by salt domes, accumulated along the Gulf Coastal Plain during the Cenozoic, and these, together with their Mesozoic predecessors, largely conceal the old Ouachita-Marathon System. Uplift of the same sort that rejuvenated the Appalachians up-arched the areas in which the Ouachitas and Marathon Mountains are exposed today, and Cenozoic erosion stripped part of the cover from these old mountains.

### Innuitian Region

In the Early Cenozoic, rocks of the Sverdrup Basin, and adjoining older strata, were compressed to form north-striking folds and thrust faults; and pluglike masses of Pennsylvanian and Permian evaporites were squeezed up into these distorted rocks. Probably much later in the

Cenozoic, the western part of the Innuitian System was broken by normal faults, most of which, like earlier Cenozoic structures, strike north or northeast. The significance of these events cannot yet be fully evaluated, but they serve to indicate that instability of the continental border, so pronounced in the Cordilleran System, also extended into Arctic regions during the Cenozoic.

Much of the Interior Lowlands is blanketed by a thick cover of glacial till, and there is clear evidence that the Canadian Shield, as well as the summits and canyons of the Cordilleran System and the New England Appalachians, has also been scoured by glacial ice. Because glacial features are the youngest ones recognized in each of these areas, we conclude that North America was subjected to extensive glaciation in the very recent past, during the Pleistocene Epoch.

### Distribution and Interpretation of the Till

North of a line approximately paralleling the Ohio and Missouri rivers, the scoured and striated bedrock of the Interior Lowlands and southernmost Canadian Shield is mantled by till. In Iowa, the till blanket is divisible into as many as four main layers, separated from one another by soils, peat deposits, and beds of loess (Figure 24-3). In Illinois, Indiana, and Ohio, however, only three till sheets are recognized. Only the youngest of these continues across the Canadian Shield, where the till is thin and patchy.

Many of the cobbles and boulders in the till of the Interior Lowlands are igneous and metamorphic rocks of types exposed at the surface today only in the Canadian Shield. Furthermore, the arrangement of end moraines (Figure 12-18), the trend of eskers and drumlins (Figure 12-19), and the alignment of grooves and scratches on the bedrock beneath the till show that the ice moved into the Lowlands from the north. The size and extent of this glacier, the **Laurentide Ice Sheet**, are shown in Figure 24-4.

The four divisions of the Lowlands till suggest that the Laurentide Ice Sheet advanced into at least some parts of the Lowlands four times, and that each advance was followed by a warm interval during which ice melted, soils and peats formed, and loess accumulated on top of the ground and end moraines. The absence of all but the youngest till sheet over the Canadian Shield, the center of ice accumulation, implies that deposits of earlier Laurentide glaciers were "bulldozed" from the shield during each successive ice advance.

## 24-3
# GLACIATION IN THE PLEISTOCENE EPOCH

**Figure 24-3**
Generalized section through the Pleistocene till of Iowa. The till is divided into four parts by soil and peat deposits (cross-hatched) and by beds of windblown sand and silt (dotted pattern). Diagonal lines indicate the depth to which each division was weathered before it was covered by a new till sheet. (Adapted from Kay and Apfel, 1929.)

The Cordilleran region of Canada and northwestern-most United States also bears evidence of extensive ice-sheet glaciation. Apparently Pleistocene ice in that region developed first as valley glaciers. These gradually grew together into the complex **Cordilleran Ice Sheet,** which covered the mountains and ultimately spilled eastward onto the plains of Alberta. The outlines of this glacier, inferred from the distribution of till and other glacial features, are shown in Figure 24-4.

South of the Cordilleran Ice Sheet, Pleistocene ice accumulated to form systems of valley glaciers in the higher parts of the Eastern and Western Cordillera. In the Western Cordillera, the Cascade volcanoes and the high summits of the Sierra Nevada were sculptured by valley glaciers; in the Eastern Cordillera, Pleistocene valley glaciers carved the spectacular landforms of Glacier National Park, Montana; the Tetons of western Wyoming; and Rocky Mountain National Park, Colorado.

## Other Effects of Glaciation

In Chapter 12, we noted some of the effects of glaciation on the patterns of pre-Pleistocene drainage in the Interior Lowlands. In addition, recall that the basins of the present Great Lakes were scooped out or enlarged by Pleistocene ice and that they were filled initially with meltwater from the last of the ice sheets. South of the glaciated areas, the effects of Pleistocene glaciation are obvious, even though there is no till. In Utah, for example, rainfall was heavier during times of glacier advance than it is now. Runoff into the Basin and Range Province accumulated to form large lakes during glacial intervals; during drier interglacial ages, these lakes shrank. Great Salt Lake is the last vestige of the largest of these lakes. Raised beaches fringe many of today's dry basins (Figure 3-2).

As ice accumulated in the north, sea level all over the world dropped. This provided streams everywhere with new, lower base levels. Thus, during times of glacial advance, streams entrenched their old valleys. During interglacial times, however, sea level rose as meltwater from glaciers returned to it, and streams filled their deeply trenched valleys with sand and gravel. Four intervals of cut and fill are evident in the Mississippi River valley, the bedrock channel of which is in some places 300 feet below the level of the river's present floodplain.

In both glaciated and nonglaciated regions of North America, the Pleistocene Epoch was clearly a time of great and repeated change. The surface and drainage were greatly modified by the ice sheets, and rainfall, meltwater, and sea-level fluctuations caused widespread alterations in the landscape south of the glacier margins. Most of the scenery with which we are familiar bears the stamp of the ice ages of the recent past.

In the last several chapters we have emphasized the concept of time, and practically all our statements have necessarily been made in the past tense. Clearly the earth's crust as we see it today has a history that stretches back into past time of almost inconceivable length (Figure 24-5). Are we to conclude from this that we stand at some dividing point in geologic time, that in some way the geologic story is "over"?

Not at all. We have seen that volcanoes, streams, and ice sheets are all as active today as ever; and their activities will continue into the future. Even the more majestic and ponderous features of the geologic past have their modern counterparts. Deep drilling for oil along the Gulf Coast of Louisiana and Texas, together with related geophysical

## 24-4
## GEOLOGIC HISTORY
## IN THE MAKING

**Figure 24-5**

Much of North America's geologic history is superbly documented in the western United States, especially in our national parks. Here are a few reminders of that fact, in the form of pictures that have appeared earlier in the book. We hope you will duplicate these views on your own trips, and will add pictures from the many other equally scenic and instructive features that constitute our geologic heritage.

All the landscapes shown on this page owe their present aspect to Neogene erosion: the work of weathering, gravity, streams, ice, and the wind.

Neogene volcanism is recorded by the lavas and pyroclastics exposed in the walls of Yellowstone Canyon. Figure 7-4, page 116.

Brightly colored strata of the Wasatch Formation, now intricately dissected by weathering and rainwash, are exposed in Bryce Canyon National Park. They were deposited in a Paleogene lake. Figure 5-7, page 96.

A feature of Cretaceous history in the western United States was the intrusion of great granitic batholiths. Among these was the Sierra Nevada batholith, a small part of which is shown here at Yosemite National Park. Figure 12-8, page 217.

The towering walls at Zion Canyon National Park are formed by the thick Navajo sandstone, of Jurassic age. A fossil dune deposit, it records desert conditions. Triassic strata are present in the slopes below the cliff. Figure 23-2, page 409.

In Monument Valley, southern Utah, remnants of a massive Triassic sandstone form spectacular buttes and spires. It is underlain by thin-bedded Permian sandstone and shale. Figure 18-9, page 332.

The record displayed in the Grand Canyon ranges from a Permian formation, the Kaibab limestone, at the rim, downward through sedimentary rocks of all the Paleozoic systems except the Silurian and Ordovician. Most of these rocks are of marine origin. At the bottom of the canyon is the Vishnu schist of Precambrian age. Figure 2-11, page 40.

surveys, has revealed a prism of sediments that is 550 miles long and at least 40,000 feet thick. Approximately parallel with the present Gulf Coast, this belt has subsided throughout the Cenozoic Era, acting as a trap for the thousands of cubic miles of sediments washed off central North America by the Mississippi River, its sister streams, and their predecessors. In short, this segment of the North American crust is behaving quite like mobile belts of the past.

Is mountain building going on today? We have only to look at the Pacific coastal ranges, whose crustal unrest is regularly revealed in such earthquakes as the one at San Fernando, California, in February 1971. Is there evidence today that whole regions are being bodily uplifted, as has happened in the past? Measurements of sea level in Scandinavia, extending over the past 125 years, have shown that the land is rising at a rate of roughly three feet per century. This is enough to drain the Baltic and raise the adjacent lands several thousand feet if continued for the geologically brief time of 100,000 years or so.

What about sea-floor spreading: Is there evidence that it is continuing? Indeed there is. Along the mid-oceanic ridges, for example, we find many shallow-focus earthquakes, an exceptionally high rate of heat loss, and volcanic activity as in Iceland and elsewhere. It is believed that North America and Europe are drifting apart at the relatively rapid rate of about one inch per year.

Thus today's deposit of sediment, flow of lava, or uplifted sea bottom is tomorrow's geologic history. We can do no better than once more to repeat the words of Hutton: "...no vestige of a beginning—no prospect of an end."

**24-5 SUMMARY**

In the Cenozoic, North America assumed its present size and shape, and may also have reached its present position with respect to other continents. Through the Paleogene Period, highlands in the Cordilleran region were reduced by erosion, and the detritus derived from them spread eastward into the Interior Lowlands and west into the Pacific mobile belt. In the early Neogene, after millions of years of erosion, the Laramide Rockies were rejuvenated, and block-faulting and extrusive igneous activity began in wide areas of the Central Cordillera. Late in the Neogene, compression in the Pacific mobile belt reached a climax in the Cascade Orogenies, and continued volcanism and block-faulting produced the Cascade Range and the Sierra Nevada, respectively. Moderate uplift along the eastern and southern borders of the continent made it possible for streams to strip the sedimentary cover from parts of the Appalachian, Ouachita, and Marathon mountains, and caused the shoreline to migrate slowly outward across the present coastal plains to its present position.

Most of the major geologic and geographic provinces of North America had developed their present form and structure by the beginning of the Pleistocene Epoch. Within the last million years, the center of the continent and northern segments of the continental border have been modified by four episodes of glacial erosion and deposition. Effects of glaciation are also recorded in nonglaciated areas, where we find evidence of repeated cut and fill in major stream valleys and indications of once-extensive lakes, which expanded during wet glacial intervals and shrank during drier interglacial times.

## SUGGESTED READINGS

Flint, R. F. 1971. *Glacial and Quaternary Geology*. New York: Wiley. 892 pp.
> A college text and reference book on glacial geology that includes detailed coverage of the glacial, climatic, and stratigraphic history of the late Cenozoic.

Schuchert, Charles. 1955. *Atlas of Paleogeographic Maps of North America,* maps 77–84. New York: Wiley.
> Eight paleogeographic maps of Cenozoic time-rock units, each with a list of major stratigraphic units represented.

Wright, H. E., and Frey, D. G. 1965. *The Quaternary of the United States*. Princeton, N. J.: Princeton University Press. 922 pp.
> A reference book that presents Quaternary geologic history of both glaciated and unglaciated areas. It also includes sections on the biogeography and archeology of the Quaternary in North America.

# 25 NATURE, ORIGIN, AND EVOLUTION OF THE BIOSPHERE

25-1   Fossils as History

25-2   Nature of the Biosphere

25-3   Organization of the Biosphere

25-4   Origin of the Biosphere

25-5   Evolution of the Biosphere

25-6   Summary

The history of the biosphere is recorded by fossils entombed in sedimentary rocks. We have already had a good
deal to say about how and where fossils are preserved,
and about what they can tell us of past climate and
environments. In addition, we have used fossils to help
reconstruct the ever-changing patterns of lands, seas, and
mountains, against which the history of life has been
enacted. With the stage thus set, we are prepared to study
fossils in a different light, that is, as relics of past stages
in the development of life.

Just as the story of continental evolution has been put
together piecemeal through laborious correlation of many
local rock sections, so the record of life has been assembled bit by bit by fitting together scraps of evidence from
all parts of the earth. Even though the evidence is widely
scattered, and in many critical places imperfectly or
indifferently preserved, the chief features of biosphere
history are well known.

### Definitions

The biosphere is the sum total of all living things, both
plant and animal. To this definition we may add the phrase
"past and present." Like most definitions, this one defines
only if we understand its significant words. These are
the terms *living, plant,* and *animal.*

Right away we hit a snag, for we must be content with
an operational definition of **living** (and thus **life**). That
is, we can only define life in terms of its capacities and
abilities, rather than in terms of what it is. Life is a process
that involves materials of very complex organization.
These materials are identified as living by their capacity
to grow, to reproduce, and to respond to various stimuli.

The living things, or **organisms,** that compose the biosphere are intricate systems of complex carbon-containing
chemical compounds such as carbohydrates, fats, and
proteins. These are termed **organic compounds** because
they occur naturally only as constituents of living organisms or as their derivatives. Inorganic compounds, including minerals, gases like carbon dioxide ($CO_2$) and
ammonia ($NH_3$), and water, may be present (often in large
quantities) in organisms, but they also occur in nature
independent of living things.

Although most of us readily differentiate plants from
animals among the familiar species, it is difficult to define
either group in a wholly satisfactory way. For our purposes it is sufficient to note that green plants build up,
or synthesize, from inorganic raw materials the organic
compounds essential to life; animals and most nongreen
plants, on the other hand, do not do this. Animals and
nongreen plants depend on green plants to make the basic
organic materials they need, although they are capable

of modifying these substances to suit their own requirements.

### Abundance and Diversity of Organisms

So much, then, for this definition of the biosphere. It is sketchy, but will fill the bill for the time being. There are two additional factors, however, that play an important role in our understanding of life and its history—the immense abundance of organisms in the modern biosphere, and the staggering diversity of living animals and plants.

There is no way of determining the number of individual organisms on earth today, or of those that have existed at any time in the past. A measure of their abundance is provided by a few recent estimates. Near Washington, D.C., for example, where more than a million humans are crowded into just a few thousand acres of space, the upper inch of soil is thought to be inhabited by more than one million animals and two million plant seeds per acre. Every acre of meadow soil in the mid-latitudes has been estimated to contain approximately 13 million animals and 34 million plant seeds, and the upper foot and a half of sediment on the bottom of a West Coast estuary contains nearly three million organisms per acre. None of these estimates includes animals or plants of microscopic size, and for good reason. There may be several hundred million bacteria in a single gram of soil!

**25-3
ORGANIZATION
OF THE
BIOSPHERE**

It would be a hopeless task to make any general remarks about the biosphere or its history if we had to consider every one of the unnumbered billions of individuals in it today. Fortunately, these many individuals may be grouped into **species** of plants and animals, which are far fewer in number than the individuals.[1] Even so, the immense diversity of the living world is indicated by the fact that the modern biosphere includes more than 300,000 named species of plants and about 1,120,000 named species of animals. And it has been estimated that the total number of animal and plant species that have existed since the beginning of the biosphere is in the neighborhood of half a billion.

The practical difficulties in discussing the biosphere in terms of its half billion or so species are immediately

[1] There are many definitions for the word *species*. Most biologists view species as groups, or populations, of morphologically distinctive organisms that are capable of interbreeding but are reproductively isolated from other such groups. Paleontologists have some trouble with this definition, for they work almost entirely with fossils. Fossil species are distinguished primarily on the basis of morphology and distribution.

apparent. For one thing, what would we call all these kinds of organisms; how would we refer to them in discussion? Further, half a billion is of course far too large a number for rational consideration.

## Classifying Organisms

A scheme for getting around both these difficulties was devised more than two centuries ago by the great Swedish botanist, Carl von Linné, usually referred to as Linnaeus. Linnaeus was disturbed not only by the lack of a systematic, logical plan for subdividing, or classifying, the biosphere, but also by the helter-skelter way in which newly recognized kinds of organisms were being named by his contemporaries. To solve both problems, he worked out a means of subdividing the biosphere and a standard way of applying names to each of the subdivisions. In addition, he compiled a thick and often revised catalog, the *Systema Naturae* (System of Nature), in which he assigned nearly all known animals and plants to one or another of his divisions, and gave names to them in accordance with the scheme he had worked out.

Linnaeus divided the biosphere into two large parts, a **plant kingdom** and an **animal kingdom.** Within each, he recognized a wide variety of organisms, related to one another in form, function, or "blood kinship," but on several different levels. For example, he noted that although leopards, tigers, lions, jaguars, and household tabbies are all distinctly different kinds, or species, of animals, they are clearly related and can all be regarded in a general way as "cats." All these cat species can then be grouped and considered to represent on a somewhat higher level a single category, or a **genus** (plural, **genera**). Cats and other catlike animals (such as extinct saber-toothed cats) can be brought together into a larger group to form a cat **family.** Families can be grouped with related families at an even more general level to form **orders,** and in a similar way, orders are grouped into **classes** and classes into **phyla** (singular, **phylum**). Each of the two biosphere kingdoms is an aggregate of several phyla.

Not only is the system of biosphere classification worked out by Linnaeus an orderly one, but also it permits discussion and study of the biosphere on several different levels. Students of cats, for example, may discuss this group on the level of species, genus, or family. Others, interested in more general matters, may be content to regard cats simply as one small group of a much larger division of animals, the Phylum Chordata, which includes all animals with backbones.

**Naming Organisms**

Recognizing natural groups of organisms on various levels of relationship is one thing; providing names for handy reference to them is another. It is fairly clear that babies learn to distinguish between mother, father, and the family dog some time before they learn what to call them. Fortunately, Linnaeus had the genius not only to classify natural specimens but also to find names for the subdivisions.

*The Linnean System*

It is immediately apparent that in the art of name-giving great confusion would arise if two different species of organisms were given the same name, or if two different names were given to the same species. Each species must have a name that is uniquely its own if any sort of sanity is to be preserved in studies of the biosphere. But there simply are not enough words to go around if each of half a billion species is to have a separate name. Linnaeus invented a system whereby each species is given a name that consists of two words, the first a noun representing the genus to which it belongs (for example, cat or dog), the second an adjective (or a word used as one) describing an outstanding attribute of the species. Because Linnaeus and other scholars of his time wrote in Latin, the two-part names were first written in that language and alphabet, and this custom has persisted.

Linnaeus's system of naming organisms is logical simplicity itself. For example, the household tabbycat is *Felis domesticus* (*Felis* means "cat" in Latin; *domesticus* is an adjective meaning "domestic"). A leopard could be termed *Felis pardus* (spotted cat). In these two examples, the noun *Felis* is the generic name; *domesticus* and *pardus* are specific (or species) names. The "scientific name" of each species consists of both noun and adjective. If we use *Felis* alone, we refer to the genus, and therefore all the species therein. Neither *domesticus* nor *pardus* has any standing without a noun, and such specific adjectives are never used alone.

*Advantages of the Linnean System*

The Linnean, or binominal, system of giving animal and plant species two-part Latin names is used all over the world, for it has several distinct advantages. First, it appreciably increases the number of times an adjective can be used in names (a spotted dog, for example, could be called *Canis pardus*). Second, relations between similar species are immediately apparent (the common generic name *Felis,* for example, tells us at once that both *Felis domesticus* and *Felis pardus* are cats). Third, Latin has long been the universal language of learned men, so the names for animals and plants look and read exactly the same in texts written in English, Chinese, Russian, German, or Swahili.

## Examples of Classification and Nomenclature

The names of families, orders, classes, and phyla are single words. All cats are included in the family Felidae. In this case, the family name is taken from that of its most typical genus *(Felis),* and to the stem is added the Latin suffix *-idae,* which signifies "family of." Thus we have Felidae as "family of cats." A complete family tree of tabby becomes:

Kingdom Animalia (all animals)
  Phylum Chordata (all animals with backbones)
    Class Mammalia (all backboned animals with
      milk-producing glands)
      Order Carnivora (meat-eating clawed mammals
        with tearing and cutting teeth well developed)
        Family Felidae (the cat family)
          Genus *Felis* (cats)
            Species *domesticus* (domestic cats)

You might be interested in comparing the above rundown with that of man. Notice the level at which our classification diverges from the other.

Kingdom Animalia (all animals)
  Phylum Chordata (all animals with backbones)
    Class Mammalia (all backboned animals with
      milk-producing glands)
      Order Primates (five-fingered mammals with flat
        nails instead of hoofs or claws)
        Family Hominidae (family of man)
          Genus *Homo* (man)
            Species *sapiens* (wise or knowing man)

The names and distinguishing features of the important plant and animal phyla are given in Tables 25-1 and 25-2. These groups and their many subdivisions are also described and illustrated in most encyclopedias and in laboratory reference books, but not always under the same names used in Tables 25-1 and 25-2. For our purposes it will be sufficient to discuss life history on the level of phylum and class; reasonable familiarity with the nature of these divisions will be assumed.

**Table 25-1  Principal Divisions of the Plant Kingdom**

| Phylum | Distinguishing Features | Other Names | |
|---|---|---|---|
| **Schizophyta** | Microscopic one-celled nongreen plants. Principal infectious-disease producers in animals, and chief agents of decay. | bacteria | |
| **Cyanophyta** **Chlorophyta** **Chrysophyta** **Rhodophyta** **Phaeophyta** | Aquatic, one-celled, colonial, or many-celled plants, whose green color is often masked by blue, yellow, brown, or red pigment. Structure varies greatly; size ranges from microscopic to 200 feet in length. Tissues little differentiated; organs less specialized than the true leaves, stems, and roots of most land plants. Includes such diverse forms as diatoms, pond scums, stoneworts, and seaweeds. | algae | thallophytes |
| **Mycophyta** | Mostly multicellular; small or microscopic; nongreen; decay- or disease-producing plants, lacking roots, stems, or leaves. Includes mushrooms, yeasts, and molds like *Penicillium*. | fungi | |
| **Bryophyta** | Small, green, leafy, multicellular plants without true roots; some with simple stems. Most inhabit damp places. | mosses and liverworts | |
| **Psilophyta** | Sparsely branched terrestrial plants with wiry, woody stems, but lacking true roots and leaves; anchored by underground stems. Nearly extinct; only two living subtropical forms. Important as fossils in Lower Devonian rocks. | whisk-ferns | woody or vascular plants, or tracheophytes |
| **Lepidophyta** | Small to large land plants with true roots, leaves, and woody stems. Spores usually in cones; adapted for wide dispersal. Modern forms (club mosses and their kin) are small, but many grew to treelike size in the Late Paleozoic. | club moss ground pine | |
| **Arthrophyta** | Small to large land plants with roots and upright branched or un-branched stem, ribbed outside and hollow in center at maturity. Leaves usually in whorls around stem joints. Spores shed from cones at stem tips. Modern scouring rushes have little woody tissue, but Late Paleozoic forms had woody stems. | horsetails  scouring rushes | |
| **Filicophyta** | Well-developed roots; large, feathery leaves; stems without much wood. Spores form in tiny sacks, usually carried on under surfaces of leaves. | ferns | |
| **Cycadophyta** | Low to tall land shrubs or trees; with fleshy roots, trunklike stems, few branches, and palmlike fronds of large, leathery leaves. Embryo protected and nourished in seeds. Roots, leathery leaves, and complex pollination mechanism insuring fertility of seeds, enable these plants to be successful in dry climates. | gymnosperms | seed plants |
| **Coniferophyta** | Mostly woody evergreen trees and shrubs, with roots, many-branched stems, broad or needlelike leaves, and seeds in cones. | | |
| **Ginkgophyta** | Woody trees with deciduous leaves. Plumlike seeds and pollen form on separate plants. Members of this phylum were common in the Mesozoic; only one species lives today. | | |
| **Gnetophyta** | Desert shrubs and trees; a few tropical vines; and an essentially stemless long-lived woody plant with only two leaves. All produce naked seeds in cones. | | |
| **Anthophyta** | Familiar flowering, fruit-, and nut-bearing plants, including crop plants, "vegetables," grasses, and ornamentals. Many are deciduous; some are woody. All have flowers, and those of many species are conspicuous and colorful. All bear seeds enclosed in an ovary, which ripens to form the fruit. | flowering plants (angiosperms) | |

**Table 25-2**   Principal Divisions of the Animal Kingdom

| Phylum | Distinguishing Features | Other Names | |
|---|---|---|---|
| **Protozoa** | Tiny, one-celled (or acellular), aquatic. Many naked; some (*Foraminiferida, Radiolaria*) with calcareous or siliceous skeletons. | | |
| **Porifera** | Multicellular, mostly marine; cells not organized into tissues or organs. Many have internal needle-, hook-, or star-shaped calcareous or siliceous spicules. | sponges | |
| **Coelenterata** | Multicellular, radially symmetrical, mostly marine. Body of two tissue layers with stinging cells: no body cavity in addition to digestive tract. *Polyps* are saclike, have tentacle ring around open end, are commonly colonial and attached, and secrete a horny or calcareous skeleton. *Medusae* are typified by jellyfish, which are not attached, have thick jellylike body walls and no mineralized skeleton. | corals<br>jellyfish<br>anemones | |
| **Platyhelminthes**<br>**Rhynchocoela**<br>**Aschelminthes**<br>**Annelida**<br>**Chaetognatha** | A diverse assemblage of both primitive and highly advanced multicellular animals, mostly aquatic, bilaterally symmetrical, and of elongate wormlike shape. They are related primarily in having a very scanty fossil record. | worms | |
| **Ectoprocta** | Tiny, colonial, multicellular, unsegmented, bilaterally symmetrical, mostly marine. Body of three tissue layers; digestive tract in body cavity; fringe of tentacles around mouth; body protected by external calcareous skeleton. | bryozoans<br>moss<br>animals | |
| **Brachiopoda** | Solitary, unsegmented, multicellular, bilaterally symmetrical; marine. Body of three tissue layers; digestive tract in body cavity; fringe of tentacles around mouth. Most are attached to sea floor by fleshy stalk; all have an external bivalved shell. In class Articulata the two calcite shell valves are locked together by teeth and sockets along their posterior margins; in class Inarticulata the two valves are chitin and calcium phosphate and lack teeth and sockets on hinge margin. | lamp<br>shells | I N V E R T E B R A T E S |
| **Mollusca** | Solitary, unsegmented, multicellular; body of three tissue layers; digestive tract in body cavity; muscular ventral foot; most with distinct head. Three classes are common: | mollusks | |
| | Gastropoda—head, ventral creeping foot; univalved external calcareous shell is cap-shaped or coiled symmetrically or asymmetrically. Marine, freshwater, land. | snails | |
| | Bivalvia—no head; foot modified for digging, or absent; external bivalved shell borne laterally; two valves locked by dorsal teeth and sockets. Marine, freshwater. | pelecypods<br>clams<br>oysters<br>scallops | |
| | Cephalopoda—head with eyes and tentacles; "foot" is a ventral tube from which water is expelled to provide movement by jet propulsion. Some with straight, curved, or symmetrically coiled external calcareous shell divided inside into chambers by cross-partitions. In others shell is internal and much reduced. Exclusively marine. | squids<br>*Octopus*<br>*Nautilus* | |
| **Arthropoda** | Bilaterally symmetrical, multicellular; with segmented body and appendages, three tissue layers, and digestive tract in body cavity. Most have a hard segmented external skeleton; in some, skeleton is largely organic material; in others it includes a substantial amount of mineral matter and is capable of fossilization. | crabs<br>lobsters<br>spiders<br>insects<br>trilobites | |

Table 25-2  (continued)  Principal Divisions of the Animal Kingdom

| Phylum | Distinguishing Features | Other Names | |
|---|---|---|---|
| **Echinoderma** | Solitary, multicellular, exclusively marine; pentamerous radial symmetry masks basic bilateral symmetry. Most with dermal skeleton of spiny calcite plates. Two main groups: | | |
| | SUBPHYLUM PELMATOZOA—attached, plantlike, with roots, stems, and bodies provided with arms or armlike structures. Only crinoids alive today. | cystoids blastoids crinoids | |
| | SUBPHYLUM ELEUTHEROZOA—vagrant, unattached echinoderms of bun-, disk-, sac-, or starlike shape. | echinoids starfish | INVERTEBRATES |
| **Hemichordata** or **Stomochorda** | Worm-, or bryozoanlike; solitary or colonial; marine. Body bilaterally symmetrical, of three tissue layers; digestive tract and body cavity. Gill slits or pores open into pharynx and a saclike structure (stomochord) projects from the anterodorsal portion of pharynx. Modern colonial forms secrete a distinctive chitinoid external skeleton, as did the extinct *graptolites*, which may be fossil representatives of this phylum. | | |
| **Conodonta** | Swimming marine invertebrates with skeletons of tiny cone-, hook-, comb-, or platform-shaped pieces composed of calcium fluorophosphate. Appeared in Cambrian; extinct by end of Triassic | conodonts | |
| **Chordata** | Multicellular, bilaterally symmetrical; with three primary tissue layers, a capacious body cavity separate from digestive tract, a distinct head, a large brain and a dorsally concentrated nervous system. In many a vertebral column (backbone) and gill slits. Marine, freshwater, land, air. | fishes amphibians reptiles birds mammals | CHORDATES |

**25-4 ORIGIN OF THE BIOSPHERE**

It is curious that few men in any generation have regarded life as eternal. As a matter of fact, most cultures agree that living things came into being, or were created, at some time in the past. Explanations of how this happened vary greatly from one time to another and from one culture to the next; but interestingly enough, most accounts assume that in the beginning masses of ordinary earth materials in some way became endowed with the capacities and patterns of behavior that we associate with life. Because most of these accounts make no pretense of being scientific, we are not concerned here with their details. Nevertheless, it is worth noting that most scientists agree that life came into being on earth, rather than arriving full-blown from outer space, and that its "coming into being" involved the accumulation of nonliving earth materials into aggregates of molecules able to grow, react to environmental stimuli, and reproduce themselves more or less faithfully generation after generation.

We should emphasize that there is no direct evidence bearing on the origin or very early history of life, and it is probable that there never will be. Thus, the scientific agreement noted above is assumption, pure and simple: a place to start, no more. Even so, scientific reconstruction of the earliest part of life history does not need to be completely speculative. Indeed, it would not be worth mentioning if it were.

## A Basic Assumption and a Problem

If we assume that life began on earth and that its beginning involved the transformation of nonliving (inorganic) materials into living things, we must also accept the conclusion that conditions on earth were such that natural organic compounds could be produced nonbiologically. The fact that this is not so today, at least on any very great scale, further implies that past conditions differed from those of the present.

Our basic assumption and the two conclusions that follow from it clearly define a problem. That is, under what conditions can the basic stuff of which living things are built be made naturally and nonbiologically? If this can be answered experimentally, the answer can be applied to understanding conditions on earth prior to the oldest fossil record of the biosphere. As geologists, we are convinced of the uniformity of process through geologic time; but we are also aware that both the scale and the rate of operation of processes have varied considerably in the past.

## Nonbiologic Synthesis: The Beginning

No one has yet created a living thing in the laboratory, but experimental attempts at making the basic building blocks of organisms have been intriguingly successful. Quite a variety of basic organic compounds (amino acids, for example, and probably sugars) have been repeatedly synthesized in the laboratory, under rather surprising conditions. Attempts to make them in an atmosphere rich in oxygen (that is, in an oxidizing environment) have generally been fruitless, whereas experiments conducted under reducing conditions [in an environment rich in such substances as methane ($CH_4$), ammonia ($NH_3$), and hydrogen ($H_2$)] have been excitingly productive. These experiments suggest that the environment in which organic compounds first came into being was a reducing rather than an oxidizing one; that is, it lacked the abundance of oxygen typical of the earth's present atmosphere.

Since relatively simple organic materials have been manufactured in the laboratory by discharging an electric spark into a mixture of methane, ammonia, hydrogen, and water, it is neither illogical nor wildly speculative to view the primitive atmosphere as a combination of some or all of these materials. The electric discharge that supplied the energy for synthesis could have been lightning, ultraviolet light, or both. (Interestingly, spectrographic studies suggest that the atmospheres of Saturn, Uranus, and Neptune are largely methane, and that of Jupiter is mostly methane and ammonia.) Quite possibly, then, in its earliest stages the earth had an atmosphere similar to those of its much larger companions in space.

## An Interval of Aggregation

It was a giant step, indeed, from nonbiologic creation of organic compounds to formation of complex aggregates of molecules with all the capabilities and abilities of life. This step has not been duplicated in the laboratory. Even so, with faith in uniformity of process and with appreciation of the vast length of geologic time, we can reasonably guess at the steps intermediate between nonbiologic synthesis and emergence of the first living things.

Without the oxygen and the protective shield of ozone ($O_3$) that occur in the present atmosphere, ultraviolet light would penetrate to the surface with much more intensity than it does now, thus providing a much greater source of energy for nonbiologic organic synthesis than is available today, except in lightning bolts. Furthermore, newly formed organic matter would accumulate in the hydrosphere, because without a biosphere, such matter would neither decay nor be eaten. Thus, the organic content of the sea probably increased as long as similar conditions prevailed.

In the thin organic soup of the ancient hydrosphere, it is likely that simpler compounds were concentrated—perhaps on clay flakes—and ultimately became banded together to form the large molecules that seem to be the building blocks of all living things. Because each organism today consists of many different kinds of these large molecules (plus a lot of water), we can guess that the next step in the development of life was an aggregation of chemically compatible large molecules that both complemented and supplemented each other's reactions. These aggregates would certainly have been more efficient and effective than individual large molecules that maintained their independence. Thus we can assume that these successful aggregates grew larger, became more complicated, and were more consistently uniform in both composition and design. Certainly those best able to form smooth-working systems would tend to dominate aggregates less efficiently and permanently organized. The successful ones probably developed into the first organisms.

## The First Organisms

The growth, reproduction, and responsiveness that characterize living systems require great expenditures of energy. Modern organisms acquire this energy either directly from the sun or by using ready-made organic food as fuel. Presumably the first organisms derived energy in one of these ways, too.

Because direct use of solar energy by modern plants depends on more sophisticated systems than the earliest organisms probably possessed, it is likely that these organisms obtained their energy by feeding on inorganically created organic matter in the hydrosphere. Fermentation, a process familiar to yeasts and home brewers, may well have played a significant role in maintenance of the earliest organisms. In fermentation, such substances as sugar are broken down to yield products like alcohol, water, and carbon dioxide. The energy liberated in such a process, often expressed as heat, may have been sufficient for growth, reproduction, and activity. Although fermentation is an inefficient process, it is a simple one that may well have tided the earliest organisms over until something better was possible. One of its side effects would have been addition of considerable carbon dioxide to the atmosphere.

## Crisis and Response

At an early date, consumption of non–biologically created organic matter may well have outstripped its production. This imbalance in supply and demand may have brought the biosphere to the first of its many times of crisis. Because we know that the biosphere survived, we must conclude that by the time food was in short supply some type of organism had developed the ability to make organic compounds itself, directly from inorganic raw materials. Self-sufficient organisms like this would rapidly have achieved dominance over more dependent types. The final achievement in this chapter of biosphere history would have been the appearance of organisms that could use carbon dioxide (a waste from fermentation) as their only external supply of carbon, and sunlight as their only external source of energy. Perhaps these were the first plants.

## Changes in the Atmosphere

While all these things were happening in the primitive biosphere, changes must also have taken place in the atmosphere. Perhaps the methane and ammonia of the

439

primitive atmosphere were mostly used up in development of an ever larger and more complex biosphere; or were changed, at least in part, into free nitrogen and carbon dioxide by the activities of ancient organisms. Early plants probably obtained most of their carbon from carbon dioxide in the atmosphere and released an increasing quantity of oxygen. By this means, the atmosphere probably changed slowly from its earlier reducing to its modern oxidizing state.

As oxygen accumulated, part of it came to form an ozone layer high in the atmosphere. Today, and probably through much of the geologic past, this layer effectively screens out much of the ultraviolet light received from the sun. Because such light is considered to have been an important source of energy in the early stages of biosphere development, organisms like plants, which depend on solar radiation for external energy, must have developed the means of using visible sunlight as an energy source. The organic pigments (chlorophylls) of modern plants serve such a purpose and probably developed early in the history of the biosphere.

Without doubt, the existence of food-manufacturing plants made possible the subsequent development of organisms that feed on them; and the appearance of these set the table for others that feed on plant eaters.

## The Geologic Evidence

One might expect some geologic evidence to support, at least indirectly, some or all of the stages in life history just outlined. It happens that certain features of very old rocks have been thought to support part of this history.

For example, if oxygen began to accumulate in the atmosphere only after development of plantlike organisms, there may have been a considerable interval in early earth history during which the atmosphere contained far less oxygen than it does now. Thus we might surmise that the whole process of chemical weathering as understood today was considerably different in the past. Whereas only mineral oxides are now especially resistant to chemical decay, even feldspars and such sulfides as pyrite would have been relatively resistant beneath an oxygenless atmosphere, and sandstones might have contained more feldspar and pyrite than those sandstones formed today. We might also expect to find ferrous iron dominant over ferric iron in old soils; few or no deposits colored red by oxidized iron; and greater quantities of unoxidized native carbon, coloring sediments black, than in rocks recording later stages of earth history.

All these characters have indeed been mentioned as

attributes of many of the oldest Precambrian rocks. However, assembling and evaluating the data necessary to support the contention that the early Precambrian atmosphere lacked oxygen are largely jobs for the future. Such evidence as is available today is much disputed, although a good deal of it seems to suggest the existence of an early Precambrian atmosphere much different in composition from the familiar oxygen-rich one of today.

Although we can be sure of only a few links in the chain of reasoning that leads to the concept of life origin we have sketched, it fits what little we know about life and the part of the solar system in which it developed. It is, at the very least, an eminently reasonable working hypothesis. However the biosphere originated, not until much later in its history did it begin to leave an objective record.

In our discussion of possible stages in the development of the primitive biosphere, change with time has been implicit. Indeed, inevitable change is built into the scheme we have outlined. Why, you may ask at this point, have we assumed the likelihood of such a sequence of changes so confidently? Why not explore the less complex (and much older) idea that every species came into being (or was created) in its present form at the beginning of things? Why, in other words, have we chosen an *evolutionary* explanation for the unrecorded early stages in biosphere history in preference to the doctrine of *special creation*?

**25-5 EVOLUTION OF THE BIOSPHERE**

### Evolution versus Special Creation

The studies of Aristotle, in the fourth century B.C., and of a few of his predecessors, resulted in establishment of a "ladder of life," with small, simple forms on the lower rungs and larger, more complex forms higher up. This progression, which was largely classificatory and not evolutionary, was interpreted to mean that organisms at every level of the ladder were independently created, separate from organisms at other levels. Each type was held to be endowed originally with the capacity to reproduce itself faithfully generation after generation. This view of the biosphere, a static one, became the doctrine of special creation.

Special creation, which influenced scientific thinking for many centuries, was called into question as careful studies revealed too many similarities between diverse types of organisms to be explained satisfactorily or easily in this manner. For example, comparison of the legs of a man with those of a dog, an anteater, a horse, and other

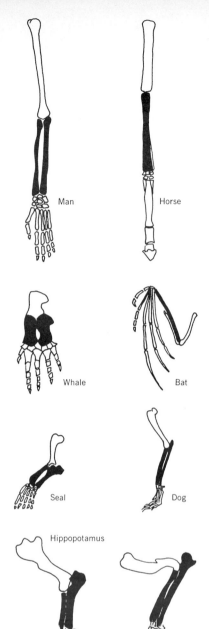

vertebrate animals (Figure 25-1) reveals that even though they differ in detail, all have virtually identical skeletal elements. Furthermore, the same arrangement of bones is found in the wing of a bat and the flipper of a seal, even though the limbs of these animals are peculiarly suited for flying in the one and swimming in the other rather than for walking or running. If all these organisms had been specially created, it seems doubtful that limbs with such different functions would have identical skeletal patterns; certainly something other than a modified walking leg would most efficiently serve as swimming or flying structures.

As early as 1760, it was concluded (by some at least) that the similarities in structure noted above are best explained by assuming that dogs, pigs, horses, bats, seals, and man are divergent descendants of a distant and long-defunct common ancestor, who had five toes and from whom they all inherit a basic limb design. The pattern is modified in many different ways, permitting success in a variety of habitats and modes of life. This view, an evolutionary one, contrasts strongly with the static system maintained by the doctrine of special creation. Which, if either, has prevailed in biosphere history?

## Evolutionary Theory and the Fossil Record

Choice between alternative explanations for the same phenomenon is made after study of all the evidence. One or the other is generally regarded as proved if it consistently provides the best or most straightforward answer to each problem. It must be thus with special creation versus evolutionary development of the biosphere. Though we cannot examine here every aspect of the biosphere in this way, one such aspect—the fossil record— is of such telling importance in the argument that it merits special attention.

Since the beginning of the Paleozoic Era, the biosphere has left a passably adequate record of itself in the form of fossils embedded in crustal rocks. Because the order of formation of these rocks can be deduced from the law of sequence, the order of appearance and range in time of species represented by fossils can also be determined. If each species had been specially created at the beginning of things, we should expect to find about the same kinds of fossils at every level in the geologic column. This is far from the case. At no two levels in the column are fossils exactly alike. Furthermore, the number and extent of differences between them increases appreciably as their separation in the column increases. Fossil Cambrian ani-

**Figure 25-1**
The forelimbs of eight types of vertebrate animals. The upper bone (humerus) in each forelimb joins the shoulder (not shown), and is followed by two bones (radius and ulna) shown in black. A cluster of wrist bones separates radius and ulna from bones of the fingers (or toes). Note that all limbs shown have these bones, but that the pattern is modified in each animal, permitting efficient running, swimming, or great manual dexterity.

mals, for example, resemble Ordovician animals more than they resemble those of the Cretaceous.

If the fossils that make up the fossil record are studied in order, from the bottom of the geologic column to the top, it is immediately apparent that change in form from layer to layer is mostly gradual; new species in successive beds are but slightly revised editions of those in older rocks beneath. If each species had been specially created at the beginning or at different times, it is quite unlikely that so much intergradation would occur. On the other hand, such a gradational relationship is just what we should expect if the biosphere had evolved in time, that is, if earlier species had gradually changed to become, or to produce, later ones.

Finally, the fossil record might be expected to verify the prediction that horses, for example, have at the root of their family tree the same sort of five-toed ancestor thought to have given rise to pigs, seals, bats, and other mammals. In other words, we might expect to find in the rocks a progression from five-toed forms to one-toed modern horses. Just such a progression has been found (Figure 30-8) and its fossil record occurs in just the order expected.

From the examples cited, and from many others like them, scientists conclude that evolution, rather than spasmodic special creation, has been a feature of biosphere history since at least the beginning of the Paleozoic. Since it seems unlikely to us that evolutionary development is an innovation of the last 570 million years, we choose to assume that the early history of the biosphere, documented by only a few small fossils, was also evolutionary.

In summary, the fossil record, useful in a variety of ways to geologists, consistently supports an evolutionary view of biosphere history. It does not tell us how evolution works (we leave this subject to more advanced courses), but it provides the most powerful proof yet discovered of the fact that it has worked. Acceptance of the evidence provided by the fossil record has given the study of animals and plants a new dimension and the collection and evaluation of fossils a new importance.

**25-6 SUMMARY**

The biosphere, which includes all living things past and present, has a long history that is recorded by fossils. Because the number of kinds, or species, of organisms is incredibly large, the Linnean system for classifying and naming them has proved very useful, for this system permits study and discussion of the biosphere on several different levels.

Evidence from fossils is lacking for the very earliest part of biosphere history. However, a working hypothesis has been developed that fits what we know of life and the part of the solar system in which it developed. If it is assumed that the biosphere originated on earth, an initial stage in its history must have involved nonbiologic synthesis of basic organic materials, probably in an environment deficient in oxygen, or lacking it completely. Subsequent stages would have involved aggregation of these materials into complex living aggregates of organic molecules, and ultimately, development of the capacity by successors of these aggregates to use sunlight as an external energy source in synthesis of new organic materials from carbon dioxide. A consequence of the latter stage in biosphere development, the appearance of photosynthetic plants, would have been the release of oxygen into the atmosphere.

An evolutionary view of biosphere history is chosen in preference to one that involves independent, or special, creation of each type of organism, since the fossil record shows a gradual succession of morphologic changes, not abrupt and erratic additions of new types of organisms. The fossil record does not tell us how evolution works, but the study of fossils adds a crucially important dimension to evolutionary theory.

## SUGGESTED READINGS

Clark, D. L. 1968. *Fossils, Paleontology and Evolution,* chaps. 3, 4, and 5, pp. 11–32. Dubuque, Iowa: W. C. Brown. (Paperback.)
The origin and evolution of life, and the classification and nomenclature of fossils, are included in this beginning-level book.

Kurtén, B. 1969. Continental drift and evolution. *Scientific American,* vol. 220, no. 3, pp. 54–64. (Offprint No. 877. San Francisco: Freeman.)
Explores possible significant effects of continental drift on the evolution of organisms.

McAlester, A. L. 1968. *The History of Life,* chaps. 1 and 2, pp. 4–38. Englewood Cliffs, N.J.: Prentice-Hall. (Paperback.)
A chronologic treatment of the evolution of life, written for beginning geology students. Chapters 1 and 2 cover the origin, early development, and diversification of life.

Newell, N. D. 1963. Crises in the history of life. *Scientific American,* vol. 208, no. 2, pp. 76–92. (Offprint No. 867. San Francisco: Freeman.)
Discussion of the reasons for extinction of entire animal groups, with emphasis on the importance of natural catastrophes in these crises.

Rhodes, F. H. T. 1962. *The Evolution of Life*. Baltimore: Penguin Books. 302 pp. (Paperback.)

A popular account of the principal groups of plants and animals and their development through time.

Rutten, M. G. 1962. *The Geological Aspects of the Origin of Life on Earth*. Amsterdam and New York: Elsevier. 146 pp.

Considers geologic features that should be found if early earth lacked an oxygenic atmosphere and if life developed as suggested in our Chapter 25.

Simpson, G. G. 1967. *The Meaning of Evolution,* rev. ed. New Haven: Yale University Press. 333 pp. (Also available in paperback. New York: Bantam Books, 1971.)

Excellent readable discussion of the history of evolution as a concept, together with a description of evolutionary mechanisms and their results.

# 26 THE PRECAMBRIAN BIOSPHERE

26-1   Early Precambrian

26-2   Middle Precambrian

26-3   Late Precambrian

26-4   Shells and Skeletons

26-5   Summary

We include in the Precambrian eras the first 4.0 billion years (or about 87 percent) of earth history. Among the rocks formed during this vast stretch of time are great thicknesses of relatively undeformed and only slightly altered sedimentary strata, of types in which we would expect to find an abundant fossil record of the Precambrian biosphere. Curiously, most of these rocks lack conspicuous fossils, and time-consuming searches for microscopic organic remains have gone largely unrewarded. Here and there in the Precambrian section, however, are beds and nodules of chert and a few thin limestones; and studies of these in the last few years have revealed a remarkably varied assortment of well-preserved microscopic fossils. These fossils shed considerable light on biosphere history for all but about the first 1.5 billion years of the Precambrian eras.

The oldest fossils thus far discovered are bacteriumlike rods and algalike spheroids, at least 3.1 billion years old, found in black chert collected from the Fig Tree Group near Barberton, South Africa (Figure 26-1). Suspiciously algalike structures have also been discovered in rocks that

**26-1
EARLY
PRECAMBRIAN**

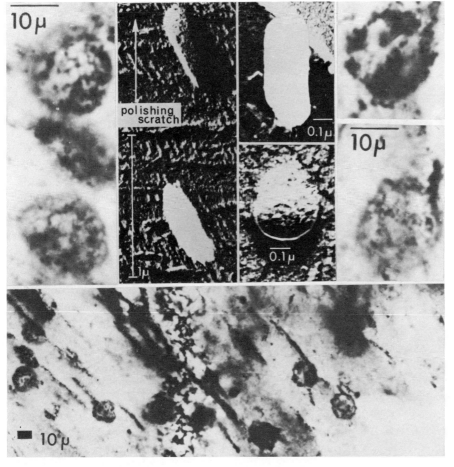

**Figure 26-1** Microfossils from the 3.1-billion-year-old Fig Tree Group of South Africa. The spherical bodies in the views with light-colored backgrounds are algalike microfossils; elongated objects in the four views with dark backgrounds are bacteriumlike cells. The symbol $\mu$ means "micron," a unit of measurement equivalent to one thousandth of a millimeter; there are 254,000 microns in an inch. (Photographs courtesy of J. W. Schopf.)

are just beneath the Fig Tree Group and about 3.2 billion years old, but it is not entirely certain that these structures were organic in origin. Comparison of the algalike microfossils with living algae suggests that the ancient organisms were photosynthetic food producers, and thus the products of a probably long period of early evolutionary development of which we have as yet no fossil record.

Additional information about the early Precambrian biosphere is provided by strange concretionlike structures, called **stromatolites.** These are known to have formed as accumulations of calcareous material on sheetlike mats formed by communities of microscopic organisms, in which filamentlike blue-green algae play the principal role. Threadlike, or filamentlike, microfossils were found many years ago in the early Precambrian Soudan Iron Formation of the Lake Superior region, but the oldest stromatolites known (Figure 26-2) are from a limestone of the Bembesi gold belt in Southern Rhodesia,

**Figure 26-2**
Stromatolites from a 3-billion-year-old limestone of the Bembesi gold belt, Southern Rhodesia. Laminated masses of calcite in specimens like this were produced by algae. Natural size. (Photo by J. M. Schopf, U. S. Geological Survey.)

which is thought to be between 2.8 and 3.1 billion years old. The organic origin of these old structures has recently been proved chemically, and they thus place a minimum date on the time of origin of the blue-green algae, on oxygen-producing photosynthesis, and on the evolution of integrated biologic communities.

**26-2**
**MIDDLE**
**PRECAMBRIAN**

Sedimentary rocks deposited in the Middle Precambrian, between about 2.5 and 1.7 billion years ago, contain a considerably better and more widespread record of the

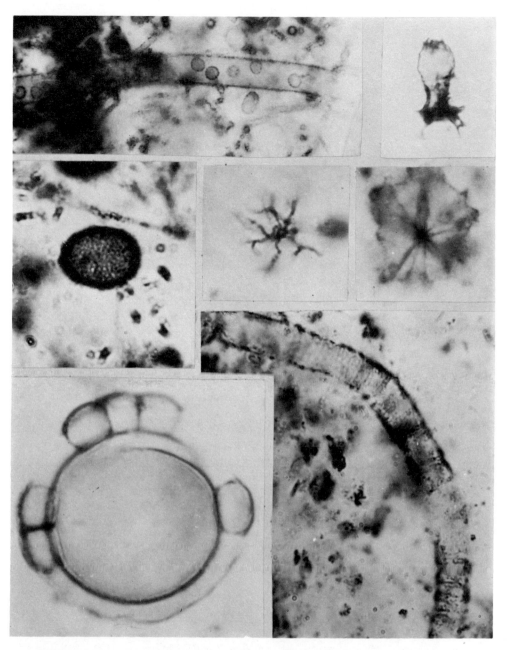

biosphere than rocks of Early Precambrian age. Middle Precambrian fossils are known not only from South Africa but also from Greenland, southern Ontario, the Belcher Islands in Hudson Bay, and several other places. Microscopic fossils from cherts of the Gunflint Formation of southern Ontario, illustrated in Figure 26-3, give an idea of the nature and diversity of the Middle Precambrian biosphere. These famous fossils, preserved organically and in three dimensions in very fine-grained chert, are primarily microscopic algae, which formed laminated mats on the bottom of a shallow, agitated basin. Less

Figure 26-3
Microscopic plant fossils, all enlarged more than 1,000 times. The specimens shown here are from the Gunflint Formation of southern Ontario, which is nearly two billion years old. (Courtesy of E. S. Barghoorn.)

449

commonly represented are forms that strongly resemble certain iron- and manganese-oxidizing bacteria, and others that may have been floating algae. Although vague dark spots within some of these one-celled fossils suggest they may have had a cell nucleus, the majority certainly lacked this important structure, as do living bacteria and blue-green algae.

**26-3**
**LATE**
**PRECAMBRIAN**

Within the past few years, both microscopic and megascopic fossils of considerable variety have been discovered in Late Precambrian sediments deposited between 1.7 billion and about 570 million years ago. This interval in biosphere history was an exceedingly important one, for it saw development of nucleate cells; the origin, diversification, and modernization of many groups of algae; and the development of multicellular organization among both plants and animals. Through the Late Precambrian, blue-green algae were the dominant members of widespread aquatic communities, but by contrast with those of earlier Precambrian times, these Late Precambrian communities also included nucleate algae and probably fungi.

**Figure 26-4**
*Brooksella canyonensis*, a supposed fossil jellyfish (phylum Coelenterata) from the Nankoweap Group, Late Precambrian, of the Grand Canyon, Arizona. Natural size. (Photo by Van Gundy, 1951.)

It is likely that unicellular animals made their debut in the Late Precambrian, but we have no certain fossil record of them. Sponge fragments, however, have been reported from Late Precambrian rocks in Brittany, Ontario, northern Arizona, and South Australia. Impressions resembling jellyfish (Figure 26-4) have been found in Late

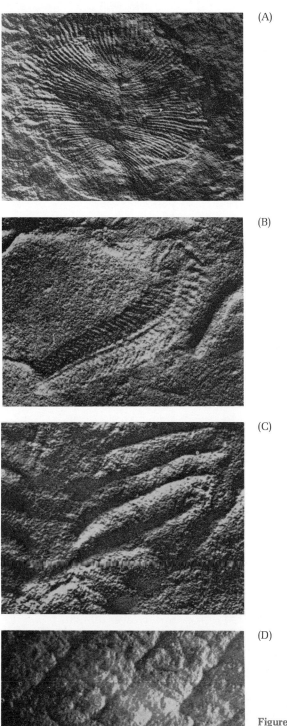

(A)

(B)

(C)

(D)

**Figure 26-5**
Late Precambrian fossils from the
Ediacara Hills, South Australia.
*Dickinsonia* (A) is similar to some
living flatworms, *Spriggina* (B), to living
annelid worms; *Rangea* and *Charnia* (C
and D), to some modern sea pens.
(Photos courtesy of M. F. Glaessner.)

451

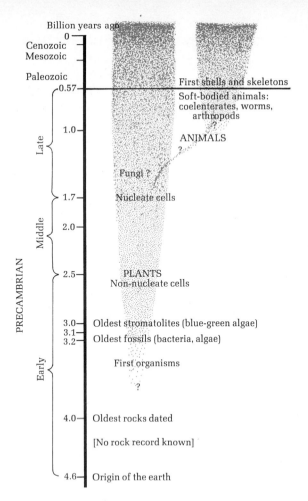

Billion years ago

Cenozoic
Mesozoic

Paleozoic
0.57 — First shells and skeletons

Soft-bodied animals:
coelenterates, worms,
arthropods
?

1.0 — ANIMALS
?

Late

Fungi ?

1.7 — Nucleate cells

2.0 —

Middle

PRECAMBRIAN

2.5 — PLANTS
Non-nucleate cells

3.0 — Oldest stromatolites (blue-green algae)
3.1 —
3.2 — Oldest fossils (bacteria, algae)

First organisms

?

Early

4.0 — Oldest rocks dated

[No rock record known]

4.6 — Origin of the earth

**Figure 26-6**
Development of the Precambrian
biosphere.

Precambrian sandstones exposed in the Grand Canyon of northern Arizona, and in South Australia. The same rocks in the latter place, as well as equivalent strata in South Africa and elsewhere, have also yielded fossils reminiscent of modern sea pens (Figure 26-5). Thus the phylum Coelenterata, to which both jellyfish and sea pens belong, was apparently represented in seas of the Late Precambrian.

Trails and burrows, thought to be the work of worms, have been discovered in Late Precambrian rocks in both Michigan and Montana; and the impressions of several types of soft-bodied organisms (Figure 26-5), interpreted as belonging to annelids, flatworms, and primitive arthropods, have recently been found in weathered quartzites deposited some 600 to 700 million years ago in South Australia. Fossils of the same types of animals, and of about the same age, are also known from southwest Africa, England, northern Sweden, the Ukraine, and Siberia, and probably southeast Newfoundland.

Thus by the end of the Precambrian eras, some 570 million years ago, the biosphere had achieved considerable diversity, and many of the groups of plants and animals important in later phases of biosphere history were clearly well established (Figure 26-6). Animals with shells and skeletons were apparently rare in Late Precambrian seas—if they existed there at all.

Although animals probably developed sometime between 1.7 billion and 570 million years ago, their Late Precambrian record is sketchy because they had apparently not yet developed the shell- or skeleton-forming capacity. Assumption of this capacity, with its dramatic effect on the fossil record, seems to have been an event of earliest Cambrian times in many different animal groups. But the abrupt and approximately simultaneous appearance of shells and skeletons in such different types of animals as snails, brachiopods, echinoderms, and arthropods near the beginning of the Cambrian Period poses an intriguing question, for which there is as yet no completely satisfactory answer.

It is possible that animals of the types just mentioned did not exist through most of Precambrian time, and that the level at which their shelly remains first appear coincides approximately with the point in time at which they came into being. Such a view virtually requires that in latest Precambrian time or very early in the Cambrian, or in an interval embracing both, a majority of the more highly organized animal phyla developed rapidly from their much simpler Precambrian ancestors. Many paleontologists have discounted such an explanation, because the later history of the biosphere seems to have unfolded much more slowly than this. However, it now appears that soft-bodied animals as highly organized as coelenterates, annelid worms (Figure 26-5), and arthropods were moderately widespread in the Late Precambrian, perhaps 100 million years or more before the beginning of the Cambrian. Such a stretch of time would certainly have been ample for "rapid" development of the even more complex organisms that developed the shell- and skeleton-forming capacity in the Early Cambrian.

Clearly, we must be cautious in equating the level of first appearance of fossil shells and skeletons with the time of origin of the animal groups represented by the fossils. There may well have been an interval of unusual innovation among animals in latest Precambrian or earliest Cambrian time, but the level at which fossils appear may mark only the point in time at which groups of previously soft-bodied organisms achieved the capacity to form inorganic hard parts. Many animals, perhaps a majority, function quite successfully today without

**26-4
SHELLS AND
SKELETONS**

mineralized hard parts, and by no means all groups that now have shells and skeletons appear as fossils for the first time in basal Cambrian rocks.

Most modern soft-bodied animals are relatively small, or they are specialized descendants of animals with hard parts, adapted for life in environments in which a shell or skeleton is of no particular advantage. Possibly, then, many types of animals had achieved the organization required to build shells or skeletons by earliest Cambrian time. The seemingly abrupt appearance of such structures in these animals may thus represent adaptation by some of the larger members of several phyla to environments in which shells or skeletons would be advantageous. That is, shells confer protection against predators and may also make life more tolerable to sedentary animals in environments swept by vigorous currents. Skeletons provide support, externally or internally, and shells and skeletons provide sites for the attachment of muscles; and muscles, in their operation against movable rigid pieces, make all sorts of complex movement possible. Development of the capacity to form shells and skeletons might thus be seen as an event in biosphere history that permitted many groups of larger animals to invade and become successfully adapted to the wide variety of environmental possibilities offered by shallow seas. Ultimately, but not until much later in the Paleozoic Era, the capacity to form a skeleton played an important role in adaptation to life on land.

**26-5
SUMMARY**

The oldest fossils known, probably representing bacteria and photosynthetic algae, are from rocks at least 3.1 billion years old. Their organization suggests that the biosphere had a long previous history, and for this we have as yet no fossil record. In younger Precambrian rocks, the variety and abundance of fossils increases, but until latest Precambrian time all represent bacteria, aquatic algae, and perhaps a few fungi. The record of animals in Precambrian rocks is limited to a few types of soft-bodied forms (sponges, coelenterates, annelids, flatworms, and arthropods), preserved under unusual conditions. The appearance of many groups of more complex animals in earliest Cambrian time is probably to be correlated with assumption of the capacity to form shells and skeletons, which may not mark the time of origin of these groups but may approximate the time at which many larger forms became adapted to specialized roles in shallow seas.

## SUGGESTED READINGS

Barghoorn, E. S. 1971. The oldest fossils. *Scientific American,* vol. 224, no. 5, pp. 30–42. (Offprint No. 895. San Francisco: Freeman.)

Discusses fossil algae and bacteria, some more than 3 billion years old, that provide evidence on the earliest stages in the evolution of life.

Barghoorn, E. S., and Tyler, S. A. 1965. Microorganisms from the Gunflint Chert. *Science,* vol. 147, no. 3658, pp. 563–77. Report on structurally preserved Precambrian fossils from the Gunflint chert of Ontario.

Glaessner, M. F. 1961. Pre-Cambrian animals. *Scientific American,* vol. 204, no. 3, pp. 72–78. (Offprint No. 837. San Francisco: Freeman.)

A report on Precambrian fossil organisms, with emphasis on those found in South Australia.

McAlester, A. L. 1968. *The History of Life,* chap. 1, pp. 4–21. Englewood Cliffs, N.J.: Prentice-Hall. (Paperback.)

Includes discussion of fossil evidence of the Precambrian biosphere.

Ross, C. P., and Rezak, Richard. 1950. The rocks and fossils of Glacier National Park: The story of their origin and history. *U. S. Geological Survey Professional Paper 294-K.* 40 pp.

This report includes information on fossils from the Precambrian Belt Series of the Glacier National Park area.

# 27 THE EARLY PALEOZOIC BIOSPHERE

27-1   The Teeming Seas
27-2   The Age of Fishes
27-3   Summary

The principal difference between Precambrian and younger rocks is that the latter are, in many places, crowded with fossils. These organic relics, besides enabling us to correlate sedimentary strata, also allow us to reconstruct the changing climates of the past and the shifting boundaries of lands and seas. They also make it possible to piece together a fairly detailed account of biosphere history for about the last 13 percent of geologic time. Following is a brief survey of the chief events in the Cambrian and later history of the biosphere.

The Cambrian, Ordovician, and Silurian periods together represent about 170 million years of Paleozoic time, or nearly a third of all the time that has elapsed since the close of the Precambrian eras. Nearly everywhere in the world this long interval was marked primarily by repeated invasion of the continents by warm, shallow seas. That these seas teemed with life is attested by the abundance of fossils entombed in sediments deposited on their floors.

Only a few rather limited tracts of *nonmarine* Cambrian and Ordovician rock are known, however, and these have yet to yield a single fossil. Indeed, neither plant nor animal seems to have inhabited the lands until very late in the Silurian Period, when a few primitive organisms developed the special capacities needed for land life. Thus, for about 170 million years after the Precambrian, all life remained confined to water.

**Aquatic Plants**

The Early Paleozoic seas were clearly the domain of aquatic plants and invertebrate animals. Bacteria, algae, and fungi are all represented in Precambrian rocks and were undoubtedly present in abundance through much of Cambrian, Ordovician, and Silurian time. Unfortunately, neither **fungi nor bacteria are especially common** as fossils, and the record of the algae is confined largely to those with the capacity to secrete calcium carbonate. Some of the latter contributed extensively to the formation of great mound-shaped reefs in Early Paleozoic seas, and much of the black color of thick Cambrian, Ordovician, and Silurian shales undoubtedly results from carbon contributed by decay of aquatic algae.

**Invertebrate Animals**

The Early Paleozoic fossil record is heavily stacked in favor of animal groups with preservable shells and skeletons. Of the twelve animal phyla that have an extensive fossil history, primitive members of eight (Protozoa, Porifera, Coelenterata, Annelida, Brachiopoda, Mollusca,

Arthropoda, and Echinoderma) were present in Early Cambrian seas. Conodonts appeared in the Middle Cambrian, the Ectoprocta (bryozoans) and the Hemichordata (graptolites) made their debut in the very early Ordovician, and the Chordata (represented by jawless "fish") appeared in limited numbers somewhat later in the Ordovician. Thus, well before the end of the Ordovician, all animal phyla with an appreciable fossil record were present (as at least four others were whose history is only sketchily documented). Since that time, no important animal phylum has appeared, and only the Conodonta has become extinct.

These facts of Early Paleozoic life may appear to contradict what we have said about progressive changes in the fossil record that permit establishment of a worldwide time scale and enable us to correlate rocks from one place to another. To explain this, we need only note that each of the animal and plant phyla includes organisms with a basically similar plan of body organization. Viewed broadly, it appears that by Mid-Ordovician time all the chief possibilities in animal structure had been achieved. The framework, however, is a broad one at the phylum level, and within it there has been room for considerable and elaborate variation since the Ordovician. But this variation has been largely at the level of order, family, genus, and species.

Figures 27-1 and 27-2 are reconstructions of the sea floor in Middle Cambrian and Silurian times. Comparison of these two figures goes far to illustrate the nature of the

**Figure 27-1**

The Mid-Cambrian sea floor in British Columbia. Seaweed (1) forms the plumelike background; the bushy colonies of tubelike stalks (2) are extinct animals closely related to sponges. Jointed trilobites (3) and several varieties of worms (4) crawl over the bottom in the foreground, and other types of arthropods (5) hover just above the trilobites. The umbrella-shaped objects (6) on and above the bottom are jellyfish. (Courtesy of Chicago Natural History Museum.)

changes that took place in the first 170 million years or so after the Precambrian Eras. We see against a background of seaweed in Figure 27-1 a marine invertebrate fauna composed largely of sponges, worms, trilobites (primitive, now extinct arthropods), brachiopods, and jellyfish (phylum Coelenterata). We might imagine the addition of a few small snails, rare bivalves and conodonts, several types of protozoans, and a variety of wholly soft-bodied wormlike animals, since all these are known as fossils from Middle Cambrian rocks, but the picture still lacks such familiar creatures as corals, ectoprocts, and fish. In addition, we should certainly note the absence of the now common crinoids, starfish, and echinoids (all members of the Echinoderma), and would miss the cephalopods, represented in modern seas by *Nautilus* and the familiar squid and octopus.

Most of the groups missing in Figure 27-1 are present in Figure 27-2. Center stage is occupied by masses of colonial coral and two types of **nautiloid** (Nautilus-like) **cephalopods;** the shells of snails are in evidence, and also those of clams, articulate brachiopods, trilobites, and cystoids. In addition, we might imagine the sands of the sea floor to contain a host of Protozoa with sandy shells (Foraminiferida), and the water above to be inhabited by a variety of swimming and floating animals such as Radiolaria, jellyfish, graptolites (phylum Hemichordata), conodonts (Figure 27-3), and a small jawless fish or two.

**Figure 27-2**
Middle Silurian sea floor in Illinois. The rock-garden-like mass (1) in the foreground is a reef composed of several species of colonial corals. Hovering just above it at the left and resting below it to the right are two types of nautiloid cephalopods (2). Trilobites (3) are in evidence in the foreground, and also several types of brachiopods (4) and snails (5). Cystoids (6; stalked, extinct members of the Echinoderma) sway gracefully in currents to right and left. (Courtesy of Chicago Natural History Museum.)

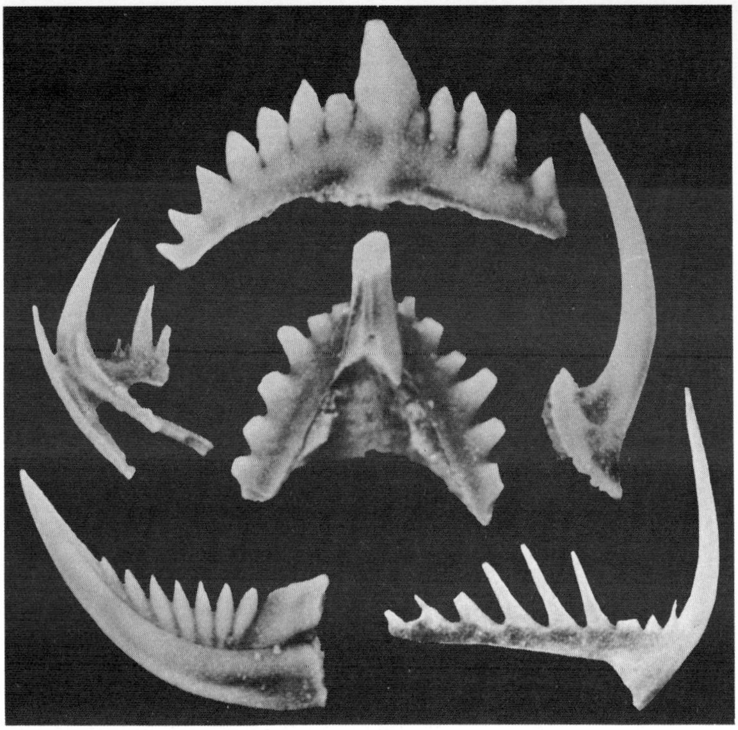

**Figure 27-3**
Conodonts were swimming marine animals, which are represented in the fossil record from Cambrian through Triassic by tiny skeletal elements like the ones shown here at about 50 times actual size. From Ordovician rocks in Kentucky.

Comparison of Figures 27-1 and 27-2 indicates that not only did whole new phyla appear between Middle Cambrian and Silurian time, but there was both expansion and contraction as well as continued variation in form among the members of even the most ancient groups. For example, the archaic trilobites, which held sway in great numbers in the Cambrian and Ordovician, were less abundant by Silurian time, but the carnivorous nautiloid cephalopods, which did not appear until near the end of the Cambrian, were clearly in their heyday in the Silurian. Indeed, the unobtrusive trilobites may have fallen prey to the cephalopods, the latter prospering on a diet of the former. However, the nautiloids were soon to lose ground to more specialized cephalopods and to fishes.

Note that changes in the invertebrate fauna have been strictly intramural. Trilobites eventually became extinct, in the Permian Period, and nautiloid cephalopods have not been especially important for many millions of years. But the phyla represented by these animals (Arthropoda and Mollusca) flourish today as never before; trilobites and nautiloids have simply been replaced by other arthropods and mollusks (or their place has been taken by a group from some other phylum). In short, since Mid-Ordovician times, the history of aquatic invertebrates has been one of constant comings and goings within the framework of some twenty invertebrate phyla. Groups have risen to prominence only to be replaced entirely or in part by others of later origin. As a whole, few principal groups (classes) have become extinct, and the net result has been gradual expansion in total numbers of different kinds of invertebrates with time.

## The Rise of Fishes

One of the strangest things about the seas of Early Paleozoic times would have been the almost complete absence of vertebrate animals. There is no record at all of Cambrian vertebrates, and few seem to have existed for the first third of Ordovician time. Many bony plates from the dermal armor of undoubted vertebrates have been found in Middle Ordovician strata in Colorado (Figure 27-4), however, and a few other scraps of bone have been discovered in still younger Ordovician rocks

**Figure 27-4**
*Astraspis desiderata.* A fragmentary bony plate, possibly from the middle of the back, of one of the oldest vertebrate animals. The fragment pictured, about three inches long, is from Middle Ordovician rocks near Canyon City, Colorado; ostracoderm remains are numerous in rocks of the same age through much of central Colorado and Wyoming. (From Bryant, 1936.)

in the western United States. Silurian strata in many parts of the world have yielded a fairly good, though tantalizingly incomplete, record of ancient vertebrates.

### Ostracoderms

All the Ordovician and Silurian vertebrate fossils thus far discovered are those of rather small aquatic animals that lacked jaws and paired fins, and had flattened fishlike bodies and tails, both enclosed in a thick bony armor.

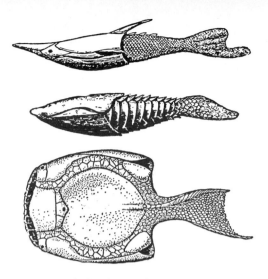

Figure 27-5
Ostracoderms. These armored vertebrates lacked jaws and paired lateral fins and apparently had no bony internal skeleton. *Poraspis* (in the middle) is from Silurian rocks on Spitzbergen. It was about 6 inches long. *Pteraspis* (top) and *Drepanaspis* (bottom) are both Lower Devonian in age; the former was some 3 inches long, whereas the latter reached a foot in length. (From A. S. Romer, *Vertebrate Paleontology,* © 1945 by The University of Chicago Press.)

Figure 27-6
Silurian eurypterids (phylum Arthropoda). Four types of these large scorpionlike predators are shown; all are similar in having 6 pairs of appendages on the head, but, as can be seen, these were variously modified for swimming, walking, and grasping prey. Some of these animals grew to be 10 feet long. (Courtesy of Chicago Natural History Museum.)

Because of the latter feature, these curious animals are commonly termed **ostracoderms** (literally, "shell-like skins"); because they lacked jaws, they are grouped with living lampreys and hagfishes in the chordate class Agnatha. Some of the better-known ostracoderms are shown in Figure 27-5.

Although their descendants came later to dominate both the seas and the freshwater bodies of the world, the ostracoderms themselves apparently never amounted to much. Their flattened shape suggests they were sluggish bottom-dwellers, and the complete absence of jaws or teeth in their slitlike mouths indicates that they fed by

grubbing for organic debris in the mud of the sea floor. They were certainly not predators, so were no competition for the larger, more agile carnivorous cephalopods that dominated Ordovician and Silurian seas. Their bony armor may have protected them from bottom-living invertebrate predators such as the giant eurypterids (phylum Arthropoda, Figure 27-6), with whom they shared the sea floor.

### Jaws and Teeth

At some time in the Silurian (we may never know when), several unusually important events took place in one or more of the several groups of ostracoderms (we may never know which ones). Slowly, through many generations of these primitive vertebrates, the most anterior of the several pairs of skeletal bars that lie between the gills on either side of the throat grew larger and swung forward toward the mouth, where they came to function as jaws. At the same time, sharp-edged denticles in the skin around the mouth became enlarged and were pressed into service as teeth. These seemingly innocuous but exceedingly important innovations enabled the groups that possessed them to become predators, feeding on larger animals and freed from the mud-grubbing existence eked out by their less progressive ostracoderm cousins.

Emancipation from the mud, however, placed a premium on the ability to dart swiftly after prey, and favored those primitive fishlike vertebrates that developed the streamlined shape that makes this mode of life most

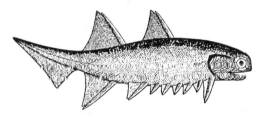

successful. Active swimming above the bottom also requires stabilizers and accessory steering organs such as paired lateral fins. These, too, appeared in the latest Silurian—not as the result of any grand design apparently, but because animals with these features were more successful than those that lacked them and were the ones that survived. A primitive fishlike vertebrate at this stage of development is shown in Figure 27-7.

**Figure 27-7**
*Climatius,* a Lower Devonian acanthodian, represents the least specialized group of placoderms. These primitive vertebrates, about 3 inches long, had an armor of bony scales and are the oldest jawed vertebrates known. (From A. S. Romer, *Vertebrate Paleontology,* © 1945 by The University of Chicago Press.)

The advent of jaws, a streamlined shape, and paired lateral fins made possible an almost explosive increase in fishes in the Devonian Period. Indeed, their fossil remains are so abundant in many Devonian rocks that

**27-2**
**THE AGE**
**OF FISHES**

this interval in earth history is often dubbed the Age of Fishes.

## Placoderms

The most primitive of the jawed fishes appeared in the Late Silurian. They are placed in the chordate class Placodermi. By Early Devonian times, these animals had developed along many different lines and were adapted to life in both fresh and salt water. Some (**antiarchs**) lived

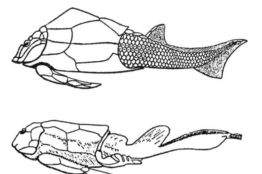

a mud-grubbing existence essentially like that of the ancestral ostracoderms; others (**acanthodians**) were river-dwellers of strikingly sharklike appearance; still another group (the **arthrodires**) grew to lengths of more than 30 feet and had powerful jaws armed with vicious sharp-edged bony projections of the skull. The mud-grubbing antiarchs (Figure 27-8) retained the bony armor, as well as the feeding habits, of the ostracoderms, but in the more agile predacious acanthodians and arthrodires (Figure 27-9) the armor was reduced to the forward part of the body.

**Figure 27-9**
Devonian shark and arthrodire. *Cladoselache* (left) was an early shark about a foot long; *Coccosteus* (right) was a jointed-necked arthrodire that grew to lengths of more than 10 feet. (Drawings by M. F. Marple.)

Successful as they were in the Devonian, the placoderms dwindled rapidly in importance after that time and were entirely extinct by the end of the Paleozoic Era. Well before their decline, however, they had successfully replaced most of the old ostracoderms, and had given

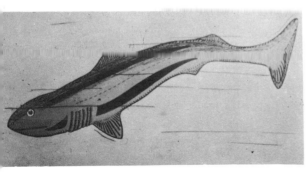

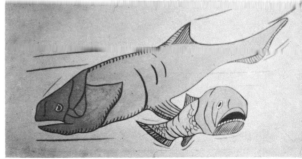

rise to the host of more advanced fishes that gradually came to replace them.

## Sharks and Bony Fishes

In the Devonian, the placoderms spawned two important groups of fishes: the sharks (chordate class Chondrichthyes) and the familiar bony fishes (chordate class Osteichthyes). The physiology of living sharks indicates quite clearly that they were originally adapted to life in the sea; the physiology of bony fishes, on the other hand, shows equally clearly that they were originally adapted to fresh water. A few sharklike fishes have invaded fresh water since the Devonian, but they have never been much at home there. Bony fishes, however, are now the dominant forms in both marine and fresh water.

Sharks and their kin have a skeleton of cartilage, and thus no bones to leave as fossils; but their skin is roughened by the inclusion of minute denticles and their well-developed jaws are studded with many razor-sharp teeth capable of preservation. Their elongate streamlined shape and capacity for rapid, powerful swimming early made them successful predators. For the most part, they have changed very little since their appearance. They probably replaced the archaic arthrodires in post-Devonian seas, and their gradual increase in size and numbers since then is probably closely connected with the increasing abundance of bony fishes, on which they feed. *Cladoselache*, a Devonian shark, is shown in Figure 27-9.

**Figure 27-10**
*Latimeria,* the only living member of the Crossopterygii, a group of bony fishes probably ancestral to all four-footed land animals. Note the stout paddle-shaped fins, which may be regarded as true limbs and represent a stage transitional between typical fins and the legs of land animals. *Latimeria* reaches a length of more than 5 feet and a weight of about 180 pounds. (Reprinted with permission. Copyright © 1955 by Scientific American, Inc. All rights reserved.)

Almost from their beginnings in the Middle Devonian, the bony fishes have been divisible into three basic groups, which were very much more like one another then than they are now. All the earliest bony fishes had lungs as well as gills, and all were covered with thick enameled bony scales. One group, however, the dominantly fresh-water **lobe-fins** (or Crossopterygii), had stout lateral fins with fleshy lobes and internal skeletal supports of bone arranged in much the same fashion as those in the limbs of land vertebrates. The fossil record indicates that lobe-fins were abundant in the Late Paleozoic; however, they have dwindled since that time and are represented today only by *Latimeria* (Figure 27-10), an inhabitant of the deep seas off the coast of South Africa.

**Lungfishes,** another group of bony fishes present in the Devonian, were (and still are) river-dwellers that pos-sessed a functional lung, and, like the closely related lobe-fins, had fleshy lobes on the fins, and nostrils opening both to the inside of the mouth and to the exterior. Lungfishes differ from lobe-fins primarily in lacking an extension of the bony skeleton into the fins. They were among the more common Devonian fishes, but like the lobe-fins, they have become less and less important. The group is represented today by only three genera, all inhabitants of streams in the Southern Hemisphere.

In the Devonian, lobe-fins and lungfishes greatly out-numbered the third basic bony-fish group, the **ray-fins** (or Actinopterygii). These animals, now the most success-ful of all the bony fishes, derive their name from the fact that the paired fins consist of a thin cover of skin spread out over a series of horny rays; the fins contain no bone, nor do they have fleshy lobes like those of the lobe-fins and lungfishes. For the most part, the ray-fins were rela-tively unimportant until late in the Paleozoic (Permian), when they began an almost explosive increase in numbers and varieties in both fresh and salt water. In the course of their development, bony scales have been gradually lost and the old bony-fish lung has evolved into a swim bladder.

**27-3 SUMMARY**

During the Cambrian, Ordovician, and Silurian periods, or for about 170 million years after the Precambrian, all life remained confined to water. Organisms were espe-cially abundant in the warm shallow seas that repeatedly invaded the continents. Aquatic plants, although not exceptionally common as fossils, were surely abundant and varied, but the fossil record is dominated by inverte-brate animals. Eight of the twelve animal phyla that had preservable shells and skeletons were present in the Early Cambrian, and the others by the end of the Ordovician. Early Paleozoic seas swarmed with sponges, trilobites, brachiopods, corals, nautiloid cephalopods, snails, clams and many other forms.

Ordovician and Silurian rocks yield fossils of primitive jawless fishes, or ostracoderms, which probably gave rise in the Late Silurian to more successful groups of primitive jawed and streamlined fishes, the placoderms. In the Devonian, often termed the Age of Fishes, more modern types of fishes became adapted to life in both salt and fresh water. Sharks, which developed in the sea, have an internal skeleton of cartilage and have probably changed little since they appeared in the Devonian. Bony fishes, with a mineralized internal skeleton, were originally adapted to fresh water but are now dominant in fresh and marine waters. All the earliest bony fishes had lungs as well as gills, and one group, the lobe-fins, had stout lateral fins with internal supports arranged much like those in the limbs of land vertebrates.

## SUGGESTED READINGS

Clark, D. L. 1968. *Fossils, Paleontology and Evolution,* chap. 7, pp. 51–87. Dubuque, Iowa: W. C. Brown Co. (Paperback.) Chapter 7 in this beginning-level reference covers the evolution of the principal groups of invertebrate animals. Although coverage is not limited to the Early Paleozoic, many of the invertebrates evolved or became dominant during this time.

Fenton, C. L., and Fenton, M. A. 1958. *The Fossil Book.* Garden City, N.Y.: Doubleday, 496 pp. A general reference book on fossils, designed especially for the layman; provides an accurate and readable account of earth history and of fossils representing each group of plants and animals.

McAlester, A. L. 1968. *The History of Life,* chaps. 3 and 4, pp. 39–81. Englewood Cliffs, N.J.: Prentice-Hall. (Paperback.) Chapter 3, "Life in the Sea," and Chapter 4, "The Transition to Land," deal with organisms that evolved mostly during the Early Paleozoic.

Millot, J. 1955. The coelacanth, *Scientific American,* vol. 193, no. 6, pp. 34–39, (Offprint No. 831. San Francisco: Freeman.) A report on a lobe-finned fish, represented today only by *Latimeria,* which until 1938 was thought to have been extinct for 70 million years.

# 28 THE BIOSPHERE OF THE LATER PALEOZOIC

28-1 Stay-at-Homes in the Sea
28-2 Colonization of Land
28-3 Summary

By Mid-Silurian time, animals and plants were both varied in structure and widespread in distribution in the seas. Representatives of the arthropods (the eurypterids) and the vertebrates (the ostracoderms) had successfully invaded fresh waters. The lands, however, remained barren and the skies empty.

Beginning slowly in the Late Silurian, but increasing rapidly in tempo in the Devonian and later Palcozoic, the lands were occupied by living things: first, almost certainly, by primitive plants, then by invertebrate animals, and finally by four-footed vertebrates. Although restless comings and goings continued among the many inhabitants of the sea, the conquest of land and sky is the principal novelty in the history of the later Paleozoic biosphere, and it is to this facet of the story that our interest is most strongly drawn.

## 28-1 STAY-AT-HOMES IN THE SEA

The interval from Devonian through Permian was a time of decline and extensive replacement in the sea. Archaic invertebrate groups, many tracing their ancestry back to the Early Cambrian, began to dwindle in the Devonian and were replaced by newcomers of more modern appearance from their own or other phyla. Trilobites and curypterids, for example, declined rapidly in numbers through the later Paleozoic periods and were both extinct by the end of the Permian. Nautiloid cephalopods, varied and successful in Ordovician and Silurian seas, became less so in the later Paleozoic. Their place was taken rapidly by the more streamlined **ammonoid cephalopods** (so named because they resemble *Ammonites,* one of the first genera named; Figure 28-1), which appeared first in the Early Devonian and rose quickly to prominence in Devonian and later seas.

Plants must have been abundant in fresh waters of the later Paleozoic, for many groups of invertebrate animals, which depend on plants for food, invaded that environment in the Devonian. Plant fossils are not numerous, however. (Why not?) Freshwater protozoans were

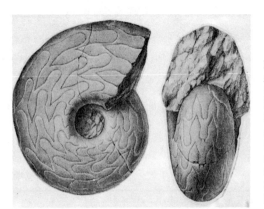

Figure 28-1
An ammonoid cephalopod (*Schistoceras missouriense*) from Pennsylvanian rocks in Missouri. Remains of the ammonoids are common in rocks of Devonian through Cretaceous age, but the group apparently became extinct at the end of the Mesozoic Era. Natural size. (Courtesy of A. K. Miller and W. M. Furnish.)

probably abundant, as undoubtedly several phyla of worms were. Many of the eurypterids were probably freshwater dwellers, and fossil freshwater clams are known from Devonian rocks in both America and Europe.

28-2
COLONIZATION
OF LAND

## Requirements of Land Life

The biosphere originated in water, and the bulk of it is still confined to the hydrosphere. The structures and capacities requisite to land life have been developed by representatives of eleven plant phyla, but by only four animal phyla. All have adapted to broadly similar conditions, but the means vary considerably. In brief, the requirements of land life are these:

1. *Obtaining and conserving water.* Obviously an adequate supply of water is no problem to aquatic plants and animals. But retention of water in body tissues is essential to active life, and rapid evaporation of water is a general feature of the terrestrial environment. Thus successful adaptation to land life requires structures that inhibit or reduce water loss to the atmosphere, as well as systems for obtaining water and conveying it to all parts of the body. Certain animals and plants have also developed the capacity to withstand prolonged periods of desiccation, and "come to life" only when conditions are favorable.

2. *Body support.* In an aquatic environment, the effects of gravity are partly compensated by the buoyant effect of water. Not so on land. Success on land requires a strong supporting skeleton for all but the smallest organisms.

3. *Reproduction independent of external water.* In a very large number of invertebrate animals and aquatic plants, sexual reproduction is accomplished by mobile sperms that swim through the water around the organisms to fertilize eggs that may themselves be shed into water. Sperms still swim to eggs of land plants and animals, but full adaptation to land required development in both types of organisms of elaborate systems of transport not dependent on external water.

4. *Adjustment to wide temperature extremes.* Temperatures vary from day to night and from season to season in water, just as they do on land. However, extremes are far greater on land than in water, which gains and loses heat less rapidly than air. Thus only those animals and plants that have developed a means of withstanding great changes in temperature have been successful on land.

## Land Pioneers

The oldest undoubted inhabitants of the land were primitive leafless plants of the phylum Psilophyta, whose remains have been found in Upper Silurian rocks in Czechoslovakia. Rocks of earliest Devonian age at many places yield fossils of psilophytes, bryophytes, and lepidophytes. The record of land plants is not good until somewhat later in the Devonian, however. Early in that period, woody plants in some abundance and diversity were present in many parts of the world.

During the Devonian, land plants gradually became common. Some reached the size of trees, and our first record of an extensive forest is preserved in rocks of Mid-Devonian age in New York (Figure 28-2). The primitive psilophytes gradually declined in the later Devonian and were largely replaced by the club mosses (lepidophytes), scouring rushes (arthrophytes), and ferns (filicophytes), all of which were abundant by the Late Devonian. Woody tissues in the stems of all these plants permit the conservation and conduction of water, and provide strong support for the body; however, their sperms still swim to eggs in at least a film of external water. These plants were a successful first step in land conquest; many of them persist in damp places today, and some have even become adapted to life in areas that are periodically deserts.

**Figure 28-2**
Reconstruction of a Devonian forest. Leafless plants in the foreground (1) are psilophytes about 2 feet tall. Others are arthrophytes (2) and filicophytes (3). The tallest plants (4, 5) are difficult to classify; they are regarded as seedless ancestors to the gymnosperms. (Reconstruction by Charles R. Knight, courtesy of Chicago Natural History Museum.)

**Figure 28-3**
*Proscorpius*, an Upper Silurian scorpionlike arthropod from New York. Invertebrate animals like this, and like others found in Upper Silurian rocks in Scotland and Sweden, may have been the first to live on land. Many paleontologists believe, however, that *Proscorpius* and its relatives spent most of their lives in water. About twice natural size. (From *Treatise on Invertebrate Paleontology*. Courtesy of Geological Society of America and University of Kansas Press.)

A few invertebrate animals may have become adapted for land life at about the same time as the earliest plants, for scattered remains of both millipede- and scorpionlike arthropods (Figure 28-3) have been found in Upper Silurian rocks in New York, Scotland, and Sweden. It is not at all certain. however, that these animals, like their living descendants, were able to live on land; both groups may well have been freshwater dwellers in the Late Silurian.

It was not until Mid-Devonian times that undoubted land animals evolved. Cherty rocks of that age in Scotland have yielded a variety of fossilized spiders, mites, and wingless insects that establishes the existence of air-breathing invertebrate animals midway through the Devonian. One of these early land animals is shown in Figure 28-4. All of them belong to a group of invertebrates, the arthropods, that were in many ways already fitted out for land life. Body support for even many aquatic arthropods is provided by a sturdy external skeleton, which also provides an external cover for body tissues and helps protect them from desiccation. Furthermore, living spiders and insects have elaborate systems for internal fertilization of eggs, hence need not depend on external water during their reproductive cycles. We may assume that spiders and mites of the Devonian were similarly equipped, though perhaps not so perfectly.

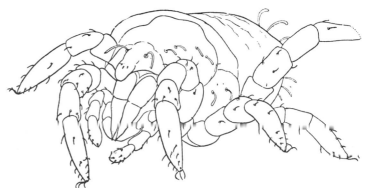

**Figure 28-4**
Reconstruction of a fossil mite (phylum Arthropoda) from cherty rocks of Mid-Devonian age in Scotland. About 170 times natural size. (From Hirst, 1923.)

## Coal Swamps

In the land flora of the Late Devonian and Mississippian there appeared a considerable variety of fernlike plants, in which the spore-bearing organs were clustered together in rather massive aggregates. These plants, commonly termed seed ferns, were much more advanced in many characters than true ferns (Filicophyta), but fall somewhat short of the organization displayed by modern seed plants. Whether they should be included with the ferns or with the seed plants is unimportant. What is important is that by the Late Devonian the immediate ancestors of the seed-bearing Cycadophyta and Coniferophyta had evolved and spread widely through the swampy places of the world.

In the Pennsylvanian Period, the several groups of primitive land plants probably reached the acme of their development, and the widespread occurrence of Pennsylvanian coal beds is a record of the abundance of these plants in swampy lowlands of the continental interior. As shown in Figure 28-5, many of these old plants grew to treelike proportions and occurred in numbers not since equalled. Among them, however, there were also many primitive seed plants of the phyla Cycadophyta and Coniferophyta: representatives of the latter were particularly conspicuous parts of the Pennsylvanian flora.

With the general uplift of continents and restriction of seas that marked the Permian and Triassic periods in most parts of the world, the environment most suitable for many of the groups of primitive plants dwindled, and broader areas of dry land emerged. In effect, this offered a challenge to those plants adapted for life in dry, harsh climates. The seed plants were fitted for life in such an environment, and apparently were adapted to it by early in the Permian Period. Their success in drier land areas was facilitated by the fact that their seeds, which enclosed and nourished the embryo, could be dormant until environmental conditions favored their continued growth.

Thus, as swampy lowlands diminished in the Late

**Figure 28-5**
A coal-forming swamp of Pennsylvanian time. The plants identified by numbers are lepidophytes (1), arthrophytes (2), ferns, or filicophytes (3), seed ferns, or cycadophytes (4), and coniferophytes (5). Note also the huge dragonfly (6) and the big cockroaches (7), both of which are insects (phylum Arthropoda). (Courtesy of Chicago Natural History Museum.)

Pennsylvanian there was wide and successful invasion of the drier uplands by Cycadophyta, Coniferophyta, and perhaps by early representatives of the little-known Gnetophyta, as well. By the end of the Paleozoic, plants were well established in a wide variety of land environments.

Late Paleozoic swamps were also the habitat of air-breathing invertebrates of great variety. By the Pennsylvanian, insects had achieved the numerical dominance they still maintain, and coal swamps of the continental interior must have buzzed with them (Figure 28-5). Several groups had developed wings and many had attained great size: a fossil dragonfly from Belgium has a wingspread of some 29 inches, and fossil cockroaches three to four inches long are not uncommon in Pennsylvanian rocks. Spiders, centipedes, and scorpions were numerous in the rotting organic debris of the coal swamps, and a few land snails (phylum Mollusca) may have existed in the same environment.

### Rise of the Tetrapods

We are more directly concerned with another part of Late Paleozoic history, and what we have said thus far merely sets the stage for it. This is the story of vertebrate development on land.

The bony fishes called crossopterygians, or lobe-fins, were abundant in Devonian streams and lakes. *Eusthenopteron,* shown in Figure 28-6, was such a fish. Animals like this, with lungs, extensions of the bony internal skeleton into the paired lateral fins, and an exterior armored with stout bony plates, already possessed most of the requirements for life on land.

Because bony fishes like *Eusthenopteron* were abundant in the Devonian, it is not surprising that the slight modifications that transformed some of them into four footed land animals **(tetrapods)** had also come about by late in the same period. Such an animal was *Ichthyostega* (Figure 28-7), whose remains have been found in terrestrial rocks of Late Devonian age in East Greenland. Indeed, *Ichthyostega* and other equally primitive amphibians are like Devonian lobe-fins in so many anatomic particulars that they have been jokingly termed "fishes with legs."

Amphibians, however, have never become fully adapted to life on land. Living amphibians (frogs, salamanders, and their kin) live in moist environments, and although air-breathers as adults, they spend much of their lives in water. All return to water to lay and fertilize their eggs, and early stages in their development mimic those of the ancient fishes from which they are remotely descended. In the absence of serious competition for life on land,

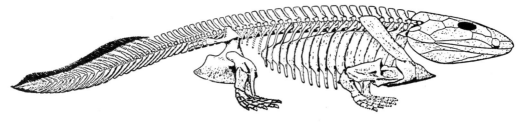

amphibians prospered in the Mississippian Period and were widespread and remarkably diversified in the dank swamps of the Pennsylvanian and Permian.

Although none of the familiar modern amphibians had appeared in the Late Paleozoic, it is appropriate to term the Pennsylvanian and Permian the Age of Amphibians, for the group as a whole has never since been so numerous or varied. *Diplovertebron,* shown in Figure 28-8 clambering out onto the bank of a Pennsylvanian swamp, was a clumsy fellow with short, sprawling legs, a large flattened head, and a still fishlike tail. The animal bears the unmistakable stamp of its lobe-fin ancestry, but it also provides evidence that, by the Late Paleozoic, amphibian legs had become long, stout, and well adapted for walking on land. The largest of the Late Paleozoic amphibians were probably no more than ten feet long; giants among them may have weighed as much as 500 pounds.

Today amphibians are not an important part of the

**Figure 28-8**

*Diplovertebron,* a small Pennsylvanian amphibian about 2 feet long. These animals probably spent most of their time in water, and may have fed on the many freshwater fish in the coal-swamp ponds. (Courtesy of American Museum of Natural History.)

land-vertebrate fauna. After their brief rise to success in the Late Paleozoic, they dwindled rapidly to their present insignificance. Their place was taken in the Mesozoic by the reptiles, a group spawned by amphibians sometime in the late Pennsylvanian and destined to complete the conquest of the land.

**28-3 SUMMARY**

Change was continuous in seas of the later Paleozoic, largely involving modernization of the existing inverte brate fauna. Groups successful in the Early Paleozoic gradually declined and were replaced by more efficiently organized descendants. Plants and animals became adapted for life on land between the Late Silurian and the Late Devonian. Such adaptation requires structures that reduce water loss to the atmosphere, strong supporting skeletons; structures that protect against great temperature changes; and systems that reduce reliance on external water during the reproductive process.

Plants, which first appeared on land in the latest Silurian, spread rapidly and gradually acquired the organization that permitted full adaptation to a wide variety of terrestrial environments. By the Pennsylvanian Period, advanced land plants were widespread and varied and their abundance in swampy lowlands is recorded by the occurrence of numerous coal beds in North America and elsewhere.

By Mid-Devonian times, invertebrates such as spiders, mites, and wingless insects had become qualified terrestrial inhabitants. Later in the Paleozoic these groups diversified greatly, and at least the insects appear to have achieved the numerical dominance they still enjoy. Vertebrate animals, in the form of fishlike amphibians, made their land debut in the Late Devonian. In the Pennsylvanian, reptiles developed from amphibians, to complete the conquest of land environments by vertebrate animals.

## SUGGESTED READINGS

Clark, D. L. 1968. *Fossils, Paleontology and Evolution,* chaps. 6, 7, and 8, pp. 33–115, Dubuque, Iowa: W. C. Brown. (Paperback.)
These chapters include information on fossil plants, and on both invertebrate and vertebrate animals, that were dominant during the Late Paleozoic.

Fenton, C. L., and Fenton, M. A. 1958. *The Fossil Book:* Garden City, N.Y.: Doubleday. 496 pp.
A general reference book on fossils designed especially for the layman; provides an accurate and readable account of earth history and of fossils representing each group of plants and animals.

McAlester, A. L. 1968. *The History of Life,* chap. 5, pp. 82–100. Englewood Cliffs, N.J.: Prentice-Hall. (Paperback.)
Chapter 5 covers the evolution of land plants from the oldest vascular plants of the Early Paleozoic through the flowering plants of the Mesozoic but concentrates on groups that were dominant during the Late Paleozoic.

# 29 GYMNOSPERMS AND REPTILES

29-1  The Mesozoic Seas
29-2  Mesozoic Land Plants
29-3  Mesozoic Tetrapods
29-4  A Time of Great Dying
29-5  Summary

In North America and much of the rest of the world, the Late Paleozoic and Early Mesozoic were times of increasing continentality. Land areas slowly enlarged through the Permian Period, and except in a few broad embayments, shallow seas were restricted to continental borders in the Late Permian and Triassic. Expanding land areas and widespread episodes of mountain making brought with them new opportunities for terrestrial plants and animals. Diminished areas of shallow sea floor, on the other hand, seem to have heightened competition among aquatic organisms. Thus late in the Paleozoic, or very early in the Mesozoic, many types of archaic marine invertebrates disappeared through extinction; others were largely replaced by groups of more modern appearance. On land, however, early semiterrestrial pioneers gave way in the Pennsylvanian and Permian to plants and animals more fully equipped for the rigors of land life, so the passage from Paleozoic to Mesozoic is marked in the fossil record of terrestrial organisms only by expanding evolutionary development. The Mesozoic Era, then, was a time of gradual modernization. In water and on land it was intermediate between an age dominated by bizarre old-fashioned organisms and the present day, with its familiar array of plants and animals.

Trilobites, the jointed denizens of Paleozoic seas, declined in the Late Paleozoic and were extinct by the end of that era. Their place may gradually have been taken over by other arthropods, including the lobsters, the oldest fossil remains of which are found in rocks of Triassic age. Solitary corals that built horn-shaped skeletons failed to survive the end of the Paleozoic, and most of the old colonial corals were replaced by more modern types early in the Mesozoic. Brachiopods, numerous and varied through the Paleozoic, were only locally abundant in Mesozoic seas, as is the case today. As they declined, their place seems to have been taken by another group of bivalved invertebrates, the clams (class Bivalvia, phylum Mollusca). Fossil remains of clams, and of other mollusks, are abundant in Mesozoic rocks, and many of them are important in correlating marine strata.

The Mesozoic was clearly the Age of Mollusks. All manner of snails and clams lived on the sea floor, and cephalopods dominated among the swimmers. Most of the archaic nautiloid cephalopods had disappeared well before the end of the Paleozoic, although a few had a brief "comeback" in the Jurassic and Cretaceous. Their place was rapidly taken in the Late Paleozoic and Mesozoic by the ammonoids (Figure 28-1), a group of efficiently streamlined, rapidly swimming cephalopods characterized by shells that display sutures of greatly complex

## 29-1
## THE MESOZOIC SEAS

479

pattern; and by the **belemnoids,** squidlike cephalopods with internal shells (Figure 29-1).

Ray-finned bony fishes were present in the Devonian, but were an insignificant part of the aquatic fauna until the Permian, when they began the great radiation that culminated in the fish fauna of today. Most Mesozoic ray-fins were transitional in many ways between their heavily plated, lung-bearing Devonian ancestors and their thin-scaled, lungless modern progeny. Only a few of these transitional forms persist today: the sturgeon and paddlefish of the Mississippi River are living remnants of an early stage in ray-fin evolution, whereas garpike and freshwater dogfish linger on as representatives of an intermediate chapter in the story. In these animals, thick bony scales are retained. The old ray-fin lung, however, has undergone a series of modifications and now serves as an internal air sac, or swim bladder.

Modern bony fishes, with thin scales, symmetrical tails, and only a few movable spines in the fins, made their appearance in the late Mesozoic. Since then, they have become remarkably diverse in shape, size, and distribution. At present, the group includes nearly every familiar freshwater and salt-water fish, from trout and salmon to sea horses, eels, and the curiously flattened flounders.

**29-2
MESOZOIC
LAND PLANTS**

The fossil record of Permian and Triassic land plants is considerably less extensive than that of the Pennsylvanian. The continents of the world were broadly uplifted, and the swampy environments favorable to the well-developed but archaic Pennsylvanian flora were severely restricted. Probably for this reason, many of the more primitive groups of land plants disappeared or diminished greatly in importance. Hardier species of mosses, rushes, and lepidophytes persisted in well-watered lowlands, and ferns continued in importance in such places.

The drier uplands, with harsh climates and limited or seasonally varied rainfall, were gradually invaded during the Permian and Triassic by advanced groups of seed plants representing the plant phyla Cycadophyta, Ginkgophyta, Coniferophyta, and Gnetophyta, and collectively termed **gymnosperms.** By late in the Triassic, members of all these phyla were firmly established and widely distributed in the uplands (see Figure 29-14). An instructive study of an upland flora of Late Triassic age can be made in the Petrified Forest of northern Arizona, where the remains of cycadophytes and coniferophytes are preserved in abundance. Conifers are recorded by silicified logs, many more than 100 feet long and 10 feet in diameter. These logs, largely devoid of bark and branches, represent giant trees that probably grew in surrounding highlands and were washed by floods into adjacent lowlands, where they were rapidly buried. Such scraps of foliage as are preserved in Triassic rocks of the Petrified Forest record cycadophytes and ferns, suggesting that these plants grew along stream courses in the lowlands themselves.

Flowering plants, the phylum Anthophyta, may have been present in limited numbers as early as the Triassic; however, they did not become abundant, as they are now, until early in the Cretaceous Period, when they spread dramatically across the continents. The rapid rise of anthophytes in the Cretaceous may well be associated with the emergence of pollinating insects, upon which most modern flowering plants depend heavily in their reproductive cycles. Unfortunately, the fossil record of insects is spotty, and to date it has not been possible to correlate their development with that of flowering plants. Bees, flies, butterflies, and moths are all known from Jurassic rocks, however, and all serve an important function today in pollinating flowering plants. Since Cretaceous times, the terrestrial flora has had essentially its modern aspect.

## Amphibians

Amphibians dominated the Pennsylvanian scene and persisted in somewhat reduced numbers into the Permian and Triassic. Since then, however, these archaic tetrapods have had only slight importance among land animals. The

**29-3
MESOZOIC
TETRAPODS**

reasons for this are plain. Although adult amphibians breathe air and can live on land, their eggs are laid and fertilized in water, and the young live a fishlike existence in the pond or puddle where they hatch. Consequently, these animals were largely unable to take advantage of the wide array of dry-land environments that spread out in bewildering variety in the Late Paleozoic and Early Mesozoic.

Some time in the Pennsylvanian Period, however, there appeared in the coal-forming swamps a group of amphibianlike tetrapods endowed with the capacity to lay their eggs on land. This event, comparable to the appearance of seeds in plants, enabled these animals (termed **reptiles**) to develop almost explosively in the Permian, when they may be said to have inherited the earth.

## Reptiles

### The Reptile Egg

The shelled egg of the reptiles was a truly remarkable innovation, and its development freed its possessors from water in much the same way that formation of seeds freed the plants. Reptile eggs are fertilized internally and the developing embryo, enclosed in a liquid-filled sac, is provided with food and a refuse sac for waste products. All this is enclosed in a tough shell whose walls are rigid enough to protect its contents, but porous enough to allow oxygen to pass into the egg and carbon dioxide to pass out of it. In effect, the growing embryo develops in its own private puddle, just as fish and amphibians do in a more public pond. Growing reptiles, however, are assured of protection and a supply of food. Furthermore, animals able to lay eggs like this can roam widely over dry land without needing to return to water to reproduce.

### Other Reptilian Characters

Living reptiles are readily distinguished from amphibians in numerous anatomic particulars. For example, the soft skin of an amphibian, kept moist by secretion of mucous and water, differs strikingly from the scaly "watertight" skin of a reptile, which requires little glandular moistening and is one of the features that makes reptiles successful in dry-land environments. Eyes, ears, and organs of smell and balance are also developed in somewhat different ways in these two groups of animals. Especially important are the many modifications in the reproductive system of reptiles that give them the ability to fertilize eggs internally, surround them with foodstuff and a shell, and lay them on land to hatch in the sun.

It is not always easy to separate the fossil remains of early reptiles from those of the amphibians from which they evolved, for both types of animals had a basically similar skeletal organization; and the features of skin, sense organs, and reproductive systems that distinguish living forms from one another disappeared by decay during the process of fossilization. One distinctive clue aids in deciding whether a fossil skull should be classified as that of an amphibian or a reptile. That is, grooves on the snout and along the sides of the skulls of fish and many amphibians mark the position of an elaborate system for detecting movement in the surrounding water. No such sensory system occurs in reptiles. Thus if a fossil skull bears traces of a lateral sensory system it represents an animal that either underwent its early development in water or lived its entire life there—and that would rule out the reptiles and their descendants, the birds and mammals.

Vertebrae of many of the more primitive reptiles are stout and fit together in such a way that up-and-down movement of the backbone must have been limited even though side-to-side movement was relatively free. Vertebrae of this sort are not characteristic of the amphibians, although they do occur in *Seymouria* (Figure 29-2), whose skull also bears traces of a lateral sensory system. Thus *Seymouria* combined amphibian and reptilian characters, and is commonly regarded as a link between the two

**Figure 29-2**
*Seymouria*, a primitive cotylosaur (reptile), was about 2 feet long and is known from many skeletons found in Permian redbeds in Texas. Rocks that yield these skeletons also produced the oldest known shelled egg. (Courtesy of American Museum of Natural History.)

groups. In recognition of its reptilian features, however, we classify *Seymouria* with the reptiles. The Permian rocks from which the remains of this ancient reptile have been collected have also yielded the oldest shelled egg known.

## Classification of Reptiles

On the basis of skull architecture, reptiles are divisible into five large groups, typical members of which are shown in Figure 29-3. The *cotylosaurs* and *turtles* of the figure make up the **Anapsida,** a main division characterized by a solidly roofed skull with no openings behind the eyes. Another main group, the **Synapsida,** includes the *pelycosaurs* and *therapsids* of Figure 29-3. In addition

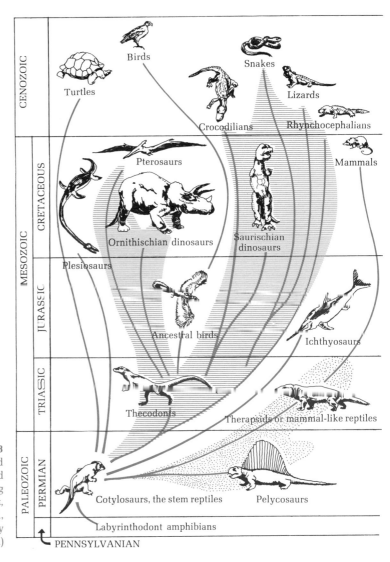

**Figure 29-3**
Family tree of the reptiles. Dotted pattern joins the Synapsida; lined pattern joins the Diapsida or "Ruling Reptiles." (From *The Dinosaur Book,* by E. H. Colbert, Copyright © 1951, McGraw-Hill Book Company. Used by permission.)

to openings for the eyes, synapsid reptiles have another opening on either side of the skull, behind and on line with the lower half of the eye opening. A third principal division, the **Parapsida** (*ichthyosaurs* of Figure 29-3), and a fourth, the **Euryapsida** (*plesiosaurs* of the figure), have a lateral skull opening behind each eye, as the synapsids do. However, in these two groups the lateral openings are on line with the upper half of the eye opening, rather than with the lower half as in the synapsids. Euryapsids and parapsids are distinguished from each other by the fact that the lateral openings are bounded by different skull bones in each group. The fifth and largest of the reptile groups, the **Diapsida,** embraces the thecodonts, dinosaurs, pterosaurs, crocodiles, snakes, lizards, and rhynchocephalians of Figure 29-3, all with two openings behind the eye on each side of the skull. Synapsid, parapsid, and euryapsid reptiles are all extinct. Living reptiles are mostly diapsids, although turtles, which date back to the Permian, represent the primitive anapsids.

*Anapsids: The First Reptiles*

*Seymouria* (Figure 29-2) and its fossil kin are commonly included among the **cotylosaurs** (Figure 29-3), an extinct group of clumsy anapsid reptiles with sprawling legs and solidly roofed skulls similar in plan to those of Pennsylvanian amphibians but not so distinctly flattened. Only a few of the cotylosaurs grew to great size; most of them were less than five feet long. They were abundant and varied in the Permian and Triassic, but disappeared before the Jurassic. Nearly every other reptile group can be traced back to them, however, so they occupy an important place in the history of life on land.

*The Permian Reptile Radiation*

Reptiles are not especially common fossils in Pennsylvanian rocks, but a host of reptile skeletons has been recovered from rocks of Permian age. These indicate that soon after they appeared, reptiles spread rapidly and widely into all the many environments offered by dry land. Although there is a certain sameness about the old cotylosaurs that gave rise to this rapid reptilian radiation, their descendants are greatly varied in size, form, and structure, and it is clear that they were adapted to many different ways of life. The general pattern of reptile radiation is shown in Figure 29-3 and the main groups are there outlined and named. We can summarize only the main features of the principal groups.

485

## Synapsids: Mammal-like Reptiles

The pelycosaurs and therapsids (Figure 29-3) belong together in a large group termed the Synapsida. Synapsids appeared early in reptile history, in the Pennsylvanian, and spread widely through North America and South America, eastern Europe, and South Africa. They were abundant in the Permian and Triassic, but were all extinct before the beginning of the Jurassic.

Many **pelycosaurs** had elongate bodies, sprawling limbs, and jaws equipped with long sharp teeth that indicate they were meat eaters. Long spines, developed from the vertebrae of Permian pelycosaurs like *Dimetrodon* (Figure 29-4), supported a saillike flap of skin down the middle of the back. The function of such a bizarre and undoubtedly cumbersome structure is not known, but it may record a primitive experiment in body-temperature regulation, in which a great area of skin was exposed to warming sun or cooling breezes.

Pelycosaurs were strange old-fashioned reptiles that were all extinct by the end of the Permian. They are of uncommon interest to us, however, for they were close relatives of an important group of late Paleozoic and Triassic reptiles, the **therapsids,** which probably gave rise in later times to the mammals.

**Figure 29-4**
Fin-backed synapsid reptiles. Those with the large heads are *Dimetrodon*, a meat-eater; the small-headed reptile left of center is *Edaphosaurus*. (From a painting by Charles R. Knight, courtesy of Chicago Natural History Museum.)

**Figure 29-5**
Mammal-like synapsid reptiles. Three
meat-eating representatives of
*Cynognathus* (left foreground and
background) have cornered a bulkier
plant-eating *Kannemeyeria*. Early
Triassic, South Africa. (Painting by
Charles R. Knight, courtesy of Chicago
Natural History Museum.)

Several groups of therapsids appeared in the Permian
and radiated widely throughout nearly every continent.
Their skulls, although modified in several noteworthy
respects, are similar in many ways to those of the pelycosaurs. However, limbs no longer sprawled out to the sides,
but were rotated in beneath the body and somewhat
twisted, so that the knees pointed forward and the elbows
backward. This arrangement made walking and running
easier, for the body was raised well off the ground.

Some therapsids were ungainly beasts of fair size but
primitive characteristics. Others, like *Cynognathus* (Figure 29-5) were smaller but had limbs that were more effectively beneath the body; an elongate, narrow skull; and
a lower jaw formed almost entirely of a single element.
Although the jaw of *Cynognathus* is reptilian in structure,
it bears highly specialized teeth, differentiated into nip-
ping incisors, piercing canines, and many-cusped post-ca-
nines useful in chewing. Furthermore, the nasal passage
of *Cynognathus* was separated from the mouth by a well-
developed secondary palate; in other words, this animal
could chew food with its mouth closed and breathe at
the same time.

*Cynognathus* and its near relatives are thought to be
close to the stock from which mammals sprang. For this
reason, they are often termed **mammal-like reptiles.** What

are the features that make the therapsids such likely mammalian ancestors?

Reptiles are scaly and cold-blooded, whereas mammals are hairy and warm-blooded. External features like hair and scales, however, rarely find their way into the fossil record, and warm-bloodedness is a capacity resulting from metabolic and physiologic organization that cannot be fossilized under any circumstances. Nevertheless there are skeletal features associated with maintaining a uniformly high body temperature, that is, with being warm-blooded, and *Cynognathus* as a mammal-like reptile has these. For example, the more advanced therapsids had a secondary palate, separating mouth from nasal passages. Such a feature, present in all mammals, suggests that in maintaining a constant body temperature, therapsids depleted oxygen in cells, blood, and lungs so rapidly that, like mammals, they had to breathe at short intervals, even while the mouth was full. In short, the secondary palate is viewed as an adaptation that permitted development of warm-bloodedness, a character by which we recognize mammals. Teeth, differentiated into incisors, canines, and molars in therapsid reptiles and in mammals, may also be viewed as useful adjuncts. Warm-blooded animals must not only breathe at short intervals, they must also process their food so that it can be swallowed quickly: the throat, which is also the principal passage for oxygen, cannot be blocked for long periods of time. Rapid chewing would be further aided by a stout lower jaw and strong

**Figure 29-6**

Ichthyosaurs. These porpoiselike marine reptiles were numerous in Mesozoic seas, but were extinct by the end of the Cretaceous Period. (Painting by Charles R. Knight, courtesy of Chicago Natural History Museum.)

muscles to operate it. Thus the tendency in *Cynognathus* and other therapsids for one of the three parts of the reptilian jaw to be more massively developed than the others is indirectly associated with other characters that can be related to warm-bloodedness. We can only guess at many other therapsid features. At least some therapsids may have given live birth or incubated their eggs; had a four-chambered heart and mammary glands; and been covered with hair, for all of these developments have something to do with warm-bloodedness.

### Parapsids and Euryapsids: Back to Sea

During the Pennsylvanian, in the earliest stages of reptile history, one group of small cotylosaurs gave up land life and returned to water. Fossils of these aquatic reptiles **(mesosaurs)**, known only from the Southern Hemisphere, indicate that they had long bodies and heads, the latter equipped with greatly lengthened jaws set with many sharp teeth. In addition, mesosaurs had a long flexible tail and paddlelike limbs.

Mesosaurs were closely related, but probably not ancestral, to **ichthyosaurs** (Figure 29-6), porpoiselike reptiles that first appeared in the Triassic and were clearly adapted to life in the sea. The ichthyosaur body was fishlike in shape and its limbs were modified to form fins. Because ichthyosaurs were completely aquatic in habitat, they could not return to land to lay eggs as other reptiles do.

**Figure 29-7**
Plesiosaurs. These long-necked marine reptiles were numerous in the sea from Triassic through Cretaceous times. They were well adapted for swimming and probably fed on fish, cephalopods, and carrion scavenged from the sea floor. (Painting by Charles R. Knight, courtesy of Chicago Natural History Museum.)

Thus it seems probable that young developed within the maternal body and were born alive.

From a stock of lizardlike late Paleozoic land reptiles, there developed in the Triassic a group of aquatic reptiles **(plesiosaurs)** with a long neck and a broad, flattened body, the limbs of which were modified to form oarlike paddles (Figure 29-7). The jaws of typical plesiosaurs were studded with sharp-pointed teeth, which suggests that they fed on fish. The shape of the body and the nature of the limbs indicate that they probably "sculled" along at the surface, or just beneath it, and that they were unable to return to land for even a short time. Fossil plesiosaurs are common in Mesozoic marine rocks, and the creatures seem to have been especially abundant in Cretaceous seas.

### Diapsids: The Ruling Reptiles

Except for turtles, an ancient group more closely related to extinct cotylosaurs than to other living reptiles, all surviving reptiles belong in a large group termed Diapsida. Living diapsid reptiles, however, are a mere shadow of the group as a whole: during the Mesozoic Era, diapsids ruled the earth.

**Thecodonts** (Figure 29-3), small carnivorous Triassic reptiles, were the rootstock of the diapsids. Unlike their four-footed ancestors, these light-bodied reptiles were bipedal, many of their bones were hollow and birdlike, and their hips were importantly modified to permit them to run on their hind legs. Although they did not last beyond the Triassic, thecodonts gave rise to a wide variety

**Figure 29-8**
*Tyrannosaurus*, a Cretaceous theropod dinosaur. This fearsome animal was 20 feet high, 50 feet long, and 8 to 10 tons in weight. (Painting by Charles R. Knight, courtesy of Chicago Natural History Museum.)

of dinosaurs, flying reptiles, crocodilians, snakes, and lizards, and were the ancestors of the birds as well.

Probably the most celebrated of the many reptiles descended from thecodonts are the ones popularly dubbed **dinosaurs** (literally, "terrible lizards"). Although not all dinosaurs were large, some grew to spectacular size; indeed some of them were the largest animals that ever roamed the land. The reptilian orders Saurischia and Ornithischia, united under the term *dinosaur* in Figure 29-3, are really quite different groups of reptiles. The former had hip bones arranged in a typically reptilian fashion, whereas the latter had hip bones arranged like those of birds. Both, however, developed from Triassic thecodonts.

Saurischian dinosaurs include both carnivores and herbivores. The bipedal pattern of the thecodonts was retained by carnivorous forms, whose development culminated in *Tyrannosaurus* (Figure 29-8). *Tyrannosaurus* stood 18 to 20 feet above the ground, was nearly 50 feet long, and probably weighed in at 8 to 10 tons. It had strong hind legs and a stout pelvis, but the forelegs were reduced to hooked claws. The skull, borne on a very short neck, was large and powerful and the massive jaws were set with daggerlike teeth. *Tyrannosaurus* and others of its ilk were clearly well-adapted big-game hunters, which probably fed largely on other dinosaurs.

Diminutive Triassic thecodonts also gave rise to herbivorous Jurassic and Cretaceous dinosaurs, some of which became the largest land animals of all time. *Brontosaurus* (Figure 29-9), a Jurassic inhabitant of the western United States, typifies this line of descent. This giant reptile was a quadruped, had a small head at the end of a long, slender neck, was nearly 85 feet long, and may have weighed 50 tons or more. Unlike the light birdlike bones of the carnivores, the bones of the large herbivores were dense

**Figure 29-9**

*Brontosaurus*, a Jurassic sauropod dinosaur. These giants, 85 feet long and 50 tons or more in weight, were the largest land animals that ever existed. (Painting by Charles R. Knight, courtesy of Chicago Natural History Museum.)

and solid to provide support for a body of such great size and weight. All seem to have been plant eaters; they probably spent much of their lives partially submerged in swamp water, where they would have been close to a source of food, protected from marauding carnivores, and rendered less clumsy by being partly buoyed up by water.

Ornithischian dinosaurs were more advanced and much more varied in form than saurischians. The **stegosaurs** (Figure 29-10), a group of Jurassic and Early Cretaceous dinosaurs, were plant-eating quadrupeds with spiked tails and prominent rows of plates along the back. **Ornithopods,** or duckbills (Figure 29-11), were mostly bipedal, and like other ornithischian dinosaurs fed almost entirely on plants. Many peculiar anatomic features suggest that these animals spent much of their time almost completely submerged in water. **Ankylosaurs** (Figure 29-11) were bulky armored quadrupeds that appeared in the Cretaceous Period. **Ceratopsians,** or horned dinosaurs (Figure 29-12), were the last major ornithischian group to evolve. These four-footed Cretaceous plant eaters had a massive beaked skull, with horns and a prominent bony frill that projected back over the neck and shoulders. This seemingly clumsy structure was evidently of service in protecting vital regions of the body, and may partly account for the fact that ceratopsians were abundant in

Figure 29-11
*Trachodon* (the three dinosaurs in the left foreground) and *Parasaurolophus* (just visible in the right background) are duckbilled dinosaurs. *Ankylosaurus* in the right foreground is an armored ornithischian dinosaur. All these reptiles are plant-eating Cretaceous species. (Painting by Charles R. Knight, courtesy of Chicago Natural History Museum.)

the Late Cretaceous and were among the last of the dinosaurs to become extinct.

**Crocodiles, snakes,** and **lizards** are familiar living diapsid reptiles, whose ancestors were also Triassic thecodonts. Crocodiles date back to the Late Triassic, as the lizards do. Snakes, which are really highly modified lizards, were the last of the major groups of reptiles to appear. Their fossil record is scanty, but it stretches back to the Cretaceous.

Our survey of the ruling diapsid reptiles would be

Figure 29-12
*Triceratops*, a Cretaceous ceratopsian dinosaur, was 20 feet long and 8 feet high. Reptiles like this were widespread and successful in the late Mesozoic and were among the last of the dinosaurs to become extinct. (Painting by Charles R. Knight, courtesy of Chicago Natural History Museum.)

incomplete without mention of the **pterosaurs,** or flying reptiles, which appeared in the Jurassic and lived through the Cretaceous. Jurassic forms were about two feet long, had an elongate skull with many sharp teeth, and bore a long tail with a diamond-shaped rudder at the tip (see Figure 29-14). Bones in the forelimbs were modified to support a wing membrane, much of which was stretched out between the body and a greatly overgrown fourth finger. Later pterosaurs were much larger; species of *Pteranodon* (Figure 29-13), for example, attained a wingspread of more than 25 feet.

### Early Birds

Birds are the rarest of fossils, but their history goes back at least as far as the Jurassic, when two of them plummeted into the sea in Bavaria and were buried in fine-grained limy mud on its bottom. Fortunately for us, these muds have since lithified to dense limestone, in which many details not ordinarily preserved in the fossil record are retained. Indeed, were it not for the fact that impressions of feathers still surround the skeletons of *Archaeopteryx* (Figure 29-14) they would undoubtedly be classified as those of early diapsid reptiles!

*Archaeopteryx* was a small animal with a long reptilian tail, hind limbs very much like those of Jurassic carnivorous dinosaurs, a pelvis similar to that of ornithischian dinosaurs, toothed jaws, and a skull reminiscent of early diapsid reptiles but somewhat enlarged to accommodate a bigger brain. Impressions in the limestone around the Bavarian skeletons show long feathers extending out from the hand and lower arm bones, and a row of feathers along either side of the tail. If the first amphibians were "fishes with legs," *Archaeopteryx* was certainly a diapsid reptile with feathers.

Important modifications in skull, pelvis, and wing structures characterize Cretaceous birds. Like modern birds, they lacked the long tail of *Archaeopteryx* and its reptile forefathers, but unlike modern species, they still had teeth.

494

The end of the Mesozoic Era brought with it a large number of puzzling extinctions. Ammonoid and belemnoid cephalopods, abundant in the Cretaceous as never before, vanished abruptly at the end of that period. Plesiosaurs also became extinct in the Cretaceous, and on land the dinosaurs and flying reptiles disappeared without issue. There have been many attempts to explain the widespread extinctions of the Late Cretaceous, but none is completely satisfactory. Because the wave of extinctions affected groups living in many different environments, but did not noticeably affect other groups living in the same places, it is not logical to attribute the "great dying" to any single cause. All the groups that disappeared, however, were highly specialized, not primitive, members of their phyla, and it is animals like this that would be most adversely affected by the changes in

## 29-4
## A TIME OF GREAT DYING

**Figure 29-14**
Four representatives of *Archaeopteryx* (1), the oldest known bird, and several flying reptiles (*Rhamphorhynchus*) (2) are the principal animals in this Jurassic scene. Two small lizardlike dinosaurs (3) and several varieties of gymnosperm plants (all cycadophytes) complete the picture. (Painting by Charles R. Knight, courtesy of Chicago Natural History Museum.)

land-sea distribution and the profound alterations in continents that took place in the Cretaceous.

As the once widespread seas retreated in the Late Cretaceous, it is probable that competition increased for living space and food on dwindling shallow sea floors. Possibly the ammonoids and belemnoids, as well as the marine reptiles, had become so specialized in diet or mode of life that they were unable to survive this more rigorous competition. They may have been replaced by the less specialized squids.

Wider land areas and new ranges of high mountains in the Late Cretaceous undoubtedly brought with them harsher climates than were characteristic of the Jurassic and Early Cretaceous, and the rapidly spreading anthophytes clothed these lands in vegetation quite different from that of earlier times. Many of the dinosaurs were clearly adapted for feeding on a limited variety of plants. Diminution in the supply of these may have spelled extinction for herbivorous dinosaurs and for the carnivorous types who fed on them. In addition, the advent of warm-blooded, well-insulated birds probably faced flying reptiles with competition they could not handle. Perhaps no single episode in biosphere history so dramatically illustrates the devious interdependence of animals and plants on one another.

Whatever the reasons for the "great dying," Late Cretaceous extinctions subtracted from the Mesozoic marine fauna most of the animals that distinguished it from the modern one, and disappearance of the dinosaurs on land left a void into which the mammals expanded rapidly in the Early Cenozoic. The marine fauna and the terrestrial flora have been virtually of modern type since the end of the Cretaceous.

## 29-5 SUMMARY

The passage from Paleozoic to Mesozoic was marked by extinctions of many types of archaic marine invertebrates; by introduction of more modernized groups; and by expanding development of terrestrial plants and animals, most of which represented groups that first appeared in the late Paleozoic. Mollusks were varied and numerous in Mesozoic seas, and ray-finned bony fishes diversified and evolved to the prominence they hold today. Mosses, rushes, lepidophytes, and ferns continued to grow luxuriantly in well-watered Mesozoic lowlands, but gymnosperms spread widely into the drier uplands, where many grew to great size. Flowering plants may have evolved as early as the Triassic, but were not abundant until the Cretaceous: their rapid development then may have been associated with the emergence of pollinating insects.

Reptiles developed from amphibians in the Pennsylvanian and inherited the earth in the Permian. Five important groups, distinguished by skull structure, were differentiated by Triassic time. Clumsy anapsid reptiles of the Permian and Triassic spawned the mammal-like synapsids, the marine parapsids and euryapsids, and the highly diverse diapsids. Synapsids evolved into mammals in the late Triassic, but the marine parapsids and euryapsids, despite their abundance in Mesozoic seas, were all extinct by the end of the Cretaceous. Diapsids, represented by pterosaurs and a great variety of saurischian and ornithischian dinosaurs, ruled the Mesozoic lands. By the end of the era, however, pterosaurs and dinosaurian diapsids had become extinct. Only crocodiles, snakes, lizards, and rhynchocephalians survive. In the Jurassic, the diapsid reptiles gave rise to the birds, a group that is represented by only a few fossils but has been highly successful since its origin.

Late Cretaceous extinctions, probably attributable to changes in land-sea distribution and Cretaceous alterations in continental architecture, subtracted from the Mesozoic biosphere most of the organisms that distinguished it from the modern one. Since the Cretaceous, the marine fauna and the terrestrial flora have changed only slightly; principal changes have been in the land fauna, for disappearance of the dinosaurs in the Cretaceous created a void into which mammals expanded rapidly in the Cenozoic.

## SUGGESTED READINGS

Clark, D. L. 1968. *Fossils, Paleontology and Evolution*, chaps. 6 and 8, pp. 33–50 and 88–115. Dubuque, Iowa: W. C. Brown. (Paperback.)

These chapters include information on those fossil plants and vertebrate animals that were dominant during the Mesozoic.

Colbert, E. H. 1949. The ancestors of mammals. *Scientific American*, vol. 180, no. 3, pp. 40–43. (Offprint No. 806. San Francisco: Freeman.)

Reports on reptiles with mammalian characteristics, the therapsids, which lived during the Permian and Triassic periods.

Colbert, E. H. 1961. *Dinosaurs: Their Discovery and Their World*. New York: Dutton. 300 pp.

A popular account of this group of animals.

McAlester, A. L. 1968. *The History of Life*, Chap. 6, pp. 101–27. Englewood Cliffs, N.J.: Prentice-Hall. (Paperback.)

Includes discussion of the evolution of reptiles and mammals.

# 30 THE AGE OF MAMMALS

30-1  The Origin of Mammals

30-2  Mesozoic Mammals

30-3  Early Cenozoic Radiation

30-4  Modernization of the Placentals

30-5  History of the Primates

30-6  Summary

No part of life history holds as much interest for us as the latest part, the chapter that unfolded during the Cenozoic Era. For the fossil record tells us that during this interval in earth history the mammals, the group to which we belong, achieved their present prominence.

Fortunately, the record of mammalian development is remarkably complete. This is partly because Cenozoic rocks, being the youngest in the geologic column, are far more widespread than older strata. Furthermore, continents have been large since the end of the Mesozoic, and erosion has not yet removed much of the thick accumulation of terrestrial sediments deposited in the last 60 or 70 million years. Because these are the sorts of rock in which remains of land plants and animals are preserved as fossils, there is a wide variety available for study. We turn now to the story reconstructed from mammalian fossils found in these rocks.

## 30-1 THE ORIGIN OF MAMMALS

Many points of similarity can be emphasized between the skeletons of mammals and those of the Triassic therapsid (mammal-like) reptiles. For example, both mammals and therapsids have a bony secondary palate that separates mouth from nasal passages; both have teeth that are differentiated into incisors, canines, and cheek teeth; and in both the limbs are in beneath the body. In addition, the skull articulates with the first neck vertebra by means of a pair of round bony knobs rather than by the single knob that occurs in nontherapsid reptiles. Thus there is little doubt that mammals descended from the therapsids (Figure 30-1).

Mammal-like therapsid reptiles were widespread and abundant in the Permian and Triassic. Probably during the height of their importance in the Triassic, the therapsids gave rise to mammals, the oldest remains of which are jaw scraps and a few tiny teeth found in rocks of Late Triassic age.

## 30-2 MESOZOIC MAMMALS

The fossil record indicates that mammals existed for more than 100 million years before they became a particularly conspicuous part of the vertebrate fauna. This is a bit surprising in view of their many advanced characters, but it is probably a result of the fact that the various habitats to which they are now so well adapted were then occupied by highly successful reptiles. Further, skeletons indicate that the first mammals were small and not very much brighter than Mesozoic reptiles. That they hung on at all through the Jurassic and Cretaceous is probably a tribute to their more efficient reproductive and physiologic systems.

Before the beginning of the Jurassic, three main groups

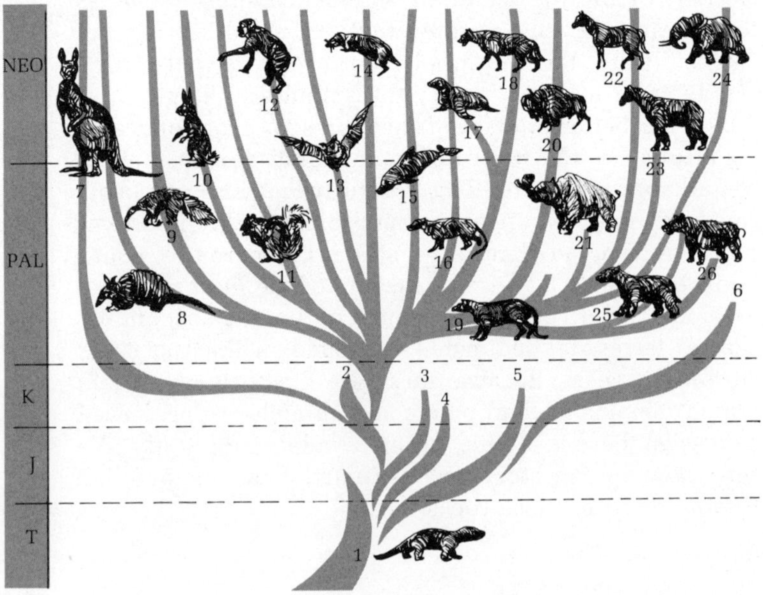

**Figure 30-1**
A family tree of the mammals. The branch numbered 1 represents the therapsid reptiles, the probable ancestors of all the mammals. Other numbered branches are:

2. pantotheres
3. symmetrodonts
4. docodonts
5. triconodonts
6. multituberculates
7. marsupials
8. armadillos
9. sloths and anteaters
10. rabbits
11. rodents
12. primates
13. bats
14. insectivores
15. whales and porpoises
16. creodonts
17. seals and walruses
18. modern carnivores
19. condylarths
20. even-toed hoofed mammals
21. titanotheres
22. horses, zebras, rhinos, tapirs
23. chalicotheres
24. elephants
25. amblypods
26. uintatheres

of mammals had appeared (Figure 30-1: groups 3, 4, 5). Two additional groups made their debut in the Jurassic (groups 2 and 6). Although these groups are named in the explanation of Figure 30-1, we do not define them here because their scanty fossil record consists largely of tooth and jaw fragments. It is sufficient to note that all were mouse- to housecat-sized quadrupeds with sharp teeth, which probably fed on the insects, worms, seeds, and eggs they found in woodlands. The groups numbered 3, 4, and 5 disappeared in the Late Jurassic and Cretaceous; seed eaters of group 6 survived into the Paleogene, when they became extinct; and the group labeled 2 lived into the Early Cretaceous, when it spawned the marsupials (group 7) and the placentals (groups 8 through 26).

**Marsupials** include such mammals as kangaroos and opossums, in which the young are born alive, but in a tiny immature state. After birth, these young are commonly kept in a pouch, or marsupium, on the lower belly of the female. **Placentals,** including man and many other familiar mammals, give birth to live young in a relatively advanced state of development and do not keep them in a pouch. The term *placental* refers to the disc-shaped organ (placenta) by means of which the unborn embryo is nourished. Marsupial skeletons are readily distinguished from those of placentals by the presence of bones to support the pouch; by a distinctive pattern of cusps on molar teeth; and by lower jaws with in-turned back corners.

The marsupials and placentals were represented in the Late Cretaceous by opossum- and shrewlike animals that were probably rather widely distributed on the several continents. Although the fossil record of this stage in mammalian history is not much better than that of earlier periods, it indicates that marsupials and placentals lived together on a more or less equal footing, because both were largely unspecialized and were still dominated by the ruling reptiles. Shortly after the end of the Cretaceous, however, the placental mammals, with their greater intelligence and more advanced reproductive system, replaced the marsupials nearly everywhere except in Australia and South America.

With the disappearance of dinosaurs near the end of the Cretaceous, mammals expanded rapidly into environments for which they were already adapted and in which there were no longer reptilian competitors. This radiation was accomplished with astonishing swiftness: close cousins of the opossumlike marsupials and the shrewlike insect-eaters multiplied quickly in numbers, grew somewhat in size, and spread like a tidal wave into every corner of all the continents. Competition soon returned full force; but this time the contestants were nearly all mammals.

Because placental mammals are advanced in many ways over the pouched marsupials, they rapidly replaced them in most of the world in the Paleogene Period. Both Australia and South America, however, were cut off from the rest of the world late in the Cretaceous or early in the Paleogene, soon after the horde of primitive mammals had invaded them. Consequently, marsupials developed on both these continents to greater heights than elsewhere. They still constitute the bulk of the native vertebrate fauna in Australia. In South America they have been largely replaced by modern placental mammals, which moved in when the Isthmus of Panama reappeared in the recent past. In North America and the rest of the world, the Cenozoic was really the Age of Placental Mammals.

## 30-3 EARLY CENOZOIC RADIATION

### Archaic Placental Mammals

In addition to shrew- and molelike **insectivores,** the mammalian tidal wave of the earliest Cenozoic contained a bizarre assortment of archaic hoofed animals (condylarths, amblypods, uintatheres: Figure 30-1, groups 19, 25, and 26), and a few primitive carnivores (creodonts: group 16). The earliest members of all these groups were much the same. For the most part, they were low-slung sheep-sized mammals with long bodies and tails, five clawed toes on each foot, and an array of teeth that suggests that they could have eaten nearly any food available. Shortly after the appearance of these archaic mammals, modifications of one sort or another began to appear, and these, sorted out by competition and adaptation, gradually produced more specialized forms in each of the ancestral placental groups.

### Condylarths

**Condylarths** (Figure 30-1, group 19) were light-bodied archaic hoofed mammals. The earliest forms were small, and their relatively primitive teeth and clawed toes indicate derivation from the insectivores. These lowly forms, however, radiated extensively in the earliest Cenozoic, grew considerably in size, and developed peculiarly specialized molar teeth. *Phenacodus* (Figure 30-2), perhaps the best known of all condylarths, had grown considerably away from its clawed, insect-eating ancestors.

**Figure 30-2**
*Phenacodus,* a sheep-sized condylarth. Note the long body and tail and the 5 hoofs on each foot. (From a painting by Charles R. Knight, courtesy of American Museum of Natural History.)

Nevertheless, it was still a clumsy animal, with a long low skull, large canine teeth, and square-crowned molars obviously adapted for chewing plants. Each foot bore five toes that ended in hoofs rather than claws.

### Amblypods and Uintatheres

Other primitive hoofed mammals, the amblypods and uintatheres, developed rapidly in Early Cenozoic time into animals of considerable size. **Amblypods,** such as *Coryphodon* (Figure 30-3), were stout massively built quadrupeds with strong limbs terminating in five hoofed toes. The shape of the molars indicates that the amblypods were browsing animals, but their large sharp canine teeth imply a close relationship to other primitive mammals, including the meat-eating creodonts.

**Uintatheres** (Figure 30-1, group 26) grew to be larger than amblypods and developed massive rhinoceros-sized bodies, and skulls from which there projected six grotesque horns. All five toes were retained on the stout, elephantlike legs of these swamp-living archaic mammals; although long canines were present, molar teeth were poorly constructed and suggest that uintatheres could eat

**Figure 30-3**
*Coryphodon*, an extinct plant-eating amblypod. Note the 5 hoofed toes and the long canine teeth. (From a painting by Charles R. Knight, courtesy of American Museum of Natural History.)

**Figure 30-4**
*Oxyaena*, an early Cenozoic creodont, feeding on a five-toed horse. (From a painting by Charles R. Knight, courtesy of American Museum of Natural History.)

nothing but soft vegetation. Neither amblypods nor uintatheres survived to the end of the Paleogene Period, nor do they seem to be the ancestors of any of the modern hoofed mammals.

### Creodonts

Meat-eating placental mammals of the Early Paleogene were little different in size or appearance from the earliest hoofed mammals. These early carnivores, collectively termed **creodonts,** were widespread and successful in the Early Cenozoic, but have been extinct since early in the Neogene Period. Creodonts, such as *Oxyaena* (Figure 30-4), were low, short-limbed mammals with clawed toes, long tails, and a pair of cheek teeth on either side specialized for slicing meat and cracking bone. The oldest members of the group were very similar to the ancestral insectivores; later creodonts were more specialized, and ranged from the size of a weasel to that of a tiger.

## 30-4 MODERNIZATION OF THE PLACENTALS

The Early Cenozoic world was filled quickly by insectivores, creodonts, amblypods, uintatheres, and condylarths. Very early in the game, however, modifications of various sorts began to appear in each of these groups. In many cases, these initiated new lines of mammalian development; in others, the archaic groups themselves gradually improved. For the most part, however, the modified descendants of the old-fashioned placentals were larger and more intelligent, with teeth and limbs better adapted for specialized competition. These animals rapidly replaced their archaic forebears, most of which were extinct before the end of the Paleogene Period.

### The Conservatives

All the placental mammals shown on the left side of Figure 30-1 are more closely related to the ancestral insectivores (group 14) than are the several groups of carnivores and hoofed mammals shown on the right side of the figure. **Bats** (group 13), for example, are insect eaters specialized for flight; **primates,** our own order (group 12), are distinguished by high mental development and great manual dexterity; but they retain five-toed limbs and rather primitive teeth. The **rabbits** and **rodents** (groups 10 and 11) are widely adaptable mammals specialized for hopping and gnawing. Although they have never been particularly big or bright, they breed rapidly and are now the most numerous of all the mammals. **Armadillos, sloths,** and **anteaters** (groups 8 and 9) belong together in a largely

South American group characterized by teeth that are much simpler than those of other mammals or are lacking entirely.

## Meat Eaters

Early in their history, the meat-eating creodonts (Figure 30-1, group 16) gave rise to two modernized groups of carnivorous placentals. One, including **whales** and **porpoises** (group 15), went back to sea, where they have developed some remarkably fishlike features. The other developed into the great host of modern flesh eaters, which includes cats, dogs, bears, wolves, weasels, raccoons, and the like (group 18). Sometime near the beginning of the Neogene Period, this branch of the order gave rise to a second seafaring generation, represented today by such creatures as seals, walruses, and sea lions (group 17).

Modern carnivores are specialized largely with respect to their teeth. In most of these animals, the canine teeth have become long sharp stabbing structures, and a pair of cheek teeth on either side (one on the upper jaw, the other on the lower) developed a shearing movement with respect to one another. These teeth aid in slicing tough meat and in cracking bone. Carnivores have also tended to become somewhat longer-legged with time, and probably somewhat brighter as well. Their bodies have not changed as much in structure, however, as those of the hoofed mammals have.

## Hoofed Mammals

The most striking feature of mammalian history in the Cenozoic has been the wide adaptive radiation of hoofed plant eaters. The Early Cenozoic wave of archaic mammals brought with it condylarths, amblypods, and uintatheres. Modernization of condylarth descendants has produced nearly all the familiar herbivorous mammals of the present. In general, this modernization has involved an overall increase in body and brain size, development of high-crowned flat-surfaced molar teeth for grinding hard, silica-rich grasses, and great modification of the limbs for speedy locomotion.

Early in their history, the low-slung five-toed condylarths sired two quite different groups of hoofed placentals. In both, the first joints of the legs became short and the second ones long, and by elongation of the bones of the feet, a functional third segment was added to each limb. The animals thus came to run on their toes, and the ancestral claws were gradually transformed into short horny hoofs at the end of each toe. Furthermore, in many hoofed forms the shorter side toes gradually lost contact with the ground, and the central toe or toes tended to become both longer and stouter than the others.

If you spread your palm flat on a table and then gradually raise the wrist, you can duplicate this development for yourself. Note that, as your wrist is raised, first your thumb then your little finger lose contact with the table; ultimately, with the wrist almost vertical, only your middle finger touches the table. Except for the earliest step in this procedure, when the thumb is above the table and the little finger is still touching it, only an odd number of fingers (5, 3, or 1) makes contact with the table. This is characteristic of the limbs in a large group of hoofed mammals to which horses, tapirs, and rhinoceroses belong, a group we term **odd-toed ungulates** (Figure 30-1, groups 21, 22, and 23).

If, as you raise your palm from the table, you twist your wrist ever so slightly outward, an even number of fingers (4 or 2) can be kept in contact with the table top. This type of limb structure is characteristic of the "cloven-hoofed" pigs, deer, cattle, and their kin, a group we refer to as **even-toed ungulates** (Figure 30-1, group 20).

**Figure 30-5**
A trio of fox-sized five-toed early Cenozoic horses. Formerly called *Eohippus* (dawn horse), these animals are now included in the genus *Hyracotherium*. (From a painting by Charles R. Knight, courtesy of American Museum of Natural History.)

## Odd-toed Ungulates

*Hyracotherium* (Figure 30-5), often termed the dawn horse, was one of the earliest of the odd-toed ungulates.

This fox-sized animal had a relatively small brain, short slender limbs, and elongate feet in which ankles were raised well above the ground. Three toes made contact with the ground on both front and hind feet; a fourth toe hung above the ground on each front foot. Teeth in *Hyracotherium* were still low-crowned and primitive, although a few showed modification for grinding.

From primitive creatures like *Hyracotherium* there diverged a wide variety of odd-toed ungulates, of which we note the characters of only three groups. One branch of the clan, the **titanotheres** (Figure 30-6), developed huge horned skulls and grew to be as much as eight feet high at the shoulder. Their great weight was supported on stocky legs that terminated in four toes in front and three in back. Although giants in their time, the titanotheres were a short-lived group. They were all extinct before the end of the Paleogene Period.

Another group, the now extinct **chalicotheres** (*Moropus* of Figure 30-7), were in many respects the odd-toed clowns of the Cenozoic landscape. These surprisingly successful animals grew to the size of modern horses, which they somewhat resembled, but their front legs were longer than their hind ones, and the three functional toes on each foot terminated in large claws rather than hoofs. The claws and the low-crowned molars of these strange creatures suggest that they browsed on roots dug up along stream courses, although occasionally they may have

**Figure 30-6**
A herd of titanotheres, extinct odd-toed ungulates, on the plains of South Dakota. (Painting by Erwin Christman and Charles R. Knight, courtesy of American Museum of Natural History.)

**Figure 30-7**
Chalicotheres and entelodonts. Two odd-toed chalicotheres (*Moropus*) browse on the trees at the right: note their horselike heads, long front legs, and clawed toes. Three entelodonts (*Dinohyus*), large even-toed relatives of modern pigs, root for food in the foreground. (From a painting by Charles R. Knight, courtesy of Chicago Natural History Museum.)

nibbled on leaves from low-hanging branches, as is being done in Figure 30-7.

A third group of odd-toed ungulates, the **horses,** developed in North America from little *Hyracotherium.* The stages in horse development outlined in Figure 30-8 show that there was considerable increase in body and brain size from Early Paleogene *Hyracotherium* to living *Equus,*

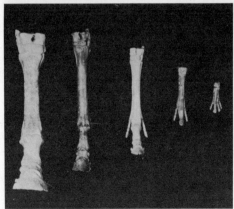

**Figure 30-8**
Development of horses. Above, a model of tiny early Cenozoic *Hyracotherium* is contrasted with the skeleton of a modern horse, *Equus.* Below, a series of skeletons shows the change in foot structure from *Hyracotherium* to *Equus.* Note that the middle toe developed more rapidly than the others and is the only one effectively retained in *Equus.* (Courtesy of American Museum of Natural History.)

that legs and feet became progressively longer, and that the middle toe grew longer and stouter as the side toes gradually disappeared. In addition, there were substantial changes in the size and shape of the teeth and jaws. The incisor teeth became broad efficient nippers, separated by a gap from the premolar cheek teeth, which were gradually converted into high-crowned molars well adapted for grinding prairie grasses.

Although odd-toed ungulates were once numerous in North America, they all became extinct here during the Pleistocene. All present-day members of the group on this continent, including horses, have been imported by man. The group as a whole reached its zenith in the Late Paleogene and Early Neogene periods and has declined steadily since that time.

## Even-toed Ungulates

Most of the hoofed mammals on the modern scene have an even number of toes in contact with the ground. The list of these is long and includes cattle, sheep, goats, giraffes, deer, pigs and peccaries, hippopotamuses, camels and llamas, muskoxen, and antelope. Although the variety is immense, the even-toed ungulates can be readily divided into two distinctly different groups, the **pigs** and their kin, and the **cud chewers.**

Pigs, peccaries, hippopotamuses, and the many extinct groups closely related to them are distinguished from cud chewers in several ways. Perhaps most important to paleontologists is the fact that molars of the "pig group" are low-crowned and surmounted by conical cusps, whereas molars of the cud chewers are both low- and high-crowned and invariably bear crescent-shaped cusps on their grinding surfaces. The earliest of the pig group were like the ancestral condylarths in being small, short-limbed animals with the full array of mammalian teeth; however, only four toes on each foot were in contact with the ground. From these modest beginnings, there radiated in Mid-Cenozoic time a wide variety of even-toed members. Some, the now extinct **entelodonts** (*Dinohyus* in Figure 30-7) developed long legs and two-toed feet and became buffalo-sized animals with immense knobby skulls. Others developed gradually into the more familiar pigs, peccaries, and hippopotamuses.

The earliest of the cud chewers may have been the now extinct **oreodonts**, a group of short-legged ungulates that roamed western North America in great herds well into the Neogene Period. Camels (Figure 30-9) appeared in Mid-Paleogene time, and like the horses, underwent the major part of their evolutionary development in North

**Figure 30-9**
Rhinoceroses and camels. The rhinos in
the foreground are odd-toed ungulates,
whereas the herd of llamalike camels in
the background represents the
even-toed ungulates. Both groups were
contemporaries of the entelodonts and
chalicotheres shown in Figure 30-7.
(From a painting by Charles R. Knight,
courtesy of Chicago Natural History
Museum.)

America. Apparently they emigrated to the Old World
and South America with the advent of the Pleistocene
ice sheets. The only indigenous camels in the Western
Hemisphere today are the llamas of South America.

All the other cud chewers are descendants of a Mid-
Paleogene mammal about the size of a jackrabbit. This
curious creature had a long tail, somewhat lengthened
feet in which functional emphasis was on the two middle
toes, and a complete complement of mammalian teeth.
Cud-chewing descendants of this early species have
become larger, the tail has become shorter, and the upper
incisor teeth have been gradually replaced with a horny
pad against which the lower incisors nip. In addition,
there has been a marked tendency toward the development of horns, some of them exceedingly spectacular, on
both the forehead and nose.

The widespread success of the cud chewers is almost
certainly a result of their complex four-part stomach.
Food, hurriedly stowed and partly digested in the forward
two sections, can be regurgitated and chewed carefully
at leisure. These animals can "eat and run" with no
harmful effects, for they can later digest in peace and
safety. In the Neogene, cud chewers multiplied rapidly
in numbers and variety, and largely replaced the odd-toed
ungulates, whose time of glory was about midway through
the Cenozoic.

## Elephants and Their Kin

Elephants, limited now to Asia and Africa, are familiar representatives of a mammalian group with a long Cenozoic history. The oldest ones, known from Mid-Paleogene deposits in Egypt, were stout pig-sized mammals with heavy legs and broad spreading feet with a hoof at the end of each toe. Although these animals probably had no trunk, the second of the three incisors on either side of the upper and lower jaws were tusklike and all but crowded out the other incisors and canines.

Later predecessors of modern elephants grew much larger than their Mid-Paleogene forebears, had massive skulls and short necks, and, through elongation of the upper lip and nose, developed the prehensile trunk so characteristic of the group (Figure 30-10). In addition, the second incisors grew prodigiously to form tusks. These developed on both upper and lower jaws in early trunked mammals, but are confined to the upper jaw in modern elephants and their immediate Neogene predecessors.

Elephants probably reached the peak of their evolutionary development and were worldwide in distribution during the ice ages of the Pleistocene (Figure 30-11). Many had shaggy coats and inhabited areas of rigorous climate near the glacier margins. These animals seem to have captured the fancy of ancient man (as living elephants fascinate us), for he decorated the walls of his caves with numerous pictures of his giant companions. They were certainly familiar to early man in America, and probably became extinct here only a few thousand years ago. No one knows why elephants have dwindled so drastically

**Figure 30-10**
A pair of "shovel-tusked" early Neogene elephants. Although clearly related to modern elephants, these animals differed from living types in several obvious ways. (From a painting by John Conrad Hansen, courtesy of Chicago Natural History Museum.)

**Figure 30-11**
Wooly mammoths and rhinoceroses of
the Pleistocene. Both types of animal
were clearly adapted for rigorous
climates at the ice-sheet margins.
(Courtesy of Chicago Natural History
Museum.)

in importance, but man may have had at least a part in
their decline.

### The Modern Fauna

By the beginning of the Pleistocene Epoch, the mammalian
faunas of the world were essentially modern in appearance, but were far richer than they are now. Buffalo,
mammoths, mastodons, and wooly rhinoceroses roamed
widely in the frigid wastelands south of the glaciers, and
saber-toothed cats, giant birds, horses, and the like were
numerous in warmer areas of the world (Figure 30-12).

But profound fluctuations in climate during the ice ages and the rise of man changed much of that.

Many formerly widespread species of mammals (elephants, camels, and horses, for example) became restricted to much smaller areas, some gradually diminished in importance and became extinct, and still others (for example, buffalo) have been reduced to extinction or to their present insignificance by the ravages of man the hunter.

The mammalian order Primates, the group to which we belong, has played a part in the overall history of the mammals far out of proportion to its size. Furthermore, man, its dominant modern representative, now has it in his power to control or markedly influence the future course of biosphere history. For these reasons, it is appropriate that we devote greater attention to man and his kin than we have been able to accord the other mammals mentioned in this chapter. Man alone knows of his past and is equipped to do something about his future.

The fossil record tells us that most mammals have undergone extensive, and in many cases extreme, changes in limb and tooth structure during their Cenozoic history, and that these have slowly made them more and more specialized users of ever more restricted parts of their environment. The theme, in short, has been specialization. Yet within the last million years or so all mammals, even

**Figure 30-12**
Pleistocene animals at the Rancho La Brea tar pits of southern California. Saber-toothed cats at the right feed on the remains of elephants mired in the tar while vulturelike carrion birds and wolves wait their turn. A small herd of Pleistocene horses is in the right distance. (Painting by Charles R. Knight, courtesy of Chicago Natural History Museum.)

the most highly specialized, have come under the domination of man, a single species of the order Primates. Oddly enough, primates are distinguished from other mammals by their lack of specialization in many features—and therein lies an especially intriguing tale.

Because the rise to dominance of a single species is without parallel in biosphere history—and more important, because we are part of that species—we devote the rest of our discussion of Cenozoic life to the story of man and his primate cousins.

## Primate Features

The order Primates, which includes lemurs, tarsiers, monkeys, apes, and man, is an assemblage of rather generalized small to medium-sized placental mammals. All have a collarbone, structurally primitive molars, and five digits on each limb. Primate skulls are distinguished from those of other mammals by a relatively large brain case, and by a small nose that is crowded in between surprisingly large eyes. The eyes are directed forward rather than laterally, and thus supply stereoscopic vision. In addition, feet and back are supple, as limbs are; digits are tipped by nails rather than claws or hoofs; and in most primates, the thumb and big toe are set well apart from the other digits. In sum, primates have highly developed brains of large size, stereoscopic vision, and grasping hands and feet. They can see things clearly and in three dimensions; they can manipulate and understand these things because of their mobile thumbs and big toes and their high degree of nervous control.

## Primitive Primates

It seems likely that the ancestral primates were long-nosed bushy-tailed insect eaters somewhat like the living tree shrews of southeast Asia: squirrellike creatures, with thumbs and big toes set slightly apart from the other fingers, and with brains relatively larger than those of the insectivores that gave rise to them. From ancestors like this, there developed very early in the Paleogene a wide variety of primitive primates, of which the lemurs of tropical Africa and Asia and the tarsiers of the East Indies are modified living descendants.

**Lemurs** are tree-living animals of modest size that eat fruit and seeds. They have foxlike faces, pointed ears, eyes directed more laterally than forward, and long tails and legs. They are clearly primates, however, for their brains are relatively larger than those of other mammals and their thumbs and big toes are widely separated from the other digits, most of which end in nails rather than claws. **Tarsiers** are tiny nocturnal treetop dwellers that have long tails, hind legs adapted for hopping, and a rounded skull with a monkeylike face in which exceptionally large eyes peer directly forward and nearly crowd out a buttonlike nose.

Tarsiers are more advanced primates than lemurs are, although fossils of both are found in earliest Cenozoic rocks. These animals apparently were numerous in the semitropical forests that spread over much of North America and Europe in the Early Paleogene. Late in the Paleogene, however, both lemurs and tarsiers disappeared from the Northern Hemisphere and became confined to tropical parts of the Southern Hemisphere, where they persist today.

## Monkeys and Apes

At some time in the Paleogene, tarsierlike Early Cenozoic primates gave rise to three distinct groups of more advanced primates. One of these, the flat-nosed South American monkeys, is typified by the organ-grinder's companion; another, the Old World monkeys, includes baboons, mandrills, and rhesus monkeys. A third group, quite distinct from the other two, is the apes, represented today by gibbons, orangutans, chimpanzees, and gorillas. Because all these advanced primate groups appeared at the same time, it is probable that each developed independently from the old tarsiers, rather than from one another. There has been no "monkey business" in man's ancestry, which is tied directly to that of the apes rather than to that of the monkeys.

A pair of small stout jaws found in Late Paleogene strata in Egypt indicates that the earliest apes were of monkeylike size and fairly generalized structure. By the Early Neogene, however, a wide variety of larger apes flourished in Europe, Asia, and Africa and left a fairly widespread fossil record. These animals are known largely from jaws and teeth; the few limb bones that have been found are slender and indicate that Early Neogene apes were probably much more agile and active than modern ones. Further, they probably spent a good deal of time on the ground; none seems to have been especially well fitted for swinging through the trees. Indeed, these Early Neogene apes may well have been a good deal more human in appearance than their modern descendants, the great apes.

Later Neogene apes represent only a handful of species, but they were clearly successful ones that roamed widely through the warm forests of Europe, Asia, and Africa, and exhibited a good deal of variation in size and limb structure. Some of the smaller, more generalized ones undoubtedly developed into gibbons, which spend most of their lives in the trees. The large size of others prevented them from taking up an arboreal life, so they spent most of their lives on the ground. Their long arms, however, fitted them to move rapidly through the forest by swinging from branch to branch. From these apes gradually evolved orangutans, chimpanzees, and gorillas.

The geographic range of the apes has shrunk as the warm forests to which they are best adapted have dwindled in area. While apes were still widespread, however, there appeared among them in both northern India and Kenya a few individuals with jaw, facial, and dental features so like those of primitive man that it is difficult to say with certainty whether they were advanced apes or early men. Whichever they were, it is clear that in the ten million years or so that elapsed between the appearance of these manlike apes and that of the first undoubted men, advanced primates gradually developed the pelvic, limb, and spinal modifications that permitted them to stand comfortably upright and to walk efficiently on their hind legs. When erect primates of this lineage put their emancipated forelimbs to work making tools and weapons, most would agree that man had finally appeared on the scene.

## Man

Man's fossil history (Figure 30-13) is confined to the latest epoch of the Neogene Period, the Pleistocene, which is thought to have begun about two million years ago and which included, in the Northern Hemisphere, a series of continental glaciations. South of the ice sheets and in much of the Southern Hemisphere, an alternation of dry and rainy periods, correlated more or less directly with ice advances and retreats in the north, conspired to produce the sand and gravel deposits, caves, and river terraces, in and on all of which man's scanty fossil record is preserved.

### The Australopithecines

The oldest human remains known come from gravels and limy cave deposits in Africa, Israel, Java, and China, which formed just before the advent of the first Pleistocene glacier. Although these remains represent several somewhat different types of primitive men, they are all very closely related and are collectively, if unfortunately, termed **australopithecines** (southern apes). These crea-

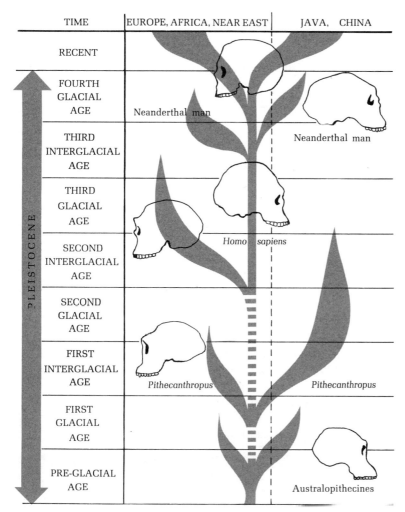

| TIME | EUROPE, AFRICA, NEAR EAST | JAVA, CHINA |
|---|---|---|

RECENT

FOURTH GLACIAL AGE — Neanderthal man — Neanderthal man

THIRD INTERGLACIAL AGE

THIRD GLACIAL AGE

SECOND INTERGLACIAL AGE — *Homo sapiens*

SECOND GLACIAL AGE

FIRST INTERGLACIAL AGE — *Pithecanthropus* — *Pithecanthropus*

FIRST GLACIAL AGE

PRE-GLACIAL AGE — Australopithecines

PLEISTOCENE

**Figure 30-13**
Geologic history of man. Three main human types are recognized: the australopithecines at the bottom of the chart, *Pithecanthropus* in the middle, and *Homo sapiens* at the top. Neanderthal man, grouped here with *Homo sapiens,* is sometimes regarded as a separate species, *Homo neanderthalensis.*

tures were short and slight of build, and most of them probably weighed less than 100 pounds. Their superficially apelike skulls (Figure 30-14), which housed a chimpanzeelike brain half the size of ours, were balanced more

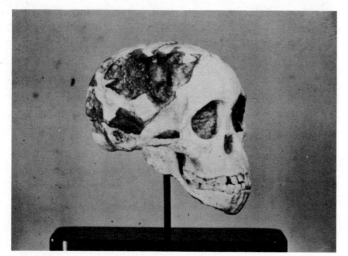

**Figure 30-14**
Restored australopithecine skull. (Courtesy of American Museum of Natural History.)

nearly on top of the spine than in the apes, and the structure of the pelvis and lower limb bones clearly indicates that they could stand and walk erect. Further, crude stone tools and crushed baboon skulls found in the same deposits with some of them are mute testimony to the fact that, despite their dull intellect, the australopithecines were the makers and users of tools, as well as hunters. In short, they were primitive men.

## Pithecanthropus

The australopithecines were succeeded in time by a highly variable species of primitive man, *Pithecanthropus* (or *Homo*) *erectus. Pithecanthropus,* or "ape man," who undoubtedly descended from the larger and brighter of the australopithecines, put in his appearance during the warm, wet interval between the first and second glaciations, or perhaps a bit earlier. He spread widely through Europe, Africa, and Asia, where he was successful until at least the latter part of the Second Interglacial Age.

*Pithecanthropus,* best known from fossils found on the island of Java and near Peking, China (Figure 30-15), was probably somewhat larger than the average australopithecine. He had a brain that ranged in size from about 775 to 1,200 cubic centimeters. The smaller of these sizes is

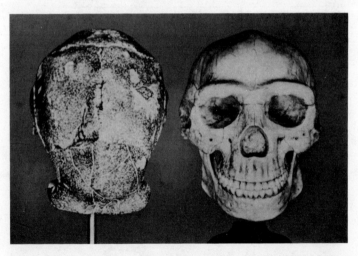

**Figure 30-15**
Skulls of *Pithecanthropus* from near Peking, China, restored by Weidenreich. Note the massive-brow ridges and the pattern of fractures in the skullcap at the left. The latter suggest that the skull had been broken by a hard blow, probably by another *Pithecanthropus.* (Courtesy of American Museum of Natural History.)

some 90 cubic centimeters larger than that of the biggest known ape brain, and the larger size is well within the range of the brain of modern man. The skull of *Pithecanthropus,* however, had a retreating forehead, prominent brow ridges, and large teeth set in a massive chinless jaw.

Interestingly, the limb bones of *Pithecanthropus* are quite similar to ours. Apparently modernization did not affect all parts at once. Rather, it seems that structures having to do with upright posture were perfected before brain, teeth, and skull reached their present degree of development. These primitive men used fire, were skilled hunters, and made crude stone tools (of a type termed Chellean), which they scattered about in Europe, Asia, and Africa.

## *Homo sapiens* Appears

During the Second Interglacial Age, while *Pithecanthropus* dominated the scene in Asia, there appeared in Europe a more advanced breed known from skull scraps found at Swanscombe, England, and from younger skull fragments discovered at Fontéchevade, France, and Steinheim, Germany. So far as it can be determined from these fragments, the men represented were similar to one another and closely related to modern man. Brain size ranged from a low of 1,100 cubic centimeters to a high of at least 1,320, well within our range but somewhat below the modern average of 1,350. However, skulls are thick and have rather prominent brow ridges, which suggests that these early men had *Pithecanthropus* in their family tree. The tendency is to regard these ancient men as the first representatives of *Homo sapiens*, the species to which we belong.

A number of fragmentary human skeletons found in central Europe and Palestine are probably younger than those from Swanscombe, Fontéchevade, and Steinheim, and are associated with tools of a somewhat more advanced culture. These skeletons indicate that by the latter half of the Third Interglacial Age *Homo sapiens* had become quite varied. Some of the more primitive individuals retained the heavy brow ridges, retreating forehead, prominent jaws, and feeble chin of earlier men, whereas others had skulls like those of the more primitive races of modern man.

## *Neanderthal Man*

Probably the best known of all early men was Neanderthal man (*Homo neanderthalensis*), who was the only type of man in Europe and North Africa during the early part of the last glaciation. Neanderthal man (Figure 30-16) was short, but heavy-bodied and of powerful build. He had a large head, slung somewhat farther forward on the neck than ours, and with a thick-walled skull, a low sloping forehead, and a flattened brain case. His jaws were massive, his chin receding, and his teeth much larger than ours. But Neanderthal was no fool. His massive head contained a brain whose volume averaged 1,450 cubic

centimeters, somewhat larger than the average brain of
modern *Homo sapiens*.

Neanderthal man was a skilled hunter, a user of fire,
and a dweller in caves. He buried his dead, fashioned
a sophisticated type of stone tools, and undoubtedly
clothed himself in the skins of the animals he slew. In
addition, he seems to have been the first collector of
fossils, for a polished Jurassic brachiopod has been found
among the effects of a Neanderthal buried near Saint
Léon, France.

Because he buried his dead, Neanderthal has left us
a far better record of himself than did earlier types of
man. Even so, his abrupt appearance has been nearly as
difficult to explain as his sudden disappearance near the
climax of the last glaciation in Europe and Asia. There
are two general explanations for this history that bear
summarizing here.

There are those who hold that at the *Pithecanthropus*
level of development man divided into two stocks, a
western one that ranged through Africa, Europe, Asia

Minor, and India, and an eastern group that lived primarily in central and eastern Asia. For much of Pleistocene time, these two groups are supposed to have been isolated from one another by the great mountain barrier that stretches east across southern Europe to northern India, then swings southward in southeast Asia. West of the barrier, man remained nomadic and naked, lived in the open, and did not learn to use fire because he could retreat to Africa during the times of glacier advance. East of this barrier, however, no such retreat was possible, and in cold weather primitive men set up housekeeping in caves, used fire for cooking and heating, and probably clothed themselves in skins. Such fossil evidence as is available suggests that "western man" may have developed the features of *Homo sapiens* early in the game but that "eastern man" changed slowly into Neanderthal.

Some time in the Third Interglacial Age, eastern (or Neanderthal) man may have spilled over the mountain barrier separating him from western man (*Homo sapiens*). For a time, the two may have lived together in Europe, North Africa, and western Asia, and highly variable early representatives of *Homo sapiens* may be the products of their mating. With the advent of the last glacier, however, *Homo sapiens* fled to warmer country, leaving icy Europe and Asia to Neanderthal, who was accustomed to life in this sort of climate.

But *Homo sapiens* may well have learned a thing or two from Neanderthal during their brief encounter in the Third Interglacial Age, for midway through the last glaciation he reappeared in Europe, fully clothed and fully familiar with fire and life in caves. His brain, though no larger than Neanderthal's on the average, is better developed in those areas controlling memory and speech, and with the hunting and homemaking arts learned early from Neanderthal, he was a competitor of considerable vigor. His appearance, in any event, marked the rapid disappearance of Neanderthal, who may well have been the first to fall before ruthless *Homo sapiens*.

Although the sequence sketched in preceding paragraphs is attractive, most students of man now believe it to be unnecessarily complicated. In what is probably the majority view, Neanderthal and modern *Homo sapiens* are two branches from the highly variable men that succeeded the Swanscombe, Fontéchevade, and Steinheim types late in the Third Interglacial Age. Neanderthal had his day of glory early in the last glacial age, whereas our branch of the family rose to prominence in the latter half of the last glacial age. Neanderthal, in this view, was a short-lived variant who failed to survive his keener cousin and progenitor, *Homo sapiens*.

*Modern Man Appears*

Fully modern man dates back about 40,000 years to the midportion of the last ice age when Europe was invaded by bands of men who, judging from their skulls and skeletons, were quite similar to modern Europeans. These men were hunters of great skill who carried razor-sharp flint knives and left paintings and crude statuary behind in the caves they inhabited. As they spread through Europe and western Asia, their toolmaking and artistic abilities became more and more refined; handsomely carved bone and ivory instruments were fashioned by later peoples, whose eyes and hands had also become coordinated enough to produce bright-colored graceful reproductions of men and other animals in murals on their cave walls.

As the last ice sheet melted away, modern man, divided by then into many races of somewhat different physique and structure, spread through Asia, into Australia, and ultimately across the Bering Straits into North America. The American pioneers were surely unaware that their return to this continent brought primate evolution full circle: some of the earliest history of the order is recorded by the lemur- and tarsierlike fossils unearthed from Paleogene rocks in western North America.

**30-6 SUMMARY** Mammals evolved from therapsid reptiles in the Late Triassic, but they were not a conspicuous part of the vertebrate fauna until early in the Cenozoic, when archaic marsupials and placentals expanded rapidly into environments without reptilian competitors. Although marsupials developed greatly in Australia and South America, which were cut off from the rest of the world, they were replaced by placentals nearly everywhere else in the Paleogene Period. Archaic placental mammals, including insectivores, condylarths, amblypods, uintatheres, and creodonts, radiated widely in the Paleogene. Through gradual changes in size, dental structure, and architecture of limbs and feet, members of these archaic groups gave rise in the late Paleogene and Neogene to the array of specialized placental groups familiar to us in the modern fauna.

Man and the modern great apes probably arose from common Mid-Neogene primate ancestors. Man's development first involved accumulation of structures permitting him to stand upright: competitive advantage then probably went to those who could make best use of their newly emancipated forelimbs. Man's later development is recorded primarily in increasing brain size, particularly by growth in the forward portions of the brain concerned with memory, forethought, speech, and manual dexterity. Associated with this has been gradual reduction in prominence of the canine teeth. Thus man's forehead has become higher, his skull rounder, and his jaws and teeth less massive with time. The oldest men known, the australopithecines, appeared just before the advent of the first Pleistocene glacier; they were succeeded by *Pithecanthropus* (or *Homo*) *erectus* in the interval between the first and second Pleistocene glaciations. *Homo sapiens* appeared during the Second Interglacial Age, but Neanderthal man was the only type in Europe and North Africa during the early part of the last glaciation. Modern man dates back about 40,000 years to the midportion of the last ice age.

## SUGGESTED READINGS

Clark, J. D. 1958. Early man in Africa. *Scientific American,* vol. 199, no. 1, pp. 76–83. (Offprint No. 820. San Francisco: Freeman.)
A general account of African discoveries that have yielded bones of man-apes and human tools, with specific reference to the author's excavations at Kalambo Falls.

Cruxent, J. M., and Rouse, I. 1969. Early man in the West Indies. *Scientific American,* vol. 221, no. 5, pp. 42–52. (Offprint No. 652. San Francisco: Freeman.)
Reveals that man may have inhabited the islands of the Caribbean as early as 5,000 years before the time of Christ.

Howells, W. W. 1966. Homo erectus. *Scientific American,* vol. 215, no. 5, pp. 46–53. (Offprint No. 820. San Francisco: Freeman.)
A report on the species that is very likely the immediate predecessor of modern man. The possibility is advanced that the transition took place some 500,000 years ago.

McAlester, A. L. 1968. *The History of Life,* chaps. 6 and 7, pp. 101–144. Englewood Cliffs, N.J.: Prentice-Hall. (Paperback.)
Chapter 6 includes a section on evolution of the mammals, and Chapter 7 covers the evolution of man and his ancestors.

Simons, E. L. 1964. The earliest relatives of man. *Scientific American,* vol. 211, no. 1, pp. 50–62. (Offprint No. 622. San Francisco: Freeman.)
Presents fossil evidence from 12 to 60 million years ago, which points to the main stages of primate evolution and singles out the stock from which the human line arose.

Simons, E. L. 1967. The earliest apes. *Scientific American,* vol. 217, no. 6, pp. 28–35. (Offprint No. 636. San Francisco: Freeman.)
Reports on the excavation of the skull of a 28-million-year-old ape in Egypt.

Simons, E. L., and Ettel, P. C. 1970. Gigantopithecus. *Scientific American*, vol. 222, no. 1, pp. 76–85. (Offprint No. 655. San Francisco: Freeman.)

An account of an extinct ape, which was the largest primate that ever lived. It is represented by a fossil tooth and four massive jawbones, one of which was recently found in India.

Weckler, J. E. 1957. Neanderthal man. *Scientific American*, vol. 197, no. 6, pp. 89–96. (Offprint No. 844. San Francisco: Freeman.)

A hypothesis on the origin of both *Homo sapiens* and Neanderthal man, and on how these contemporaries came into contact with each other.

**Abrasion** Mechanical wear of rock particles against each other or against bedrock, as in streams, glaciers, and waves.

**Absolute age** The length of time in years since the occurrence of a given geologic event, such as the deposition of a bed of rock or the intrusion of a batholith.

**Abyssal hills** Groups of irregular, rocky hills that stand 150 to 300 feet above abyssal plains.

**Abyssal plains** Smooth, sediment-veneered submarine plains in deeper parts of ocean basins.

**Age** In geochronology, the time of occurrence of an event. See *Absolute age, Relative age.* In stratigraphy, an interval of geologic time, a subdivision of an epoch.

**Alluvial fan** A cone- or fan-shaped deposit of poorly sorted alluvium, formed where a heavily loaded stream emerges from a narrow valley and deposits most or all of its load on a land surface.

**Alluvial stream** A stream, generally large, that flows on a broad valley floor across unconsolidated stream-deposited sediment, or alluvium.

**Alluvium** Unconsolidated stream-deposited sediment.

**Amblypod** A stout-bodied, four-footed placental mammal with strong limbs each terminating in five hoofed toes.

**Ammonoid cephalopod** Cephalopod with straight or coiled external shell. When shell dissolves, the edges of internal cross-partitions are exposed as highly sinuous or fluted lines around the core of sediment that fills the shell.

**Anapsida** An order of reptiles characterized by a solidly roofed skull with no openings behind the eyes.

**Angle of repose** The maximum angle or slope at which loose materials such as sand, soil, or rock fragments remain stable under gravity.

**Angular unconformity** An unconformity below which the strata are not parallel to those above.

**Anhydrite** A mineral, calcium sulfate, $CaSO_4$.

**Animal kingdom** A division of the biosphere that includes all animals.

**Ankylosaur** Bulky, armored four-footed dinosaur.

**Anthracite** A nonfoliated metamorphic rock produced by the very slight metamorphism of bituminous coal.

**Anticline** An up-fold in rocks.

**Aquifer** Any body of regolith or rock that contains and transmits ground water.

**Arête** A knife-edge ridge of rock produced when cirques are cut into a divide from opposite sides.

**Artesian well** Any well in which the water rises above the level at which it was encountered. May be flowing or nonflowing.

**Ash** Fine-grained pyroclastics produced in explosive volcanic eruptions.

**Astrogeology** Systematic study of the solid bodies that can be observed in space, especially the moon.

**Atmosphere** The gaseous envelope that surrounds the lithosphere.

**Atom** The smallest portion of an element that has the properties of that element.

**Augite** A silicate mineral containing aluminum, calcium, iron, and magnesium, with a basic structure of single-chain silica tetrahedra. A ferromagnesian mineral.

**Australopithecine** Collective term for several different but closely related types of primitive man, who were short, slight, and had superficially apelike skulls that housed a brain half the size of modern man's.

**Avalanche** A mass of snow and rock fragments that moves very rapidly down a mountain slope.

**Barchan** A crescent-shaped sand dune, with the horns pointing downwind.

**Barrier island** A low ridge of sand extending parallel to a low-plains shoreline, separated from the mainland by a lagoon.

**Basal slip** The motion of a mass of glacier ice as it moves bodily across a rock floor.

**Basalt** A dark, heavy, fine-grained igneous rock, commonly extrusive, which is made up chiefly of plagioclase and augite.

**Base level** The downward limit of stream erosion, which is ultimately sea level. Local base levels along a stream's course may be imposed by such features as lakes or exceptionally resistant rock layers. For a major drainage system, base level is an imaginary plane beneath the continent, sloping gently upward from sea level.

**Basement, basement complex** Precambrian rocks beneath the sedimentary-rock veneer of the Interior Lowlands.

**Batholith** A body of intrusive rock, generally granitic, with dimensions

# GLOSSARY

in the scores of miles and no known floor. Batholiths are characteristic of crustal belts that have been deformed by earth movements.

**Bay barrier** Sand bar extending across the mouth of a bay, sealing it off from the open sea.

**Beach** Sand or other clastic sediment, produced by hydraulic action and abrasion, that accumulates along the water's edge on a shoreline.

**Bed** See *Stratum.*

**Bed load** Relatively coarse particles that are pushed, dragged, and shoved along the bottom of a stream.

**Bedded** See *Stratified.*

**Bedding plane** Surface separating different layers or strata in sedimentary rocks.

**Belemnoid** Extinct squid-like cephalopod with an internal shell.

**Biosphere** The earth's living matter in all its forms.

**Biotite** A ferromagnesian mineral, the black mica; a hydrous silicate of potassium, aluminum, iron, and magnesium.

**Bituminous coal** A black sedimentary rock formed from plant material by compaction. It is typically finely banded and breaks with a blocky fracture.

**Block** Coarse pyroclastic fragment blown from a volcanic vent in explosive eruption.

**Bomb** Gob of lava blown from a volcanic vent while liquid, and solidifying before falling to the surface.

**Braided stream** A stream that flows in many channels among a maze of sand or gravel bars, as a result of receiving so much sediment that normal stream flow is obstructed.

**Breakwater** A large-scale stone structure built along a shoreline to retard wave erosion.

**Breccia** Rock made up largely of coarse clastic fragments that are sharp and angular.

**Brown clay** Deposit in deep ocean basins consisting of wind-blown silt and clay, volcanic matter, meteoritic dust, and minor quantities of material derived from land.

**Butte** A small isolated remnant of formerly extensive flat-lying rocks.

**Calcite** A mineral, calcium carbonate, $CaCO_3$.

**Caldera** A large depression, as much as several miles across, which marks the site of a former volcanic cone that has exploded or collapsed.

**Carbon-14** A radioactive isotope of the element carbon, with mass number 14, which is produced by bombardment of atmospheric nitrogen by cosmic rays.

**Carbonification** The conversion of plant matter to coal, in which moisture and gases are gradually driven off, and the proportion of carbon increases.

**Cast** See *Natural cast.*

**Cementation** The process of lithification in which mineral matter is precipitated around particles of sediment so as to bind them together.

**Ceratopsian** A horned dinosaur; four-footed plant-eater with massive, beaked skull equipped with horns and a prominent bony frill that projected back over the neck and shoulders.

**Chalcopyrite** A mineral, a sulfide of copper and iron, $CuFeS_2$. An ore mineral of copper.

**Chalicothere** Extinct odd-toed ungulate about the size of a modern horse; front legs longer than hind ones; three functional toes on each foot terminating in large claws rather than hoofs.

**Chert** A hard, dense sedimentary rock made up of extremely fine-grained silica. Black varieties are known as flint.

**Chlorite** A metamorphic mineral; a hydrous silicate of aluminum, magnesium, and iron.

**Cinder cone** Relatively small, steep-sided volcanic cone, composed mainly of pyroclastics.

**Cinders** Sand- and pebble-size pyroclastics produced in explosive volcanic eruptions.

**Cirque** The scoop-shaped depression in bedrock at the head of a valley glacier.

**Class** A group of related orders.

**Clastic** Fragmental, broken; term applied to those sedimentary rocks that consist of fragments, grains, and particles of mineral and rock material.

**Clay** The finest sediment, consisting of particles less than 1/256 mm in diameter. Composition is chiefly of the clay minerals.

**Clay minerals** A group of extremely fine-grained hydrous silicate minerals, which form as a result of chemical weathering. Kaolinite is a common variety.

**Cleavage** In minerals, the property of breaking along smooth plane surfaces that are a function of the internal atomic structure of the mineral. In rocks, chiefly slate, the property of splitting into thin sheets as a result of alignment of platy and flaky minerals during metamorphism.

**Coal** See *Anthracite, Bituminous coal.*

**Coastal plain** Low, almost flat area developed on nearly horizontal beds of sand, clay, and limestone along the continental margin; merges seaward with continental shelf.

**Compaction** The process of lithification by which mud is converted into shale, and peat into coal. Weight of overlying sediments forces water out of the mud or peat and packs the solid particles closely together.

**Composite volcano** The assemblage of lava flows and layers of pyroclastics that builds up around an explosive volcanic vent. Commonly has steep slopes and a markedly symmetrical profile.

**Compound** A substance formed by the chemical combination of two or more elements.

**Concretion** A sharply defined mass of mineral matter in a sedimentary rock, which was precipitated from solution around a nucleus.

**Condylarth** Light-bodied, archaic, hoofed placental mammal.

**Confining pressure** Pressure that is uniform in all directions.

**Conglomerate** A clastic sedimentary rock consisting of rounded gravel-size sediments and coarse sand, held together by a mineral cement.

**Contact metamorphism** Alteration that takes place in rocks at and near their contact with intrusive magma.

**Continental drift** A term describing the break-up and dispersion of the super-continent Pangaea in the Mesozoic and Cenozoic eras.

**Continental rise** A broad area of very gentle gradient that extends seaward from the base of the continental slope for hundreds of miles to merge with abyssal plains.

**Continental shelf** Shallowly submerged margin of a continent, extending seaward to the edge of the continental slope.

**Continental slope** Narrow belt of abruptly steeper sea floor, seaward from the outer edge of a continental shelf.

**Coprolite** Fossilized animal excrement.

**Core** The central part of the lithosphere, about 2,150 miles in radius, which has a very high density and presumably a composition of iron and nickel. The outer part is thought to be liquid and the inner part solid.

**Correlation** Establishing chronologic equivalence of rocks in different areas, on the basis of fossil assemblages.

**Cotylosaur** Member of an extinct group of clumsy anapsid reptiles.

**Country rock** The rocks invaded by intrusive igneous rocks.

**Crater** Bowl-shaped depression at the top of a volcanic cone.

**Creep** Imperceptibly slow downslope movement of regolith.

**Creodont** A primitive, carnivorous placental mammal, with clawed toes, a long tail, and a pair of cheek teeth on either side specialized for slicing meat and cracking bone.

**Crevasse** A deep crevice or crack in glacier ice.

**Cross section** Illustration in the form of a vertical cut or slice through a body of rocks or a part of the crust.

**Cross-stratification** Bedding or layering in sediments that is inclined at an angle to the top and bottom of the bed in which it occurs. The result of deposition from moving currents, as in dune sands and stream gravels.

**Crust** The outermost thin shell of the lithosphere, 3 to 20 miles thick, bounded at its base by the Moho.

**Cycle of regional reduction** The changes that a region undergoes as a result of stream erosion.

**Cyclothem** A set of strata recording a cycle of nonmarine and marine deposition. The term is customarily used only for Pennsylvanian coal-bearing sequences.

**Debris flow** A water-saturated mass of broken-up rock and regolith of all particle sizes, flowing down a slope.

**Decomposition** Collective term for processes that produce breakdown of rock by chemical means; chemical weathering.

**Deflation** A process of wind erosion in which fine material is blown away.

**Delta** A deposit of sediment built up at a stream's mouth when the stream enters a body of standing water, loses kinetic energy, and deposits its load.

**Dendritic drainage pattern** The branching, treelike pattern formed by streams that flow on massive or uniform rock.

**Density** See *Relative density*.

**Desert pavement** Mosaic of rock fragments on the surface of a desert, left behind when finer material is blown away by the wind.

**Diamond** A mineral composed of carbon, C. Its internal structure produces crystals of extreme hardness.

**Diapsida** An order of reptiles with two openings behind the eye on either side of the skull.

**Differential weathering** Differences in relief of an exposed rock surface, produced by uneven resistance to the processes of weathering.

**Dike** A sheet of intrusive igneous rock that cuts across the structure of the enclosing rocks.

**Dinosaur** Popular name for extinct diapsid reptiles belonging to either the order Saurischia or the order Ornithischia.

**Diorite** A coarse-grained igneous rock composed chiefly of plagioclase, hornblende, augite, and biotite. It is intermediate in composition between granite and gabbro.

**Dip** Inclination of bedding or other surface in rocks. Includes *direction,* toward which the surface is inclined, and *angle,* by which the surface departs from a horizontal plane.

**Directed pressure** Pressure that is more intense in one direction than in others; nonuniform or unbalanced pressure.

**Discharge** The volume of water in a stream that moves past a point in a unit of time. Generally expressed in cubic feet per second, or cfs.

**Disconformity** An unconformity above and below which the strata are parallel.

**Discontinuity** A surface or thin zone at depth within the earth where an abrupt change in velocity of earthquake waves marks a change in relative density and in type of earth material.

**Disintegration** Collective term for processes that produce breakdown of rock by mechanical means; mechanical weathering.

**Divide** The area of higher elevation

separating one valley from the next.

**Dolomite** A mineral, calcium-magnesium carbonate, $CaMg(CO_3)_2$.

**Dolomitization** A process of replacement in which calcite is converted to dolomite by substitution of magnesium ions for some of the calcium ions.

**Dolostone** A sedimentary rock made up of 50 percent or more of the mineral dolomite.

**Dome** An anticlinal uplift that is roughly equidimensional in plan view.

**Dome mountains** Roughly circular uplifts of the crust, from which much of the sedimentary cover has been removed by erosion.

**Downslope movements** Collective term for all movements of rock and regolith down slopes under gravity.

**Dripstone** General term for stalactites, stalagmites, and other cave deposits.

**Drumlin** A hill in an area of ground moraine, made of till, streamlined, and elongated parallel with the direction of former ice movement.

**Dune** A heap of loose sand formed by the wind.

**Early maturity** Stage of regional reduction in which the drainage network is well developed, divides are narrow, and the land is mostly in slope.

**Earthquake** A series of vibrations or shock waves passing through rock and regolith.

**Economic geology** The study of ore deposits, oil pools, and other earth materials of value.

**Electric log** A record, plotted to scale on a strip of paper, of the rocks encountered in a well, as recorded by instruments that determine the rocks' electrical properties.

**Electron** An electrically negative subatomic particle that revolves in spherical orbit about the nucleus of the atom.

**Element** A substance that cannot be changed into simpler kinds of matter by ordinary chemical processes.

**End moraine** A ridge or belt of low hills, composed of till, that marks the former position of the ice front of a glacier.

**Energy** The capacity for doing work. See *Geothermal energy, Kinetic energy.*

**Entelodont** An extinct even-toed ungulate. One group of this buffalo-sized animal developed long legs,

two-toed feet, and an immense knobby skull; another was ancestral to pigs, peccaries, and hippopotamuses.

**Epicenter** The point on the earth's surface directly above the focus of an earthquake.

**Epoch** An interval of geologic time; a subdivision of a period.

**Era** The longest recognized unit of geologic time, for example, the Paleozoic Era.

**Erosion** The wearing away of soil and rock by weathering, downslope movement, and the action of streams, glaciers, waves, wind, and underground water.

**Erratic** A boulder that is transported by glacier ice far from its parent outcrop.

**Esker** Narrow winding ridge of sand and gravel, marking the former course of a stream that flowed in a tunnel beneath a slowly moving or stagnant ice sheet.

**Euryapsida** An order of reptiles with lateral openings in the skull behind and on line with the upper half of the eye opening; distinguished from Parapsida by fact that lateral openings are bounded by different skull bones.

**Evaporite** A sedimentary rock, such as gypsum or rock salt, that is produced by evaporation of a water body.

**Exfoliation** A process of disintegration in which thin concentric shells of weathered rock are separated from the rock exposure.

**Extrusive rocks** Those igneous rocks that are formed from material that is poured out, or extruded, on the surface of the lithosphere.

**Fabric** The shape and arrangement of mineral grains in a rock.

**Family** A group of related genera, including all the species of those genera.

**Fault** A fracture along which the rocks on one side have been displaced with respect to those on the other.

**Fault-block mountains** Blocks of the crust that form mountains because they have been uplifted along faults.

**Feldspars** A group of silicate minerals, the most common in the earth's crust. Orthoclase, a potassium-bearing aluminosilicate, and plagioclase, a group of sodium- and calcium-bearing aluminosilicates, are the common varieties.

**Felsite** General name for fine-grained, extrusive igneous rock that is made up chiefly of quartz, orthoclase, and biotite.

**Ferromagnesian minerals** Silicate minerals that are rich in iron and magnesium, especially biotite, hornblende, augite, and olivine.

**Field relations** The way in which rock masses occur in relation to each other in an area or region.

**Fiery cloud** Turbulent mass of burning gases and bits of incandescent lava, ejected horizontally near the summit of an explosive volcano and traveling down the slopes of the cone with great speed and destructiveness.

**Fissure** A crack in the earth's crust from which basaltic lava may be quietly extruded.

**Flint** Black chert.

**Floodplain** Flat land along alluvial streams, subject to flooding. Underlain by unconsolidated stream-deposited sediment.

**Focus** The point within the lithosphere where the rock slippage occurs that produces the shock waves of an earthquake.

**Foliation** The alignment of platy minerals in many metamorphic rocks, in parallel sheets, bands, and streaks.

**Footwall, footwall block** The rocks that lie below an inclined fault.

**Formation** A rock unit that is distinguished from others on the basis of rock type, or lithology.

**Fossil** Any direct indication of life in the geologic past.

**Fossilization** The preservation of organisms as fossils.

**Fracture** Breaking of a mineral along irregular surfaces.

**Frost wedging** A process of mechanical weathering in which rock is split apart when an irregular crevice becomes filled with water that freezes.

**Gabbro** A coarse-grained igneous rock composed mainly of plagioclase and augite.

**Galena** A mineral, lead sulfide, PbS. An important ore mineral of lead.

**Garnet** A mineral; a complex silicate of aluminum, calcium, magnesium, and iron.

**Gastroliths** Highly polished pebbles, the "gizzard stones" of extinct Mesozoic reptiles.

**Genus** (pl. genera) A group of related species.

**Geochemistry** The application of certain principles of chemistry to problems of the earth.

**Geochronology** Determination and study of the age of the earth's rocks, the earth itself, and the solar system.

**Geode** A hollow, rounded body lined with crystals that project into the central cavity.

**Geologic map** A map that shows by means of distinctive colors or patterns the distribution, age, and attitude of the bedrock units in a particular area or region.

**Geology** Systematic study of the materials, processes, origin, structure, and history of the earth.

**Geomorphology** Systematic study of the nature and origin of landforms.

**Geophysics** The application of certain principles of physics to problems of the earth.

**Geothermal energy** Energy derived from underground steam that is obtained from wells. Can be utilized for generation of electricity.

**Geyser** A boiling spring that intermittently erupts a column or fountain of water and steam.

**Glacier** A thick mass of ice that moves slowly on a land surface. Varieties are valley glaciers in high mountains, and ice sheets that blanket whole regions.

**Gneiss** Medium- to coarse-grained metamorphic rock with crude foliation, in which the minerals are in bands and streaks.

**Gradient** The slope of a stream's channel, expressed in feet per mile.

**Granite** A coarse-grained igneous rock composed chiefly of orthoclase, quartz, and biotite or hornblende.

**Graphite** A mineral composed of carbon, C. The carbon atoms are arranged in sheets, which are easily separated; hence graphite is a soft mineral.

**Gravel** Sediment consisting of pebbles and other clastic fragments larger than sand size.

**Groins** Low walls built at right angles to a beach in order to arrest longshore drift of sand.

**Ground moraine** Deposit of till spread widely across bedrock when an ice sheet melted away.

**Ground water** Water underground.

**Gymnosperm** An advanced seed-plant of the plant phyla Cycadophyta, Ginkgophyta, Coniferophyta, or Gnetophyta.

**Gypsum** A mineral, hydrous calcium sulfate, $CaSO_4 \cdot 2H_2O$. Also, a rock made of this mineral.

**Halite** A mineral, sodium chloride, NaCl. The chief constituent of rock salt.

**Hanging valley** A tributary valley that joins a deep U-shaped glaciated valley far above its floor.

**Hanging wall, hanging-wall block** The rocks that lie above an inclined fault.

**Hardness** The resistance that a mineral offers to scratching.

**Headwall** The steep to vertical upper end of a cirque.

**Hematite** A mineral, iron oxide, $Fe_2O_3$. A common red pigment in rocks, and an important ore of iron.

**Historical geology** Study of the history of the earth and of living things from earliest times to the present.

**Horizon** In soil science, one of the several zones that are commonly present in soil from the surface down to the unaltered parent material.

**Horn** High, rocky spire produced when three or more valley glaciers push their cirques headward into an isolated mountain.

**Hornblende** A ferromagnesian mineral; a complex silicate of aluminum, calcium, iron, and magnesium, with a basic structure of double-chain silica tetrahedra.

**Hot spring** Water naturally emerging from underground at a temperature well above the normal surface temperature. Source of the heat is generally magmatic.

**Humus** Decomposed organic matter present in soil.

**Hydraulic action** Removal of loose material by direct impact of moving water.

**Hydrologic cycle** The cycle through which water passes at and near the earth's surface. It includes evaporation and transpiration; condensation, cloud formation and precipitation; infiltration and runoff; and accumulation in bodies of standing water.

**Hydrosphere** The earth's water in all its forms—gaseous, liquid, and solid.

**Hydrous** Containing chemically combined water.

**Ice front** The lower end of a glacier.

**Ice sheet** A broad glacier, with a maximum thickness of several thousand feet, that blankets a whole region.

**Icefield** Area of permanent ice in

high mountains; the larger ones are sources of valley glaciers.

**Ichthyosaur** An extinct, porpoise-like aquatic reptile.

**Identification** Establishing lithologic equivalence of rocks in different areas, on the basis of rock type.

**Igneous rocks** Rocks resulting from the cooling and solidification of molten silicate fluid.

**Impression** Imprint or surface replica of an organism in fine sediments, often carbonified, as a leaf impression.

**Infiltration** The sinking-in of surface water.

**Insectivore** A shrew- or mole-like placental mammal adapted primarily to feed on insects.

**Insoluble residue** Materials such as clay, quartz grains, and iron oxide, that do not readily dissolve in ground water, and accumulate when soluble material is removed.

**Intensity** Of an earthquake, a measure of severity. Commonly given on the Mercalli scale, ranging from I to a maximum of XII.

**Internal flow** Movement of a glacier that results from microscopic slippage along planes within individual grains of ice.

**Intrusive rocks** Those igneous rocks formed when rising magma stops, cools, and solidifies within the crust before reaching the surface.

**Ion** An atom that carries a positive or negative electric charge.

**Island arc** A festoon of volcanic islands along the landward side of a sea-floor trench, for example, Aleutians, Japanese Islands, Philippines.

**Isopach map** A contoured map on which the contours, or isopachs, join points at which the rock record of a chosen interval is of the same thickness.

**Isostatic equilibrium** A descriptive expression for segments of the crust that are in a state of floating balance.

**Isotope** An alternative form of an element, produced by a variation in the number of neutrons in the atomic nucleus.

**Kame** Small hill of sand and gravel, which originated as a pocket of these sediments in glacier ice and was left behind when the ice melted away.

**Kaolinite** A clay mineral, hydrous aluminum silicate.

**Kettle** A depression in ground moraine or outwash, formed when an isolated block of ice melted away.

**Kinetic energy** The energy of motion. The kinetic energy of an object is dependent on its mass and velocity.

**Lagoon** The long, narrow body of shallow water that separates a barrier island from the mainland.

**Lamina** (pl. laminae) A sedimentary layer less than one-half inch thick.

**Laminated** Formed of laminae.

**Landslide** General term for any perceptible downslope movement of earth materials.

**Lava** An informal term that applies to both extruded molten fluid and the solid rock formed from it.

**Law of original horizontality** Law stating that sedimentary strata are horizontal or nearly so when deposited.

**Law of sequence** Law stating that the record of an event in the geologic past surrounds, overlies, or is impressed upon the record of earlier events.

**Law of superposition** The statement that in any undisturbed series of sedimentary rock the oldest strata are at the bottom and the youngest at the top. An example of the general law of sequence.

**Life** A process that provides materials of very complex organization (organic compounds) with the capacity to grow, reproduce, and respond to various stimuli.

**Lignite** A brownish-black coaly material, intermediate between peat and bituminous coal.

**Limb** Side or flank of an anticline or syncline.

**Limestone** A sedimentary rock consisting of 50 percent or more of calcite. There are many varieties. Most limestones are of marine origin.

**Limonite** A general term for several hydrous iron-oxide minerals. A common yellowish-brown pigment in rocks.

**Lithification** Consolidation into rock.

**Lithified** Consolidated into rock.

**Lithofacies map** A map that shows the distribution of the kinds of rock deposited during a chosen interval of time.

**Lithologic log** A record, plotted to scale on a strip of paper, of the rocks encountered in a well, as determined from examination of cores or cuttings brought up during drilling.

**Lithosphere** The solid earth (as contrasted with atmosphere, hydrosphere, and biosphere).

**Lobe-fin** A bony fish (member of the Crossopterygii) with stout lateral fins containing internal skeletal supports of bone arranged in much the same fashion as those in the limbs of land vertebrates.

**Local base level** A local limit on the downward erosion of a stream, as produced by a lake or an exceptionally resistant rock layer in the stream's course.

**Loess** Wind-deposited silt, buff in color, unstratified, and capable of standing in steep or vertical cuts.

**Long profile** Profile of a stream channel from source to mouth.

**Longshore bar** A submerged ridge of sand parallel to a low-plains shoreline and some distance seaward from it.

**Longshore current** Water moving slowly parallel to a shoreline, as a result of deflection of water from waves that advance shoreward diagonally.

**Longshore drift** Gradual shifting of sand and pebbles along a beach, in a sawtooth or zigzag path.

**Lungfish** A group of bony fishes with lobed fins and functional lungs.

**Magma** Molten silicate fluid within the earth's crust, before loss of gases or other constituents. The parent material from which the igneous rocks are derived.

**Magnetite** A mineral, iron oxide, $FeO \cdot Fe_2O_3$. An ore of iron.

**Magnitude** Of an earthquake, a measure of the energy released. Magnitudes are given on the Richter scale and generally range between 3 and 8.

**Mammal-like reptile** See *Therapsid*.

**Mantle** That part of the lithosphere that extends from the Moho to a depth of about 1,800 miles. It makes up more than 80 percent of the earth's total mass.

**Marble** A nonfoliated metamorphic rock consisting mostly or entirely of calcite or dolomite.

**Marsupial** A mammal, such as the kangaroo and the opossum, in which the young are born alive, but in a tiny immature state; after birth, young are commonly kept in a pouch on the lower belly of the mother.

**Mass number** The sum of the number of protons and the number of neutrons in the nucleus of an atom.

**Massive** Occurring in a large and relatively uniform body, as a massive granite.

**Mature valley** A valley whose floor is near base level. Characterized by a pan-shaped cross-profile, a floodplain wider than the stream channel, and a laterally-swinging stream with a low gradient.

**Maturity** See *Early maturity, Regional maturity.*

**Meander** A wide-looping curve characteristic of large alluvial streams on floodplains.

**Mesa** A broad, flat-topped topographic feature that stands above the surrounding countryside.

**Mesosaur** An extinct aquatic reptile.

**Meta-igneous rocks** Metamorphic rocks derived from igneous rocks.

**Metamorphic rocks** Rocks formed from pre-existing rocks by heat, pressure, and the action of hot vapors or solutions within the crust.

**Metamorphism** Collective term for the alteration of rocks by heat, pressure, and the action of hot vapors or solutions within the crust.

**Metasedimentary rocks** Metamorphic rocks derived from sedimentary rocks.

**Micas** A group of silicate minerals with well-developed cleavage that allows them to be split into thin sheets. Biotite, the black ferromagnesian mica, and muscovite, the light-colored potash-bearing mica, are the most common varieties.

**Mid-oceanic ridge** A long, broad ridge in an ocean basin, rising as much as 10,000 feet above abyssal plains; crest fractured parallel to its length by normal faults (for example, Mid-Atlantic Ridge).

**Mineral** A naturally occurring solid inorganic substance that has an orderly internal structure and characteristic chemical composition, crystal form, and physical properties.

**Mineralogy** The systematic study of minerals.

**Mobile belt** A greatly elongate, slowly-sinking strip of crust, blanketed with sediments and volcanic rocks as it sank, which formed early in the history of a continental mountain system.

**Moho** The discontinuity between the crust of the lithosphere and the underlying mantle. Named for the seismologist Mohorovičić.

**Mold** See *Natural mold.*

**Mud** An informal term for a mixture of water, clay, and silt.

**Mud cracks** Cracks formed in mud when it shrinks in drying.

**Mudflow** A mass of mud flowing down a slope.

**Muscovite** A mineral, the light-colored mica. A hydrous potassium-aluminum silicate.

**Native element** An element that may occur in nature uncombined with other elements. Sulfur, copper, and gold are examples.

**Natural cast** Replica of a fossil, formed by new mineral matter that fills a natural mold.

**Natural levee** Ridge of fine sediment on each side of the channel of a large alluvial stream that is subject to flooding.

**Natural mold** Cavity left in rock when a fossil is removed in solution and its surface irregularities remain imprinted on the walls of the cavity.

**Nautiloid cephalopod** A cephalopod with a straight, curved, or coiled external shell. When the shell dissolves, the edges of internal cross-partitions are exposed as straight or only slightly sinuous lines around the core of sediment that fills the shell.

**Neck** See *Volcanic neck.*

**Neutron** Electrically neutral particle present in the nucleus of atoms.

**Nonconformity** An unconformity on which sedimentary rocks lie upon older eroded massive rocks such as granite or gabbro.

**Normal fault** A fault on which the hanging wall has apparently moved downward with relation to the footwall.

**Nucleus** The central part of an atom, consisting chiefly of protons and neutrons. The nucleus is relatively dense and constitutes most of the atomic mass.

**Obsidian** Volcanic glass with about the same composition as felsite. Commonly dark; breaks with conchoidal fracture.

**Olivine** A silicate mineral containing iron and magnesium. A ferromagnesian mineral.

**Ooze** Calcareous and siliceous deposits of the deep ocean floor, formed from the shells of microscopic animals that inhabit the water above.

**Order** A group of related families.

**Ore deposit** A body of rock from which a metal or metals can be obtained commercially.

**Oreodont** An extinct short-legged cud-chewing even-toed ungulate.

**Organic compound** A complex carbon-bearing chemical compound such as carbohydrate, fat, and protein.

**Organism** A living thing, plant or animal.

**Original horizontality** See *Law of original horizontality.*

**Ornithopod** Duckbilled dinosaur; most were two-footed plant-eaters.

**Orogenic cycle** The sequence of events involved in orogeny.

**Orogeny** The processes by which linear belts of thick sediments and volcanic rocks (mobile belts) are converted into systems of folded mountains.

**Orthoclase** A mineral, the potassium-bearing member of the feldspars; formula $KAlSi_3O_8$.

**Ostracoderm** A primitive fishlike chordate (Class Agnatha) that lacked jaws and paired fins and had a flattened body and a tail enclosed in a thick bony armor.

**Outwash** Sand and gravel spread out below the end of a glacier by meltwater.

**Outwash plain** A broad deposit of outwash sloping gently away from the edge or the end moraine of an ice sheet.

**Paleogeographic map** A map that shows the inferred distribution of major geographic features at some time in the past.

**Paleogeography** Arrangement of lands and seas, and other earth features, as they were at some time in the geologic past.

**Paleontology** Systematic study of fossils and of the history of life on earth.

**Parallel stratification** Arrangement in which rock layers are parallel to each other.

**Parapsida** An order of reptiles with lateral openings in the skull behind and on line with the upper half of the eye opening; distinguished from Euryapsida by the fact that lateral openings are bounded by different skull bones.

**Peat** A brown porous spongy material made up of partly decayed plant remains. The parent material of coal.

**Pegmatite** An igneous rock, generally with the composition of granite, which is exceptionally coarse-grained. Commonly occurs in dikes associated with batholiths.

**Pelycosaur** An extinct synapsid

reptile with an elongate body, sprawling limbs, and jaws equipped with long sharp teeth. Some had a saillike flap of skin down the middle of the back.

**Perched water table** A water table that exists above the general level of ground water.

**Period** The standard subdivision of the geologic time scale, during which a system of rocks was formed; for example, the Devonian Period.

**Permeability** The ability of a rock to transmit ground water.

**Permineralization** Process of fossilization in which the pore spaces in bone, shell, or wood are filled in by mineral matter.

**Petrology** The systematic study of rocks.

**Phenocryst** A mineral grain that is appreciably larger than the surrounding grains; applied to igneous rocks. A rock with many phenocrysts is said to be porphyritic.

**Phyllite** A metamorphic rock with fine wavy foliation and a texture intermediate between the texture of slate and that of schist.

**Phylum** (pl. phyla) A group of related classes.

**Physical geology** Study of the materials of the earth's crust and the processes that form and modify them.

**Placental** A mammal that gives birth to live young in a relatively advanced stage of development; the name of the group refers to the disc-shaped organ, or placenta, by means of which the unborn embryo is nourished.

**Placoderm** The most primitive of the jawed fishes.

**Plagioclase** A mineral series; general name for those feldspars that range in composition from sodium aluminum silicate to calcium aluminum silicate.

**Plant kingdom** A division of the biosphere that includes all plants.

**Plate-tectonic theory** A theory that explains present and historic features of both continental and oceanic crust as the results of interaction between large mobile crustal segments, or plates.

**Plateau basalts** A thick, region-wide series of basalt flows that were extruded from fissures in the crust over a long period of time.

**Plesiosaur** An extinct aquatic reptile with a long neck and a broad, flattened body, the limbs of which were modified to form oarlike paddles.

**Plucking** A process of glacier erosion in which the ice freezes to fragments of the bedrock on and against which it rests, plucking these out and pulling them along when the glacier moves.

**Plunge pool** Depression immediately below a waterfall, which has been scoured out by abrasion and hydraulic action.

**Point bar** Crescent-shaped bar of sand or other alluvial sediment, formed on the inside of a meander bend.

**Porosity** Portion of the total volume of a rock that is not occupied by solid matter. Commonly expressed in percent.

**Porphyritic** Term applied to igneous-rock texture in which phenocrysts are present.

**Pothole** A cylindrical depression in bedrock in a stream channel, caused by abrasion by sand and pebbles caught in a whirlpool.

**Principle of uniformity of process** The concept that geologic processes of today are the same as those of the geologic past; familiarly stated, "the present is the key to the past."

**Proton** Electrically positive particle present in the nucleus of atoms.

**Pterosaur** An extinct flying reptile.

**Pumice** Highly porous, cellular volcanic glass.

**Pyrite** A mineral, iron sulfide, $FeS_2$. "Fool's gold."

**Pyroclastics** Collective term for fragments of shattered rock blown from a volcanic vent in explosive eruption.

**Quartz** A common silicate mineral, $SiO_2$.

**Quartzite** A rock, of either sedimentary or metamorphic origin, consisting of quartz grains cemented by quartz.

**Radial drainage pattern** Pattern formed by streams that flow outward from a hill, volcanic peak, or other isolated high area.

**Radioactive** Term applied to elements that undergo radioactive disintegration.

**Radioactive disintegration** Process of slow breakdown or decay of certain elements into simpler, more stable elements.

**Radiocarbon** See *Carbon-14*.

**Rainwash** The impact of raindrops and the washing action of the water as it trickles over the surface immediately after falling.

**Ray-fin** Member of a group (Actinopterygii) of bony fishes with paired fins consisting of a thin cover of skin spread out over a series of horny rays.

**Reaction series** The order in which the rock-forming silicate minerals crystallize from molten magma.

**Recrystallization** Solution and re-precipitation of calcite or other mineral matter by percolating waters. A common process in altering the fabric of limestones.

**Regional erosion surface** A land surface near sea level, produced by stream erosion in the late mature stage of regional reduction.

**Regional maturity** Middle and late stages of erosion of a land surface by streams, from early maturity to development of a regional erosion surface.

**Regional metamorphism** Metamorphic alteration of rocks over a large area, as a result of intensive deformation within the crust.

**Regional reduction** See *Cycle of regional reduction*.

**Regional youth** Early stage in erosion of a land surface by streams, characterized by wide undissected interstream areas and small youthful valleys.

**Regolith** Loose, unconsolidated material, including soil, that rests on solid rock or bedrock. See *Residual regolith, Transported regolith*.

**Rejuvenated stream** A stream provided with renewed energy as a result of regional uplift of its drainage area.

**Relative age** The age of a geologic feature in relation to the age of associated features.

**Relative density** Ratio of the weight of a given volume of any material to the weight of an equal volume of water. Synonymous with specific gravity.

**Relief** Difference in elevation between the lowest and the highest points on a land surface. Also used as a general term for roughness of an area.

**Replacement** The simultaneous removal of one type of material and the substitution of another, on a volume-for-volume basis. A common process in the earth's crust. Limestone is especially susceptible to replacement.

**Residual mountains** Mountains produced by deep erosion of thick masses of elevated stratified rocks.

**Residual regolith** Clay, sand, and other loose materials that are derived from rock decomposition in place.

**Reverse fault** A fault on which the hanging wall has apparently moved upward with relation to the footwall.

**Ripple marks** Parallel forms produced on the surface of a bed of sand that is agitated by waves or currents.

**Rock** An aggregate of minerals. Among the few exceptions to this definition are pumice, obsidian, and coal.

**Rock cleavage** See *Cleavage.*

**Rock flour** Ground-up rock material of silt size, produced by glacier abrasion.

**Rock salt** A sedimentary rock made of the mineral halite.

**Rockfall** The instantaneous descent of a mass of rock from a steep or vertical face.

**Rockslide** Rapid downslope movement of a mass of rock, generally on bedding, foliation, or other surface of weakness within the rock mass.

**Sand** Sediment consisting of clastic grains ranging in diameter from 1/16 mm to 2.0 mm. The dominant mineral is generally quartz.

**Sandstone** A clastic sedimentary rock consisting of particles 1/16 mm to 2.0 mm in diameter, held together by a mineral cement.

**Schist** A foliated metamorphic rock made up largely of platy or flaky minerals coarse enough to be identified with the unaided eye.

**Scoria** Highly cellular basalt, honeycombed with gas-bubble holes, or vesicles.

**Sea-floor spreading** A hypothesis postulating that oceanic crust is generated at, and later spreads away from, mid-oceanic ridge crests.

**Sea-floor trench** A narrow, elongate trench depressed at least 6,000 feet below the adjacent sea floor.

**Sedimentary rocks** Rocks formed by consolidation of clay, sand, shell fragments, or other sediment; or from the precipitation of dissolved salts from a water body that dries up.

**Seismograph** An instrument that picks up and records earthquake vibrations.

**Sequence** Informal term for all the rocks formed during an era of geologic time, for example, the Paleozoic sequence. Also, chronologic order; see *Law of sequence.*

**Series** A time-rock unit, constitut-

ing all the rocks formed during an epoch of geologic time.

**Serpentine** A metamorphic mineral; a hydrous magnesium silicate.

**Shale** A clastic sedimentary rock consisting chiefly of clay-size mineral grains. These grains are tightly compacted and more or less parallel, giving shale a characteristic thin bedding.

**Sheet structure** Flat or curved tabular masses of rock, 2 to 25 feet or more in thickness, that split off some bodies of massive rock. Probably the result of relief of confining pressure on removal of overlying rocks by erosion.

**Shield volcano** Volcanic cone formed when highly fluid lava erupts from a central pipe or vent and spreads widely, building a broad cone with very gentle slopes.

**Silica tetrahedron** The basic unit of structure in silicate minerals. It consists of one silicon ion surrounded by four equidistant oxygen ions.

**Silicification** A type of replacement in which silica-bearing waters percolating through a rock dissolve a part of the rock and precipitate silica in its place.

**Sill** A sheet of intrusive igneous rock that lies parallel with the structure of the enclosing rocks.

**Silt** Sediment consisting of clastic grains ranging in diameter from 1/256 mm to 1/16 mm. Intermediate in grain size between clay and sand.

**Siltstone** A clastic sedimentary rock consisting of particles 1/256 mm to 1/16 mm in diameter, held together by clay or a mineral cement. Intermediate in grain size between shale and sandstone.

**Sink** Closed depression in a limestone area, formed by solution of the bedrock.

**Slate** An extremely fine-grained metamorphic rock, derived from shale, which possesses excellent rock cleavage.

**Sliderock** The loose blocks of rock that constitute a talus.

**Slip face** The leeward, or downwind, face of a sand dune.

**Slump** Downslope movement of a mass of regolith, moving as a unit, that is bounded on its upper edge by a sharp break and on its downhill side by a bulge.

**Soil** That part of the regolith that supports plant life.

**Soil profile** The succession of soil zones, or horizons, from the surface down to the unaltered parent material.

**Solifluction** Imperceptibly slow downslope movement of water-saturated regolith in arctic regions.

**Sorting** A process of deposition of clastic sediments, typically shown in deltas, in which the coarse, heavy grains tend to be dropped first and the finer and lighter grains progressively farther out from the stream mouth.

**Species** (pl. species) A group, or population, of morphologically distinctive organisms that are capable of interbreeding but are reproductively isolated from other such groups.

**Specific gravity** See *Relative density.*

**Sphalerite** A mineral, zinc sulfide, ZnS. An important ore mineral of zinc.

**Spit** A sand bar extending from the end of a beach into the mouth of an adjacent bay.

**Spring** Any natural surface outflow of ground water.

**Stage** A time-rock unit, constituting all the rocks formed during an age of geologic time.

**Stalactite** Icicle-shaped deposit of dripstone extending downward from the ceiling of a cavern.

**Stalagmite** Mound or dome of dripstone, built upward from the floor of a cavern.

**Stegosaur** Member of a group of extinct plant-eating four-footed dinosaurs, with spiked tails and prominent rows of plates along the back.

**Stratified** Arranged in beds or layers.

**Stratified drift** General term for glacier-derived sediments that have been sorted and layered by meltwater.

**Stratigraphy** Systematic study of layered or stratified rocks.

**Stratum** (pl. strata) A layer or bed of sedimentary rock.

**Strike** Direction of the line formed by the intersection of a dipping bed or other rock surface with a horizontal plane. The strike is always at right angles to the direction of dip.

**Strike-slip fault** A fault on which the movement has been dominantly horizontal and parallel with the strike of the fault.

**Stromatolite** A concretion-like structure that formed as an accumulation of calcareous material on a sheetlike mat formed by a community of microscopic organisms, in which filamentous blue-green algae play the principal role.

**Structural geology** The systematic

study of folds, faults, and other aspects of the architecture of rock masses in the earth's crust.

**Submarine canyon** Submerged extension of a major river valley onto or across the continental shelf; or a canyon cut into the outer edges of a continental shelf and slope by masses of water-laden sediment sliding down the slope toward the deep-sea floor.

**Superposition** See *Law of superposition.*

**Suspended load** Particles held and carried in a stream current.

**Synapsida** An order of reptiles distinguished by skulls that have openings for the eyes and another opening on either side of the skull, behind and on line with the lower half of the eye opening.

**Syncline** A downfold in rocks.

**System** A time-rock unit, constituting all the rocks formed during a period of geologic time; for example, the Devonian System. The fundamental subdivision of the rock record.

**Taconite** Siliceous iron-bearing rock of the Lake Superior region, now used as an ore of iron.

**Talc** A metamorphic mineral; a hydrous magnesium silicate.

**Talus** A heap of loose blocks of rock at the foot of a cliff or steep canyon wall, produced by frost wedging on the ledges above and accumulated under gravity.

**Tarn** Pond or small lake in a cirque from which the ice has disappeared.

**Tetrahedron** See *Silica tetrahedron.*

**Tetrapod** A four-footed land animal of the Phylum Chordata.

**Texture** The size of the individual mineral grains in a rock.

**Thecodont** A small two-footed bipedal carnivorous diapsid reptile; extinct.

**Therapsid** Member of a group of synapsid reptiles that includes many forms with mammal-like characters; often termed mammal-like reptile.

**Thrust fault** A reverse fault with a low angle of dip.

**Till** Mixture of unsorted rock debris deposited directly by glacier ice.

**Till plain** A large region underlain by till.

**Tillite** Lithified till.

**Time-rock unit** A body of rock that was deposited during a specified interval of past time.

**Time scale** The geologic calendar, on which the units of geologic time are arranged in chronologic sequence and are each assigned a length of time on the basis of absolute age.

**Titanothere** Member of a group of extinct odd-toed ungulates that developed huge horned skulls and grew to be as much as eight feet high at the shoulder.

**Tombolo** A sand bar that connects two islands, or an island and the mainland.

**Transported regolith** Clay, sand, and other loose materials brought into an area from elsewhere by natural agents, such as streams and glaciers.

**Trellised drainage pattern** Right-angle pattern of drainage formed by streams in regions of tilted sedimentary rocks of varying resistance to erosion.

**Tributary** A stream that flows into a larger one.

**Tuff** Consolidated volcanic ash.

**Turbulence** The pattern of eddies and crosscurrents characteristic of water in streams.

**Uintathere** A massive-bodied four-footed placental mammal with a horned skull and elephantlike legs retaining five toes.

**Ultimate base level** Sea level. See *Base level.*

**Unconformity** General term for a surface of erosion or nondeposition that became buried by younger sedimentary rocks.

**Underclay** Clay that directly underlies most coal beds. It is probably the much-altered soil in which the coal-forming plants grew.

**Ungulate** A hoofed placental mammal.

**Uniformity of process** See *Principle of uniformity of process.*

**Valley glacier** A glacier in a valley in high mountains.

**Valley train** Deposit of outwash sloping gently downstream from the end of a valley glacier.

**Vein** A tabular body of mineral matter, formed in an open crack or fissure by precipitation from hot watery solutions that are generally of magmatic origin.

**Velocity** Of a stream, the rate of flow, in feet per second or miles per hour.

**Ventifact** A stone that has been

abraded by wind-driven sand, with development of faceting and a high polish.

**Vesicle** Small cavity produced by escaping gas during cooling of an extrusive igneous rock such as basalt.

**Vesicular** Term applied to extrusive igneous rock that contains many small cavities, or vesicles, produced by escape of gas during cooling and solidification.

**Volcanic breccia** See *Breccia.*

**Volcanic neck** A spire or cylindrical tower of igneous rock, which represents the feeder pipe of a volcano that has been removed by erosion.

**Volcano** See *Composite volcano, Shield volcano.*

**Water gap** Segment of a stream valley where the stream cuts through a ridge of resistant rock.

**Water table** The upper boundary of the zone of ground-water saturation.

**Water-table well** A well in which the water stands at the level at which it was encountered.

**Waves** Oscillatory forms that pass through water as a result of frictional drag of wind across the water surface.

**Wave-built terrace** Deposit of sediment seaward from the wave-cut bench along mountainous shore lines.

**Wave-cut bench** Shelf or platform at sea level, cut on bedrock by wave erosion. Typical of mountainous coasts, especially headlands and promontories.

**Wave-cut cliff** A seaward-facing cliff along a mountainous shoreline, produced by wave erosion and downslope movements.

**Weathering** The mechanical and chemical changes that rocks undergo on exposure to air, water, and organic matter.

**Young valley** Valley with a steep gradient, a V-shaped cross-profile, and a floor occupied entirely by the stream.

**Youth** See *Regional youth, Young valley.*

**Youthful valley** See *Young valley.*

**Zone** A unit of sedimentary rock that is defined on the basis of the fossils that it contains.

**Zone of aeration**  The rock or regolith between the ground surface and the water table.

**Zone of saturation**  The rock and regolith below the water table, in which all pore spaces are filled with water.

Abrasion, 117, 525
  by glaciers, 214–15, 223–24, 232
  by streams, 117–19, 142
  by waves, 131, 142
  by wind, 139, 142
Absaroka Mountains, 204
Absolute age, 58, 59, 347, 525
Absolute time scale, 61, 347
Abyssal fans, 289
Abyssal hills, 287, 525
Abyssal plains, 287, 525
Acadia National Park, 131
Acadian Mountains, 391–92, 401
Acadian Orogeny, 366, 403
Acanthodians, 464
Accretion
  continental, 353, 356
  planetary, 343, 356
Adirondack dome, 378
Adirondack Mountains, 224, 233, 257, 324, 378
Agassiz, Louis, 10
Age
  as part of time scale, 362, 525
  of the earth, 49, 50, 339, 525
Age of Amphibians, 475
Age of Fishes, 464
Age of Mammals, 498
Age of Mollusks, 479
Aggregate, for concrete, 227, 232
Agnotozoic era, 55
Agricola, 178
Algae, 434, 447–52, 454
Algoman Mountains, 350
Algoman Orogeny, 349–50
Allegheny Orogeny, 366, 402
Allegheny Plateau, 325, 327
Alluvial fans, 125–28, 142, 525
Alluvial streams, 122, 142, 525
Alluvium, 122, 525
Alps, 261, 307
Amblypods, 501, 502–504, 522, 525
Ammonoid cephalopods, 469, 479, 495, 496, 525
Amphibians, 436, 474–76, 477, 481–83, 497
Analysis, process of, 10, 16
Anapsida, 484, 497, 525
Anchorage, Alaska, earthquake at, 104, 265
Ancient history, 56
Andean System, 309
Andes Mountains, 261, 307, 318
Angle of repose, 106, 525
Angular unconformity, 268, 271, 525
Anhydrite, 31, 33, 36, 147, 158, 525
Animal kingdom, 431, 435–36, 525
Animals, 429–30, 435–36, 452
  in weathering, 90
Ankylosaurs, 492, 525
Annelida, 435
Antarctica, ice sheet on, 211, 222, 232
Anteaters, 504
Anthracite, 281–82, 283, 525
Antiarchs, 464

Anticlines, 253–56, 270, 277, 525
Antler Mountains, 392, 403
Antler Orogeny, 366, 402
Apes, 514, 515–16, 522
Apollo missions, 53
Appalachian Basin, 396
Appalachian Mountains, 261, 295, 307, 309, 310, 313
Appalachian-Ouachita mobile belt, 364
Appalachian System, 307, 312, 327–28, 335, 336
Aquifer, 236, 238–39, 246, 247, 525
Archaeopteryx, 494
Arches, and basins, 365
Arêtes, 214, 232, 525
Aristotle, 176, 441
Armadillos, 504
Artesian wells, 238–39, 247, 525
Arthrodires, 464
Arthropoda, 435, 453, 454
Ash, 69, 73, 525
Astrogeology, 4, 5, 525
Astronomy, 4
Astrophysics, 4
Aswan High Dam, 129, 130
Atlantic Coastal Plain, 418, 420
Atmosphere, 19, 38, 525
  primitive, 437–41, 444
Atom, 25, 26, 41, 51, 525
Augite, 30, 31, 32, 35, 36, 525
Australopithecines, 516–18, 525
Avalanches, 108, 525
Azoic era, 55


Bacteria, 434, 450, 454
Banff National Park, Canada, 218
Barchans, 140, 525
Barrier islands, 136, 142, 525
Basal slip, 212, 232, 525
Basalt, 65, 68, 73, 84–85, 87, 525
  in crust, 23–25, 290–94, 320
  lunar, 341–44, 356
  weathering of, 89
Base level, 196–98, 208, 209, 525
Basement complex, 326, 525
Basin and Range Province, 330
Basins, and arches, 365
Batholiths, 80, 82, 87, 274, 282, 525
Bats, 504
Bauer, George, 178
Bay barrier, 134, 142, 525
Beach, 134, 525
Bed load, 120, 142, 525
Bedding planes, 34, 76, 277, 525
Beds, 145, 165, 525
Belemnoids, 480, 495, 496, 525
Bembesi gold belt, 448
Biochemistry, 4
Biology, 4
Biosphere, 20, 429–44, 525
  Early Paleozoic, 457–67
  Later Paleozoic, 469–77
  Precambrian, 447–54

Biotite, 30, 32, 36, 188, 525
  weathering of, 93, 94, 95, 98
Birds, 436, 494, 496, 497, 512
Bituminous coal, 147, 154–155, 164, 166, 281, 396–98, 525
Black Hills, 257, 325, 345
Black shale, Devonian-Mississippian, 387–88
Blastoids, 436
Blocks, 69, 73, 525
Blue Ridge, 328
Bombs, 69, 73, 525
Bonanza ore deposits, 260
Brachiopoda, 435
Braided streams, 122, 221, 232, 525
Breakwaters, 136, 525
Breccia, 146, 525
  volcanic, 68, 69, 73
Bright Angel Shale, 189–90
Brines, 19, 246, 247, 248
Brontosaurus, 491
Brown clays, 289, 526
Bryce Canyon, 330
Bryozoans, 435
Buffalo, 512, 513
Burrows, 176, 180
Butte, Montana, copper at, 86
Buttes, 204, 526


Calcite, 31, 33, 35, 36, 149, 165, 526
  metamorphism of, 273, 276
  solution of, 94, 95, 98
Calderas, 72, 73, 87, 526
Cambrian Period, 56, 57
  biosphere of, 453–54, 457–61, 466
  physical history of, 371–75, 381–82, 384
  record of, in Grand Canyon, 188–90, 191
Camels, 509–10, 513
Canadian Shield, 323, 324–25, 335, 345–46, 365, 421
Capitan Reef, 401
Carbon-14, 52, 526
Carbon dioxide, 19, 241, 243
Carbonates, 31, 33, 35, 36
Carbonic acid, 90, 93, 97, 98
Carbonification, 154, 172, 173, 180, 526
Carlsbad Caverns, 241, 401

INDEX

Carnivores, 504, 505
Cascade Mountains, 318, 328, 331
Cascade orogenies, 366, 418–19, 426
Casts, 174, 180, 526
Catskill Delta, 389–90, 391, 403
Catskill Mountains, 204, 389
Causal connection, 11
Caverns, 241, 243, 245, 248
Cementation, 147, 149, 526
Cenozoic Era, 55, 57
 biosphere of, 501–23
 physical history of, 417–21, 426
Central Cordillera, 329
Central Province, Canadian Shield, 352
Central Valley, of California, 328
Cephalopods
 ammonoid, 469
 nautiloid, 459–60
Ceratopsians, 492–93, 526
Chalcopyrite, 31, 526
Chalicotheres, 507–508, 526
Chemical rocks, 147, 156
Chemical weathering, 91, 92–95, 98
Chemistry, 4
Chert, 147, 156, 166, 526
Chloride, 31, 33, 36
Chlorite, 30, 31, 35, 36, 275, 526
Chlorophylls, 440
Chordata, 436
Chromite ores, South Africa, 86
Churchill Province, 352, 353
Cincinnati Arch, 378, 389
Cinder cone, 70, 73, 87, 526
Cinders, 69, 73, 526
Circum-Pacific belt, 73–74, 266, 270–71
Cirques, 214, 232, 526
Clams, 435
Class, of organisms, 431, 433, 526
Classification, process of, 11, 16
Clastic rocks, 146–51, 165
Clay, 89, 93, 94, 98, 120, 142, 147, 149, 165, 276, 526
Clay minerals, 31, 33, 93, 94, 276, 526
Cleavage, 27, 526
 in slate, 277, 526
Climates, ancient, 179, 181
Climatic zones, 382
Clock
 radioactive, 51
 sedimentary, 50
Coal: see Anthracite; Bituminous coal
Coal swamps, 472–74
Coast Ranges, 328
Coastal Plains, 323, 331, 334, 336, 526
Coelenterata, 435, 452, 453, 454
Colorado Mountains, 398–99, 403–404
Colorado Plateaus, 330
Colorado River, 40, 49, 330
Columbia River Plateau, 66, 330
Compaction, 147, 151, 165, 526
Composite volcanoes, 70–73, 87, 526

Composition
 of the crust, 22–25
 of minerals, 27–32
 of rocks, 32–38, 68
Compound, 25, 526
Concentrative generalizations, 16
Concretions, 160–62, 166, 245, 248, 526
Condylarths, 501, 502, 504, 505, 522, 526
Confining pressure, 276, 282, 526
Conglomerate, 146, 147, 165, 526
Conodonts, 436
Contact metamorphism, 273–74, 282, 526
Contamination, of ground water, 246, 248
Continental crust, 308–21, 354, 356
Continental drift, 296–303, 353–54, 526
Continental rise, 287, 312–13, 318, 321, 526
Continental shelves, 23, 285, 286, 312–15, 321, 526
 of North America, 323, 334
Continental slope, 23, 287, 526
Continents, 285, 293, 294, 295, 303
Convection currents, 319–20
Copper, 31, 86, 355, 357
Coprolites, 176, 180, 526
Corals, 435
Cordilleran Ice Sheet, 422
Cordilleran System, 308, 328–31, 335, 336, 418–20
Core, of the earth, 21, 41, 526
Core holes, in ocean basins, 24, 25
Correlation, 184, 186–88, 190–91, 360, 367, 526
Cotylosaurs, 485, 526
Country rock, 75, 80, 81, 87, 526
Crabs, 435
Crater Lake, 72
Craters, 65, 526
 lunar, 341–44
Creep, 101, 105, 111, 526
Creodonts, 501, 502, 504, 505, 522, 526
Cretaceous Period, 57
 biosphere of, 481, 490–97, 500–501
 physical history of, 407, 409–15
Crevasses, 213, 232, 526
Crinoidal limestone, 388
Crinoids, 436
Crocodiles, 491, 493, 497
Cross sections, 22, 360, 526
Cross-stratification, 139, 140, 166, 526
Crust, 21, 22–25, 41, 526
 continental, 308–21, 354, 356
 lunar, 344, 356
 materials of, 25–38
 oceanic, 290–94, 303
Cryptozoic era, 55
Crystalline limestone, 152
Crystallization, 147
Cud chewers, 509–10
Cumberland Mountains, 204

Cumberland Plateau, 325, 327
Currents, 131, 132, 142
Cycle of regional reduction, 200–204, 209, 526
 interruptions in, 204–206, 209
Cycle of valley development, 193
Cyclothems, 396–98, 526
Cystoids, 436

Debris flows, 101, 103, 111, 526
Decomposition, 91, 92–95, 98, 526
Deductive reasoning, 12, 16
Deep Sea Drilling Project, 24, 25
Deflation, 138, 142, 527
Deltas, 124, 125, 142, 527
Dendritic drainage pattern, 208, 209, 527
Density, relative, 20
Deposition
 by glaciers, 219–22, 224–28, 232–33
 by streams, 121, 128, 142
 by waves and currents, 132
 by wind, 139, 142
Desert pavement, 138, 139, 142, 527
Desiccation, 171, 180
Devonian Period, 57
 biosphere of, 463–66, 467, 469–74
 physical history of, 387–93, 403
Diamond, 31, 527
Diapsida, 485, 490–94, 497, 527
Differential weathering, 97, 99, 527
Dike, 76, 87, 527
Dinosaurs, 491–93, 495–97, 527
Diorite, 68, 79, 527
Dip, 251–52, 255, 270, 527
Discharge, 115, 117, 141, 142, 208, 527
Disconformity, 266–68, 271, 527
Discontinuities, 21, 527
Disintegration, 91, 92, 98, 527
Distributive generalizations, 16
Divides, 200, 201, 209, 527
Dolomite, 31, 33, 35, 36, 158, 276, 527
Dolomitization, 158, 527
Dolostone, 147, 158, 159, 166, 241, 527
Dome mountains, 257, 270, 325, 527
Domes
 anticlinal, 256, 270, 527
 and basins, 365
Downslope movements, 101–11, 527
Drainage patterns, 208, 209
Drainage, rearrangement of, 229, 233
Dripstone, 243–45, 248, 527
Drumlins, 225, 233, 527
Duckbills, 492
Dunes, 139, 140, 142, 527

Early maturity, 201, 209, 527
Earth
 age of, 49, 50, 339
 as a planet, 19
 energy of, 38, 39
 origin of, 343

Earthquake waves, 20, 21, 23, 41
Earthquakes, 101, 102, 104, 262–66, 270, 293–94, 527
East Greenland System, 333, 335, 336
Eastern Cordillera, 328–29
Echinoderma, 436
Echinoids, 436
Economic geology, 5, 527
Ectoprocta, 435
Ejecta, lunar, 342–44
Electric logs, 185, 527
Electrons, 26, 27, 29, 527
Element, 25, 28, 41, 527
Elephants, 511–12, 513
End moraines, 219, 224–25, 232, 233, 527
Energy, 38–42, 439, 527
    geothermal, 83, 87
Entelodonts, 509, 527
Epicenter, 263, 270, 527
Epoch, 362, 527
Eras, 55, 56, 57, 527
Erosion, 39, 41, 527
    by glaciers, 214–19, 221, 222–24
    by streams, 117, 142, 194, 195, 197, 198
    by waves, 131, 142
    by wind, 138, 142
Erratics, 219, 225–26, 232, 527
Eskers, 226, 233, 527
Eugeosyncline, 310n
Euryapsida, 485, 489–90, 497, 527
Evaporites, 157, 166, 379, 384, 401, 527
Even-toed ungulates, 509–10
Evolution, organic, 179–80, 191, 441–43, 444
Exfoliation, 91, 98, 527
Explosive eruptions, 66
Extrusive rocks, 63, 73, 87, 527

Fabric, 32, 34, 35, 37, 527
Family, of organisms, 433, 527
Fault scarps, 262
Fault-block mountains, 262, 314, 527
Faults, 257–62, 270, 527
Feldspars, 30, 527
    weathering of, 98
Felsite, 68, 73, 87, 528
Fermentation, 439
Ferns, 434
Ferromagnesian minerals, 30, 31, 95, 98, 528
Field observations, 14, 16
Field relations, 33, 528
Fiery clouds, 72, 104, 528
Fig Tree Group, 447–48
Finger Lakes, 223, 233
First forest, 471
Fishes, 436, 461–67, 480
Fissures, 66
Flash floods, 127
Flint, 156, 166, 528
Floodplains, 124, 142, 201, 209, 528
Flow, volcanic, 66, 79, 80

Flying reptiles, 491, 494
Focus, of earthquakes, 263, 270, 294, 528
Folds, 253–56, 270
Foliation, 35, 37, 212, 275, 282, 528
Fool's gold, 31
Footprints, 176
Footwall block, 259, 260, 270, 528
Formations, 183, 184, 190, 359–60, 528
Fossil fuels, 164
Fossilization, 169–76, 528
Fossils, 34, 169–81, 360, 363, 367, 429, 442–44, 528
    Precambrian, 447–54
Fracture, 27, 528
Franklin mobile belt, 364, 381, 390, 393, 403
Franklin Mountains, 329
Frost wedging, 91, 98, 528
Fungi, 434, 450, 454

Gabbro, 68, 79, 528
Galena, 31, 528
Garnet, 30, 31, 35, 36, 528
Gas, natural, 164
Gastroliths, 176, 180, 528
Genus, 431–33, 528
Geochemistry, 4, 5, 528
Geochronology, 4, 5, 528
Geodes, 162, 166, 245, 248, 528
Geologic maps, 360, 362, 367, 528
Geologic time, 49, 54, 55, 56, 58
    and probability, 59
    fossils and, 179
Geology, 4, 16, 528
Geomorphology, 5, 528
Geophysics, 4, 5, 528
Geosynclinal theory, 300, 310n
Geosynclines, 310n
Geothermal energy, 83, 87, 528
Geothermal Steam Act, 84
Geysers, 83, 87, 528
Glacial theory, 10
Glaciation, 211–33, 335, 336, 354–55, 357, 421–23, 427
Glacier motion, 212, 232
Glacier National Park, 218, 231, 329, 422
Glaciers, 211–28, 232, 331, 354, 528
*Glomar Challenger,* 24, 25
*Glossopteris,* 298
Gneiss, 279, 283, 528
Gold, 31, 86
Gondwanaland, 300
Gradient, 115–17, 142, 528
Grand Canyon, 40, 49, 196, 330
    Cambrian rocks of, 188–90, 191
Grand Coulee, 66
Grand Coulee Dam, 109
Grand Teton National Park, 103, 231
Granite, 68, 79, 87, 279, 528
    in crust, 23
    quarried, 84–85
    weathering of, 92–94

Granitization, 82, 279
Graphite, 31, 36, 273, 275, 528
Graptolites, 436
Gravel, 120, 146, 147, 148, 528
Gravity, 38, 42, 101, 110, 114, 293
Great Basin, 330
Great Bear Lake, uranium at, 86
Great Salt Lake, 423
Great Sand Dunes National Monument, 140
Great Smoky Mountains, 328
Green Mountains, 224
Greenland, ice sheet on, 211, 222, 232
Grenville Province, 352, 353
Groins, 137, 528
Gros Ventre River, 103, 108
Ground moraine, 225, 233, 528
Ground water, 235, 247, 528
Guadalupe Mountains, 401
Gulf Coast mobile belt, 425–26
Gulf Coastal Plain, 334, 418, 420
Gunflint Formation, 449
Gymnosperms, 481, 496, 528
Gypsum, 31, 33, 36, 147, 157, 158, 166, 528

Halite, 27, 31, 33, 36, 528
Hanging valleys, 215, 528
Hanging-wall block, 259, 260, 270, 528
Hardness, 27, 528
Hawaiian Islands, 60, 63
Headwall, 214, 232, 528
Heat, 39, 40, 41, 319–20, 321, 344
Hebgen Lake, 102
Hematite, 31, 33, 36, 149, 165, 355, 528
Hemichordata, 436
Herodotus, 50, 177
High Plains, 325, 329
Highlands, lunar, 341–44, 356
Himalayas, 307, 309
Historical geology, 4, 5, 528
*Homo sapiens,* 519, 521
Hoofed mammals, 504, 505–10
Hooke, Robert, 178
Horizons, of soil, 98, 99, 528
Horizontality, law of original, 159, 166, 251
Horn, 214, 232, 528
Hornblende, 30, 31, 32, 36, 275, 528
Horses, 506, 508–509, 512, 513
Hot springs, 83, 87, 528
Humus, 97, 99, 528
Huronian Province, 348, 352
Huronian System, 350, 354, 356
Hutton, James, 48, 49, 50, 53, 61, 426
Hydraulic action, 528
    by streams, 117, 142
    by waves, 131, 142
Hydrologic cycle, 39, 114, 528
Hydrosphere, 19, 528
Hydroxyl, 30
Hypothesis, 10, 12, 16
*Hyracotherium,* 506–507, 508

Ice
  in glaciers, 212, 232
  as a mineral, 31
  in weathering, 90, 98
Ice front, 219, 232, 528
Ice sheets, 211, 222–28, 232, 354–55, 383, 528
Icefields, 214, 232, 528
Ichthyosaurs, 485, 489–90, 529
Identification, 184–86, 187, 190, 529
Igneous rocks, 32, 33, 41, 63–83, 529
Illinois Basin, 396
Impression, 172, 529
Inductive reasoning, 12, 16
Infiltration, 235, 247, 529
Innuitian Orogeny, 366
Innuitian System, 308, 332–33, 335, 336, 420–21
Insectivores, 501, 504, 514, 522, 529
Insects, 435, 474, 477, 481, 497
Insoluble residue, 94, 529
Intensity, of earthquakes, 266, 270, 529
Interior Lowlands, 323, 325–26, 331, 335, 336
Internal flow, 212, 232, 529
Intrusive rocks, 63, 74, 87, 529
Ions, 26, 27, 29, 41, 529
Iron ore, 165, 355, 357
Iron oxide, 94, 98, 149, 165
Island arcs, 287, 312–13, 317–18, 321, 529
Isopach maps, 363, 364, 367, 529
Isostatic equilibrium, 313, 319, 367, 529
Isotopes, 26, 51, 529

Jasper National Park, Canada, 218
Jellyfish, 435, 450, 452
JOIDES, 24
Jurassic Period, 57
  biosphere of, 491–97, 500
  physical history of, 407, 409–15

Kames, 226–27, 233, 529
Kaolinite, 30, 31, 33, 36, 529
Katmai, volcano, 72
Kelleys Island, 224
Kettles, 231, 233, 529
Keweenawan System, 351, 353, 355
Kinetic energy, 38, 41, 114, 251, 529

Lagoons, 136, 142, 529
Lake basins, 230–31, 233
Lake Huron, 354
Lake Nasser, 130
Lake Superior region, 165, 324, 348–52, 353
Lakes, 230–31
Laminae, 159, 529
Land plants, 471–74, 476
Land snails, Hawaiian, 60
Landslides, 101, 529

Laramide orogenies, 366, 414, 415
Lassen Peak, 328
*Latimeria,* 466
Laurasia, 300
Laurentian Highlands, 224
Laurentide Ice Sheet, 421
Lava, 63, 64, 73, 79, 80, 529
Law of original horizontality, 159, 166, 251, 529
Law of sequence, 45–48, 53, 529
Law of superposition, 46, 529
Laws, natural, 3, 16
Layers of geology, 269–70, 271
Lemurs, 514–15
Leonardo da Vinci, 178
Life, 429, 529
Lignite, 154, 529
Limb, of fold, 253, 529
Limestone, 147, 151, 152, 165, 166, 529
  metamorphism of, 273, 274, 276
  solution of, 94, 95, 98, 240–42, 247–48
Limonite, 31, 33, 36, 94, 149, 529
Linnaeus, 431
Linné, Carl von, 431
Linnean system, 432–33, 443
Lithification, 147, 529
Lithofacies maps, 362, 364, 367, 529
Lithologic logs, 185, 529
Lithosphere, 19, 20, 21, 39, 41, 529
Lizards, 491, 493, 497
Load, of streams, 120, 142
Lobe-fins, 466, 467, 474, 529
Lobsters, 435
Local base level, 197, 208, 529
Loess, 140, 141, 142, 529
Long profile, 115, 116, 529
Longshore bars, 136, 529
Longshore currents, 131, 529
Longshore drift, 132, 529
Los Angeles region, 110
Lungfishes, 466, 529

Mackenzie Mountains, 329
Madison Canyon, 102, 108, 263, 265
Magma, 63, 75, 82, 87, 316–18, 320, 529
Magnetic field, of earth, 291–93, 302, 383
Magnetite, 31, 86, 291, 355, 529
Magnitude, of earthquakes, 265, 270, 529
Mammal-like reptiles, 486–89, 497, 499
Mammals, 436, 488–89, 499–513, 522
Mammoth Cave, 241
Mammoths, 171, 512
Man
  and downslope movements, 108–10
  and ground water, 245, 248
  and the Nile, 124, 129
  and shorelines, 136
  history of, 513, 516–23

Mantle, 21, 23, 41, 63, 291, 293, 319–20, 321, 529
Mantle rock, 290, 292, 313–14
Maps, 360–67
Marathon Mountains, 332, 402
Marble, 273, 274, 275, 276, 280–81, 283, 529
Maria, lunar, 341–44, 356
Mariana Trench, 21, 286
Marquette Range, 355
Marsupials, 500–501, 522, 529
Mass number, 51, 529
Massanutten syncline, 254, 255
Mastodons, 512
Mature valleys, 198–200, 208, 209, 530
Maturity, regional, 201–204, 209
Mauna Loa, 65
Maximum relief, 21, 41, 201, 209
Meanders, 122, 123, 209, 530
Mechanical load, 120, 121, 142
Mechanical weathering, 91, 98
Mercalli scale, 266, 270
Mesabi Range, 350, 355
Mesas, 204, 530
Mesosaurs, 489, 530
Mesozoic Era, 55, 57, 407
  biosphere of, 479–82, 489–97, 499–501
Meta-igneous rocks, 273, 282, 530
Metamorphic rocks, 35, 37, 38, 41, 212, 273, 276–82, 530
Metamorphism, 273–76, 282, 530
Metasedimentary rocks, 273, 282, 530
Meteorites, 343, 356
Micas, 30, 530
Michigan Basin, 378, 379, 381, 384, 396
Mid-Atlantic Ridge, 288, 292, 301
Mid-oceanic ridges, 287–88, 290–94, 303, 320, 321, 426, 530
Milwaukee Deep, 287
Mineral, 27, 28, 30, 41, 530
Mineral-age provinces, 347–48
Mineral resources, 84, 87, 164–65, 227–28, 260, 355–56
Mineralogy, 4, 5, 530
Miogeosyncline, 310n
Mississippi River, 115, 124, 229
Mississippi Valley, 50
Mississippian Period, 56, 57
  biosphere of, 472, 475
  physical history of, 387–95, 399, 403
Mobile belts, 309–15, 321, 530
Modified Mercalli scale, 266, 270
Moho, 22, 23, 24, 41, 530
Mohole, 24
Mohorovičić, 22
Molds, 174, 180, 530
Molecules, organic, 438, 444
Mollusca, 435
Monkeys, 514, 515
Mont Pelée, 72
Monument Valley, 330, 401
Moon, 53, 339–45, 356
Morrison Formation, 411–12, 413, 415

Mosses, 434
Mother Lode, 86
Mount Baker, 328
Mount Everest, 21
Mount Hood, 328
Mount Rainier, 328
Mount Shasta, 328
Mountain systems, 307–309, 315–19, 320–21
Mountains
 continental, 307–21, 357, 367
 dome, 257
 fault-block, 262
 residual, 204
 volcanic, 65, 70
Mt. Agung, 72, 104
Muav Limestone, 189–90
Mud, 120, 150, 530
Mud cracks, 160, 166, 530
Mudflows, 101, 103, 104, 111, 530
Muscovite, 30, 35, 36, 275, 276, 530

Native elements, 31, 36, 530
Natural casts, 174, 180, 530
Natural levees, 124, 142, 530
Natural molds, 174, 180, 530
Nautiloid cephalopods, 459–60, 530
*Nautilus*, 435, 459
Neanderthal man, 519–21
Neck, volcanic, 76, 87, 113
Necking, thermal, 314, 321
Neogene Period, 56, 57
 biosphere of, 505, 509–22
 physical history of, 418–20
Neutrons, 26, 51, 530
Nevadan Orogeny, 366, 413–14, 415
Nickel, 86, 355
Nile Delta, 130
Nile River, 50, 124, 129
Nitrogen, 19
Nodules, 156
Nonconformity, 269, 271, 530
Nonsilicate minerals, 31, 35, 41
Normal faults, 259–60, 270, 530
North America
 geologic history of, 335–36, 345–57
 provinces of, 323–34
Nucleus, atomic, 26, 51, 530

Observation, process of, 10, 14, 16
Obsidian, 68, 69, 73, 87, 530
Ocean basins, 24, 41, 51, 285–89, 303, 316, 321
Oceanic crust, 290–96, 303, 308, 312, 316, 320, 321
*Octopus*, 435, 459
Odd-toed ungulates, 506–509
Odds, of dice, 59
Ohio River, 193, 229
Oil, 164, 246
Oklahoma Mountains, 398–99, 403–404
Oldest rocks known, 52
Olivine, 28, 30, 31, 32, 35, 36, 530

Ontarian System, 349
Oozes, 289, 530
Order, of organisms, 431, 433, 530
Ordovician Period, 56, 57
 biosphere of, 457–63, 466–67
 physical history of, 371–84
Ore deposits, 85, 87, 260, 355–56, 530
Oreodonts, 509, 530
Organic compounds, 429, 437–41, 444, 530
Organic rocks, 147, 151
Organisms, 429–41, 530
 in weathering, 90, 98
Ornithopods, 492, 530
Orogenic cycles, 315, 353–54, 356, 530
Orogeny, 315–20, 321, 352, 367, 530
Orthoclase, 30, 32, 36, 530
 weathering of, 93, 95, 98
Ostracoderms, 461–63, 467, 530
Ouachita-Marathon System, 308, 331–32, 334, 335, 336
Ouachita Mountains, 331–32, 402
Ouachita Orogeny, 366
Outwash, 221, 232, 530
Outwash plains, 226, 233, 530
Oxides, 31, 33, 36
Oxygen, 19, 28, 29
 in weathering, 90, 98
Oysters, 435
Ozark dome, 378
Ozark Mountains, 257, 325, 345, 378
Ozark Plateau, 325

Pacific mobile belt, 413, 419, 426
Pacific Ocean, volcanic belt around, 73–74, 266, 270–71
Pacific Ranges, 331
Painted Desert, 408
Paleogene Period, 56, 57
 biosphere of, 500–15, 522
 physical history of, 418–19
Paleogeographic maps, 363, 364, 365, 367, 530
Paleogeography, 179, 180, 531
Paleomagnetism, 291–93, 302, 383
Paleontology, 4, 5, 180, 181, 530
Paleozoic Era, 55, 56, 57
Palisades disturbance, 366, 407
Panama Canal, 109
Panamint Mountains, 262
Pangaea, 297–302, 303
Parallel stratification, 159, 166, 530
Parapsida, 485, 489–90, 497, 530
Parícutin, 70
Peat, 147, 154, 530
Pecos Valley, 246
Pedology, 97
Pegmatite, 81, 87, 530
Pelecypods, 435
Pelycosaurs, 484, 486, 530
Pennsylvanian Period, 57
 biosphere of, 473–77, 481–82, 485–86, 497
 physical history of, 395–404
Penokean Orogeny, 350–51

Penokee Range, 351
Perched water table, 239, 247, 531
Periods, of time, 56, 57, 362, 531
Permeability, 235, 247, 531
Permian Basin, 401
Permian Period, 57
 biosphere of, 469, 473, 475, 480–81, 484, 485–87, 497, 499
 physical history of, 395–404
Permineralization, 172–73, 180, 531
Petrified Forest, 481
Petrology, 4, 5, 531
Phenocrysts, 69, 531
Photosynthesis, 448
Phyllite, 279, 283, 531
Phylum, 431, 433, 531
Physical chemistry, 4
Physical geology, 4, 5, 531
Physics, 4
Piedmont Province, 391
Piedmont Upland, 328
Pigs, 509
Piles, 137
*Pithecanthropus*, 518–19, 520
Placentals, 500–502, 504–505, 522, 531
Placoderms, 464–65, 467, 531
Plagioclase, 30, 32, 36, 531
 weathering of, 95
Plant kingdom, 431, 434, 531
Plants, 429–30, 434, 439, 452, 531
 appearance on land, 471
 of coal swamps, 472–74
 Mesozoic, 480–81, 496
 in weathering, 90
Plateau basalts, 66, 73, 87, 330, 531
Plateau Central, Mexico, 330
Plate-tectonic theory, 294–96, 302, 303, 314–15, 316, 320, 353–54, 531
Pleistocene Epoch, 56, 57
 biosphere of, 509, 511, 518–23
 physical history of, 421–23, 427
Pleistocene ice sheets, 228, 229, 233, 421–23, 427
Plesiosaurs, 485, 490, 495, 531
Plucking, 214, 223, 232, 531
Plunge pool, 119, 142, 194, 531
Point bars, 123, 142, 209, 531
Pollution, 246
Pompeii, 73
Porifera, 435
Porosity, 235, 247, 531
Porphyritic texture, 68, 69, 75, 531
Porpoises, 505
Post-Keweenawan sediments, 351
Potassium-argon method, 52
Potassium salts, 387
Pothole, 119, 142, 194, 531
Precambrian Eras, 55, 57
 biosphere of, 447–54
 physical history of, 339, 345–57
Precipitation, chemical, 147, 157
Preservation, of fossils, 171
Pressure, 275, 276, 282
Primates, 504, 513–23
Principle of uniformity of process, 48–49, 60–61, 531

Probability, 59
Protons, 26, 51, 531
Protozoa, 435
*Pteranodon,* 494
Pterosaurs, 494, 497, 531
Puddingstone, 146
Pumice, 68, 69, 73, 87, 531
Pyrite, 31, 531
Pyroclastics, 69, 87, 531

Quartz, 27, 28, 32, 33, 36, 149, 531
    weathering of, 93, 95, 98
Quartzite, 279–80, 283, 356, 531
Quaternary Period, 56, 57
Queenston Delta, 377, 378–79, 381, 384
Quiet eruptions, 64

Radial drainage pattern, 208, 209, 531
Radioactive clock, 51
Radioactive disintegration, 40, 42, 51, 52, 320, 321, 344, 531
Radioactive elements, 51, 52
Radiocarbon, 52
Rainwash, 97, 531
Ranges, 307
Rapids, 194
Ray-fins, 466, 480, 496, 531
Reaction series, 75, 87, 89, 531
Recent time, 56, 57
Recharge, 246
Recrystallization, 37, 152, 159, 274, 276, 280, 531
Reefs, 152, 387, 391, 401
Regional erosion surface, 204–206, 266, 531
Regional maturity, 201–204, 209, 531
Regional metamorphism, 275, 282, 531
Regional reduction, cycle of, 200–206, 209
Regional youth, 200–201, 209, 531
Regolith, 97, 99, 228, 531
    lunar, 341–44, 356
Rejuvenated streams, 206, 531
Relative age, 59, 531
Relative density, 20, 41, 531
Relative time scale, 61
Relief, 21, 41, 201, 531
Repeated glaciation, 228, 233
Replacement, 245, 248, 531
    forming chert, 156, 166
    forming dolostone, 147, 158, 166
    forming ore deposits, 86, 87
    in fossilization, 174, 180
Reptiles, 436, 482–97
Residual mountains, 204, 209, 531
Residual regolith, 97, 99, 532
Reverse faults, 260–61, 270, 532
Rhinoceroses, 171, 512
Rhynchocephalians, 497
Richardson Mountains, 329
Richter scale, 265, 270

Ripple marks, 160, 166, 532
Rock cleavage, 277, 282
Rock cycle, 282
Rock flour, 218, 532
Rock-forming minerals, 36
Rock salt, 147, 151, 166, 532
Rock Springs anticline, 254
Rockfalls, 101, 103, 108, 111, 532
Rocks, 32–38, 41–42, 532
Rockslides, 101, 103, 108, 110, 195, 532
Rocky Mountain National Park, 329, 422
Rocky Mountain trough, 413–14, 415
Rocky Mountains, 218, 325, 329
Rodents, 504
Roots, of mountains, 308, 318
Rubidium-strontium method, 52
Running water, work of, 113–30
Runoff, 235

Saber-toothed cats, 512
Salt: *see* Rock salt
Salt domes, 418
San Andreas fault, 258, 263, 270
San Fernando earthquake, 263–65, 266
San Francisco earthquake, 263, 265
Sand, 120, 142, 147, 532
Sandstone, 34, 147, 149, 165, 532
Sanitary landfill, 246, 247
Sawatch Range, 231
Scallops, 435
Schist, 36, 278, 283, 532
Science, 3, 4, 5, 16
Scientific methods, 5, 10, 16
Scoria, 65, 68, 532
Sea-floor spreading, 289–94, 303, 316, 320, 345, 532
Sea-floor trenches, 286, 291, 293–94, 303, 316–19, 321, 532
Sea pens, 452
Sedimentary clock, 50
Sedimentary rocks, 33, 34, 40, 41, 50, 145–66, 308, 532
Sediments, 145, 146
Seed plants, 434, 472–73
Seismographs, 21, 532
Sequence
    determination of, 46, 61
    law of, 45–48, 53, 529
Sequences, major rock divisions, 53, 56, 57, 532;
    *see also* Law of sequence
Series, 362, 532
Serpentine, 30, 31, 35, 36, 532
*Seymouria,* 483–84, 485
Shale, 147, 150, 165, 532
Sharks, 465, 467
Sheet structure, 92, 98, 532
Shelf ice, 222
Shells, first appearance of, 453–54
Shield volcanoes, 65, 73, 87, 532
Shiprock, 76, 79
Shorelines, 132–37, 142

Sierra Madre Occidental, Mexico, 330
Sierra Nevada, 80, 218, 231, 262, 270, 328
Silica, 28, 68, 93, 149, 165, 166, 356
Silica tetrahedron, 28, 29, 30, 41, 532
Silicate minerals, 28, 32, 33, 36, 41
Silicification, 156, 532
Silicon, 28, 29
Sills, 76, 79, 80, 87, 532
Silt, 120, 142, 147, 150, 532
Siltstone, 147, 150, 165, 532
Silurian Period, 57
    biosphere of, 457–63, 466–67
    physical history of, 375–84
Sinks, 242, 247–48, 532
Slate, 276–78, 282, 532
Slate Range, 47–48, 262, 270
Slave Province, 352, 353
Sliderock, 106, 111, 532
Slip face, 139, 142, 532
Sloths, 504
Slump, 101, 104, 111, 195, 532
Smith, William, 54, 178
Snails, 435
Snake River plateau, 330
Snakes, 491, 493, 497
Soil, 97–99, 532
Solar energy, 38, 439
Solar system, 19, 343
Solifluction, 101, 105, 111, 532
Solution
    by ground water, 240–42, 247–48
    by streams, 120
Solution load, 120, 121, 142
Sorting, 125, 138, 142, 145, 532
Soudan Iron Formation, 448
Space biology, 4
Special creation, 441–43, 444
Species, 430–33, 443, 532
Sphalerite, 31, 532
Spiders, 435
Spit, 134, 142, 532
Sponges, 434, 450, 454
Springs, 239–40, 247, 532
    hot, 83, 87
Squids, 435
Stage, 362, 532
Stalactites, 243–44, 248, 532
Stalagmites, 244, 248, 532
Starfish, 436
Stegosaurs, 492, 532
Steno, Nicolaus, 45, 48, 50
Strata, 145, 532
Stratification
    cross-, 139, 140, 166
    parallel, 159, 166
Stratified drift, 221, 226–28, 232, 233, 532
Stratigraphic procedures, 359–64, 367
Stratigraphy, 4, 5, 359, 367, 532
Streams, 113–40, 141–42, 206
Strike, 251–52, 255, 270, 532
Strike-slip faults, 258, 270, 532

Stromatolites, 448, 532
Structural geology, 4, 5, 532
Submarine canyons, 287, 312, 533
Subsidence, in mobile belts, 313–19, 321, 363
Sudbury, Ontario, nickel deposit at, 86
Sulfates, 31, 33, 36
Sulfur, 31
Sun, 19, 38, 42
Superior Province, 347–48, 352, 353, 354, 356
Superposition, 46
Suspended load, 120, 142, 533
Sverdrup Basin, 403, 409, 412–13
Synapsida, 484–85, 486, 497, 533
Synclines, 253–56, 270, 277, 533
Synthesis, process of, 11, 16
*Systema Naturae,* 431
Systems
  of mountains, 307–309, 315–19, 320
  of rock, 56, 57, 362, 533

Taconic Mountains, 382
Taconic Orogeny, 366, 382, 384
Taconite, 165, 355, 533
Talc, 30, 31, 35, 36, 356, 357, 533
Talus masses, 101, 105, 106, 111, 195, 533
Tapeats Sandstone, 189–90
Tarns, 218, 533
Tarsiers, 514–15
Temperature, in decomposition, 95, 99
Tertiary Period, 56, 57
Tethyan trench, 300
Teton Range, 218, 262, 270, 422
Tetrapods, 474, 533
Texture, 32, 34, 35, 37, 533
  of igneous rocks, 68, 74, 75, 81
"The Geysers," California, 84
Thecodonts, 490–91, 533
Theophrastus, 176
Theory, 12, 16
Therapsids, 484, 486–89, 499, 522, 533
Thorium, 356
Thrust faults, 261, 270, 533
Till, 219, 224–26, 228, 232, 233, 421, 533
Till plain, 225, 533
Tillites, 231, 232, 298, 533
Time
  fossils and, 179
  geologic, 49–61
Time-rock units, 362, 364, 367, 533

Time scale, 53, 56, 57, 58, 61, 533
Time significance, of unconformities, 267, 268, 271
Timiskamian System, 349
Titanotheres, 507, 533
Tombolo, 134, 533
Tracks, 176, 180
Trails, 176
Transcontinental Arch, 393–95, 398
Transportation
  by glaciers, 218–19
  by streams, 120
  by waves and currents, 132
Transported regolith, 97, 99, 533
Trellised drainage pattern, 208, 209, 533
Triassic Period, 57
  biosphere of, 479–81, 485–86, 490, 491, 493, 496, 497, 499, 522
  physical history of, 398, 402, 407–409, 414
Tributaries, 115, 533
Trilobites, 435
Tuff, 68, 69, 73, 87, 533
Turbulence, 114, 117, 141, 142, 533
Type sections, 360, 367
*Tyrannosaurus,* 491

Uintatheres, 501, 502–504, 522, 533
Ultimate base level, 197, 204, 208, 533
Umbrella effect, 245–46
Unconformities, 205, 266–70, 271, 365, 371, 533
Underclay, 155, 396–98, 533
Underground rivers, 242
Ungulates, 533
  even-toed, 509–10
  odd-toed, 506–509
Uniformity of process, 48–49, 60–61
Ural Mountains, 295, 307
Uranium deposits, 86, 355–56, 357
Uranium-lead method, 52
U-shaped valleys, 215, 232

Vaiont Dam, Italy, 103, 108
Valley and Ridge Province, 327, 328
Valley development, cycle of, 193
Valley glaciers, 211, 213–22, 232, 331, 533
Valley train, 221, 232, 533
Variables, 15, 16
Veins, 86, 87, 260, 533
Velocity, 116, 117, 142, 533
Venetz, 10

Ventifacts, 139, 142, 533
Vesicles, 65, 533
Vesuvius, 72
Virginia, 60
Volcanic neck, 76, 87, 113, 533
Volcanoes, 65, 70–73, 87
V-shaped valleys, 107, 193, 208

Water
  in atmosphere and hydrosphere, 19
  in downslope movements, 101, 111
  underground, 235–49
  in weathering, 90–95, 98–99
Water gaps, 208, 209, 533
Water table, 236, 247, 248, 533
Waterfalls, 119, 142, 194
Water-pressure surface, 238, 247
Water-table wells, 237–38, 247, 533
Waterton National Park, Canada, 218
Watkins Glen, 196
Wave-built terrace, 134, 533
Wave-cut bench, 134, 135, 142, 533
Wave-cut cliff, 134, 135, 142, 533
Waves, 130, 131, 142, 533
Weathering, 89–99, 533
  of till, 228
Wells, 237–39, 247
Western Cordillera, 328
Whales, 505
White Mountains, 224
Wichita-Arbuckle Trough, 364
Williston Basin, 378, 379, 380
Wind, 137, 142
Wind River Range, 218
Worms, 435, 453, 454

Xanthus, 177
Xenophanes, 177

Yellowstone National Park, 83, 102, 193
Yellowstone River, 116, 193, 195, 200
Young valleys, 193–96, 208, 533
Youth, regional, 200–201

Zion Canyon, 330
Zircon, 354
Zone of aeration, 236, 247, 248, 534
Zone of saturation, 236, 247, 534
Zones, 183, 184, 190, 360, 362, 367, 533

3 4 5 6 7 8 9 10